Role of Agriculture in Global Economy

Role of Agriculture in Global Economy

Edited by **Jamie Hanks**

SYRAWOOD
PUBLISHING HOUSE

New York

Published by Syrawood Publishing House,
750 Third Avenue, 9th Floor,
New York, NY 10017, USA
www.syrawoodpublishinghouse.com

Role of Agriculture in Global Economy
Edited by Jamie Hanks

International Standard Book Number: 978-1-68286-112-7 (Hardback)

Contents

Preface

This book has been a concerted effort by a group of academicians, researchers and scientists, who have contributed their research works for the realization of the book. This book has materialized in the wake of emerging advancements and innovations in this field. Therefore, the need of the hour was to compile all the required researches and disseminate the knowledge to a broad spectrum of people comprising of students, researchers and specialists of the field.

Since the emergence of human civilization, agriculture has played a critical role in maintaining livelihood and economy. Agricultural products play a major role in determining national economy as well as food security for nations in modern times. The topics covered in this book are aimed at providing a critical overview about the theory and practice of agriculture and its impact on global economy. It provides significant information of this discipline to help develop a good understanding of agricultural marketing, industrial and production technology and related fields. This book will serve as a reference to a broad spectrum of readers.

At the end of the preface, I would like to thank the authors for their brilliant chapters and the publisher for guiding us all-through the making of the book till its final stage. Also, I would like to thank my family for providing the support and encouragement throughout my academic career and research projects.

Editor

The impact of climate change towards Malaysian paddy farmers

R. B. Radin Firdaus[1]*, Ismail Abdul Latiff[2] and P. Borkotoky[2]

[1]School of Social Sciences, Universiti Sains Malaysia, Pulau Pinang, Malaysia.
[2]Faculty of Agriculture, Universiti Putra Malaysia, Serdang, Malaysia.

Ricardian model was used to estimate the impact of climate change on granary level net revenue of paddy sector in Malaysia. Panel data from eight granaries was used in this study and the annual net revenue was regressed against climate variables and other controlled variables. The results indicate that climate is one of the determinants of Malaysian paddy granaries' profitability. Given future predicted climate scenario indicates that farmers are predicted to lose an average of about 67% of revenue within the period of 2020 to 2029, 88% within 2050 to 2059, and 127% within 2090 to 2099. The losses might affect the livelihood of 296,000 paddy farmers in Malaysia. Thus, immediate adaptation strategies are critical in order to deal with such predicted effects.

Key words: Climate change, Malaysia, paddy, Ricardian model.

INTRODUCTION

Paddy is mainly planted in Peninsular Malaysia (PM), contributing 85.5% of Malaysia's total paddy production. The area involves eight major granaries, which have been reserved and designated purposely for wetland paddy cultivation by the government in the National Agricultural Policy. Most of the government support programs and interventions in paddy sector are concentrated in these areas, which also makes it an important area for research and development activities. As shown in Table 1, these eight granaries are located in seven states of PM. The designation is based on their record as large producer of paddy for the country. In order to supervise and monitor this intervention programs, a dedicated government authority is mandated to help the paddy farmers to improve their productivity, output, earning and technology practices.

The agricultural sector in Malaysia can be divided into three sub-sectors comprising agro-industrial, food and the miscellaneous group sub-sector. Paddy can be considered as the most important crop under the food sub-sector simply for two reasons. Firstly, Malaysian per capita consumptions accounted for 500 to 799 of calorie intake per day, which make rice the main staple food in Malaysia (Nguyen, 2005). Secondly, the crop provides a source of income for small-scale farmers that depend solely on paddy cultivation. With approximately 296,000 paddy farmers in Malaysia, almost 40% of them cultivate on full time basis (Elenita and Ema, 2005). Nonetheless, the local supplies still lag behind to cater for local demand, as Malaysia is still dependent on imported rice. Malaysian Padiberas Nasional Bhd (BERNAS), which holds exclusive right to import rice, is importing about 30% of rice from Vietnam, Thailand and Pakistan (Department of Agriculture, 2009b). Although Malaysia continues to be a net importer of rice, yet, in the wake of the global food crisis in 2008, the government effort to boost up local rice production is indispensable, as the crisis has indicated that cheaper imported rice would not always be available to cater local demand.

For instance in the early 2008, China has introduced export restrictions in order to safeguard the stocks of rice for domestic consumption due to the abnormally cold weather that affected production. The disastrous flooding in 2010 has forced Indonesia to import a large quantity of rice, as paddy fields across the country suffered almost a year of excessive flooding due to heavy rain (Deutsch

*Corresponding author. E-mail: radinfir@gmail.com.

Table 1. Location and average planted area of the granary areas in 2008.

State	Granary areas	Average planted area (ha)	Share to total paddy area in Malaysia (%)
Kedah and Perlis	Muda Agricultural Development Authority (MADA)	96553	38.4
Kelantan	Kemubu Agricultural Development Authority (KADA)	24965	10.0
	Kemasin Semarak Integrated Agricultural Development Project	2797	1.1
Terengganu	Northern Terengganu Integrated Agricultural Development (KETARA)	4293	1.7
Pulau Pinang	Pulau Pinang Integrated Agricultural Development Project	10305	4.1
Perak	Krian Sg. Manik Integrated Agricultural Development Project	26959	10.7
	Seberang Perak Integrated Agricultural Development Project	7272	2.9
Selangor	Barat Laut Selangor Integrated Agricultural Development Project	18301	7.3

Source: Department of Agriculture (2009a).

and Hidayat, 2011). Recent massive flood that hits Thailand in October, 2011 had destroyed approximately 10% of the nation's rice crop (Thanyarat, 2011). As a consequence, international rice trade volume in 2012 could drop to 33.8 million tons from projected 34.3 million [Food and Agriculture Organization (FAO), 2011]. Prior to the projection, the Asia-Pacific region has been expected to experience the worst effect on rice and wheat yields worldwide as result of climate change and decreased yields could threaten the food security of 1.6 billion people in South Asia [International Food Policy Research Institute (IFPRI), 2009].

The future climate scenario of PM is expected to experience a worrying trend. In every decade, the temperature is predicted to signify an increasing pattern, whereas rainfall is expected to have an opposite direction. Based on figures by Malaysian Meteorological Department (MMD), both west coast and east coast of PM has shown an increasing annual mean temperature trend (Figure 1) from 1968 to 2007 and a reduction in rainfall (Figure 2) from 1975 to 2005 (line A) as compared to those of 1951 to 1974 (line B). Temperature, precipitation, sea level rise, higher atmospheric carbon dioxide (CO_2) content and the incidence of extreme events are the main climate change related factors, which impact agricultural production (McCarl et al., 2001).

Several studies in South East Asia have confirmed that climate change can produce detrimental effects on rice production. Among the earliest studies that probe the

impact of climate change towards paddy production in South East Asia was carried out by Amien et al. (1996). The study predicts the impacts of climate change in the decades of 2010, 2030, and 2050 and found that an increase in maximum and minimum temperatures, rainfall, and solar radiation leads to reduction of about 1% of rice production annually in East Java. In Malaysia, study to assess the economic impact of climate change on paddy sector has only started in recent years. Chamhuri et al. (2009) asserted that paddy production could be affected by climate change as changes may cause shortage in water and other resources, which eventually would affect soil fertility and lead to pest and disease outbreaks. They imply that this climate crisis is highly expected to hit states of Malaysia where the poverty level is high especially in Kelantan, Terengganu, Perlis, Kedah and Perak. Recent study by Al-Amin et al. (2010) forecast that rice yields are expected to decline between 4.6 and 6.1% as a result of 1°C increase in temperature at the current CO_2 concentration level which could reduce rice productivity by 34.8% per hectare in 2060. They indicate that rise in temperature above 26°C will increase plant respiration and shorten the grain-filling periods. The periods become shorter as the plant growth rate increase and the growth duration decrease.

None of the above discussed studies especially those in Malaysia apply the Ricardian method in assessing the impact of climate change on agriculture. Actually, the application of Ricardian method has been primarily used

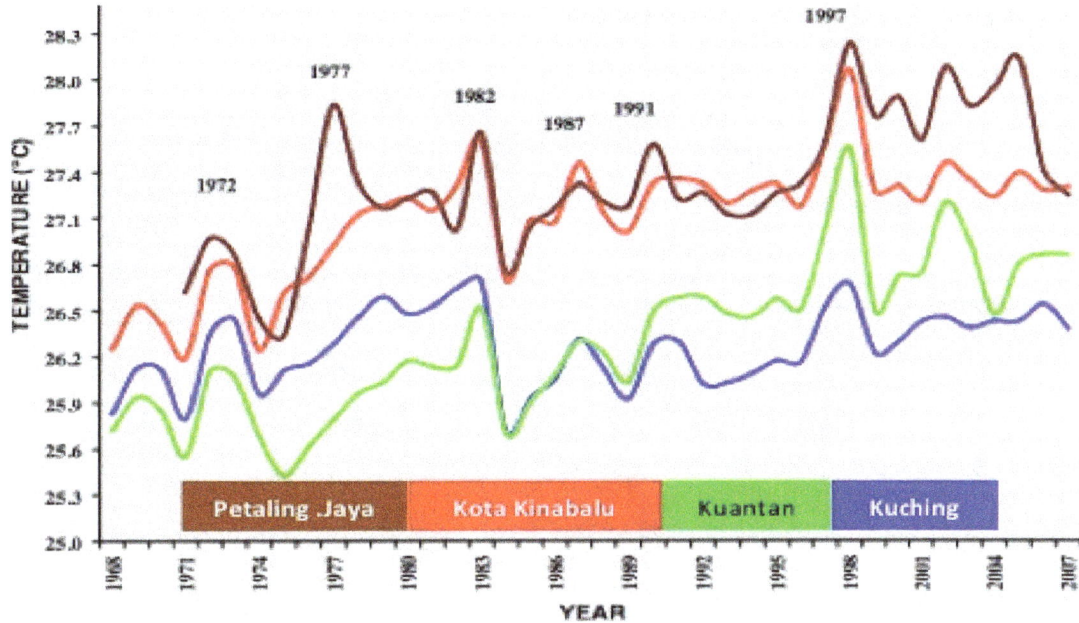

Figure 1. Annual mean temperature in Malaysia from 1968 to 2007.

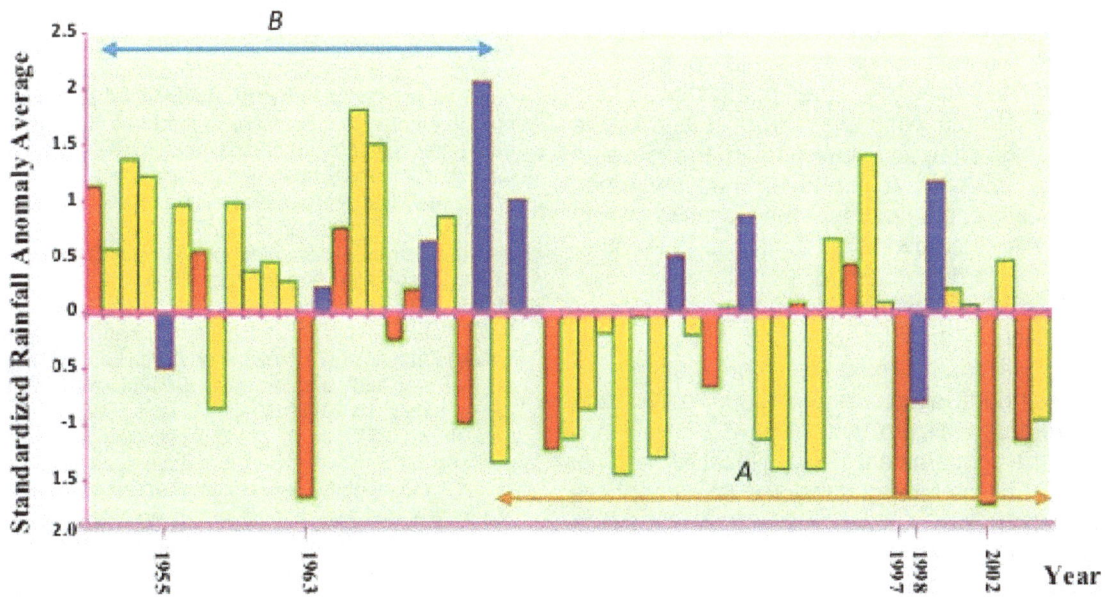

Figure 2. Standardized annual rainfall anomaly in Malaysia from 1951 to 2005. Source: MMD (2009).

in the US to predict the damages from climate change (Mendelsohn et al., 1994; Mendelsohn, 1996; Schelenker et al., 2004). There have also been a number of studies in India (Dinar et al., 1998; Kumar and Parikh, 2001), India and Brazil (Sanghi and Mendelsohn, 2008), Canada (Reinsborough, 2003), China (Liu et al., 2004), Sri Lanka (Seo et al., 2005; Kurukulasuriya and Ajwad, 2007), South African (Gbetibouo and Hassan, 2005; Behnin,

2008), Israel (Fleischer et al., 2008), Cameroon (Molua, 2008) and Nigeria (Ajetomobi et al., 2011). By and large, the literatures suggest that the impacts of climate change would be favourable to some cropping areas in America, eastern Mediterranean and some regions in China, but the impacts will be deleterious in tropical and semi-tropical countries, where most of them fall under the group of developing and low-income countries. Inevitably,

Table 2. Descriptive Statistics: Variables for net revenue regression model (Main Season and Off Season from 1999 - 2008).

Parameter	Paddy seasons in PM			
	Main season		Off season	
	Mean	SD	Mean	SD
Net revenue per ha (RM)	1218.21	681.10	1500.39	713.12
Rain (mm)	1682.94	649.52	1411.26	450.92
Temperature (°C)	27.14	0.3954	27.58	0.4539
Planted Area (ha)	24059.30	28631.46	24023.14	29007.84

Source: Author's simulation.

relevant adaptation strategies need to be taken as the negative impact could lead to insufficient food supplies in the country, especially in the case of 'staple' food crops such as paddy.

The main advantage offered by the Ricardian method is that, it allows a number of controlled variables to be included so that a measure of adaptation and adjustment strategy to mitigate the effects of climate change on farmers' productivity level can be taken into account. In this study, two different adaptation strategies were taken into account. In spite of its popularity, nevertheless the model has its own limitations. Mendelsohn (2007) revealed that the Ricardian approach could not measure CO_2 fertilization because they are relatively uniform across the world. He pinpoints this failure as one of the weaknesses in this analysis. However, in the case of Malaysian paddy, al-Amin et al. (2010) has confirmed that increase in CO_2 could not offset the negative impact brought by the temperature increase. Another shortcoming of Ricardian method is the use cross-sectional analysis. Since changes in climate are related to dynamic impacts, Kurukulasuriya and Ajwad (2007) suggest that panel data would be more reliable to measure inter temporal effects. In this study, panel data was used to overcome this problem.

Undeniably, the government has allocated a vast amount of budget in the form of subsidies and incentives to encourage production, but this may not be enough to retain the farmer's interest, if changes in future climate could substantially affect production and income. Furthermore, with competition by other agriculture crops such as palm oil and rubber that perhaps could offer better income, this may force farmers to leave the paddy cultivation and shift to the other lucrative agriculture crops. In fact, the grave long-term challenges spear-headed by changes in climate may dampen the effort of government in boosting rice production in Malaysia since climate is one of the primary determinants of paddy productivity. Consequently, the nation's food security will be under threat. Thus, this study attempts to estimate a relationship between climate variables and granary-level net revenue and project the impacts in a predicted future climate scenario with a view to find potential

strategy of adapting.

METHODOLOGY

Data and variable specifications

The granary and climate data were obtained from the Department of Agriculture and MMD (Table 2). Granary level data on net revenue requires data on paddy production in metric tons per hectare, total area planted per season, total farmers cost from planting to harvesting and the selling floor price of paddy (RM per ton) including the price subsidy. These data are seasonal data, which involves period of 1999 to 2008. The price subsidy from 1999 to 2008 is at a rate of RM248.10 per metric ton, which is actually given as an incentive by the government to encourage farmers to produce more. The data are in the form of seasonal granary level data (main season and off season), which covers farmers' activities starting from land preparation to harvesting.

Seasonal data on temperature and precipitation are based on the location of the weather station, not on granary. This climate data was taken from six stations namely Alor Star, Bayan Lepas, Ipoh, Kuala Terengganu Airport, Kota Bharu and Petaling Jaya. Alor Star station will represents climate for MADA; Bayan Lepas station for IADP Pulau Pinang; Ipoh station for Krian Sg Manik and Seberang Perak; Petaling Jaya station for Barat Laut Selangor; Kota Bharu station for KADA and Kemasin Semarak; and Kuala Terengganu Airport station for KETARA.

As shown in Table 2, the statistics indicate that the off-season period is generally warmer than main season due to lower level of rainfall and slightly higher temperature. Although more planted areas are devoted during the main season, however, the net revenue per hectare is slightly better during the off-season period. This suggests that the planted area or the size of the granary may hold a non-linear relationship (a u-shaped or inverted u-shaped) with farmers' profitability.

Net revenue

Based on the Ricardian model, theoretically the net revenue per hectare (NR/ha) for this study, which is the dependent variable in Equation 1 is defined as:

$$NR/ha_{it} = [P_{it} * Q_{it}(X_{it}, E_t) + S_{it} * Q_{it}(X_{it}, E_t)] - \Sigma C_{it}(Q_{it}, X_{it}, E_t)] / \Sigma L_{it} \quad (1)$$

where, S_{it}, P_{it} and Q_{it} are, respectively the price subsidy, selling price and total paddy production for granary i in period t. The symbol of t indicated the series of time for each variable, since this

Table 3. Average forecasted annual mean rain and temperature changes in the all granary areas for period 2020 to 2029, 2050 to 2059 and 2090 to 2099.

Annual mean rain	Temperature
2020 to 2029	
-13.4	+1.3
2050 to 2059	
+2.7	+1.9
2090 to 2099	
+10.3	+3.1

Rain is in % changes; temperature is in °C changes.
Source: Malaysian Meteorological Department (2009).

study applies pooled time series cross-section regression analysis instead of individual cross-sectional analysis. C( ) is the function of all purchased inputs (X) other than land, E is a vector of climate variables and L represents the planted area. Cost of fertilizers, which is subsidized by the government, is assumed to be borne by the farmers. Most of the paddy land in the granary areas is privately owned by the farmers. Some of them leased the land on a short-term basis to other farmer or private enterprise but the data pertaining on the number of land tenants are not available. Thus, cost of renting the land will not be included in this analysis. It is assumed that all farmers are landowners.

Temperature and rainfall

In this study, temperature and rainfall was used as a proxy to measure the variations in climate towards net revenue. The average temperature and total rainfall will be computed based on two cultivation seasons in Malaysia, namely the 'Main season' and 'Off-season' from 1999 to 2008. The main or off-season's period in one granary is not the same between other granaries. The main or off-season's period in one granary also varies from 1 year to the next. For example, the period for main or off-season in MADA is not the same as in other seven granary areas. MADA's main or off-season period varies from 1998 to 1999, 2000 and the subsequent years. However, normally paddy is planted in main season between the month of August and February of the following calendar year, while for off-season starts from the end of March until July in a year.

Most of the Ricardian method uses an average of 30 years climate data, while Kumar and Parikh (2001) computed 20 years average climate data. Instead of climate, this study however relies on seasonal average temperature and total rainfall computed from annual weather of 1999 to 2008. Same approach was applied by Kurukulasuriya and Ajwad (2007). In fact, year 1998 to 2007 was considered as a long-term average climatic in terms of temperature and precipitation in Malaysia (MMD, 2009). It is worth to note that decreasing seasonal rainfall trend for PM can be observed in 1998 to 2007 as compared to those in 1961 to 1990 and there is also no significant difference in temperature between period 1961 to 1990 and 1998 to 2007.

In order to examine a wide range of climate outcomes, the approach relies on the 'Scientific Report on Climate Change' published by MMD. As shown in Table 3, the average forecasted annual mean rain and temperature changes reflects climate scenario that might happen in north-west, north-east and central region in Malaysia within 2020 to 2029, 2050 to 2059 and 2090 to 2099. The north-west PM covers granary areas in MADA, Pulau

Pinang, Seberang Perak and Krian-Sungai Manik; Central PM represents Barat Laut Selangor and north-east PM represents Kemasin Semarak, Ketara and KADA. The simulation of annual temperature and rainfall in this report is based on ProvidingRegional Climates for Impacts Studies (PRECIS) simulation driven by HadCM3. PRECIS is the regional climate model (RCM), while HadCM3 is part of the Atmosphere-Ocean General Circulation Model (AOCGM). In practice, the AOCGM simulation model has been widely exerted by most of Ricardian method studies as in the paper by Reinsborough (2003), Liu et al. (2004), Seo et al. (2005), Kurukulasuriya and Ajwad (2007), Sanghi and Mendelsohn (2008), Fleischer et al. (2008), Kurukulasuriya and Mendelsohn (2008), and Ajetomobi et al. (2011).

Non-climatic variables

Total planted area was used as a controlled variable as number of studies have used farm's size or size of cropping area as a factor to measure net revenue gained through the concept of economy of scale (Kurukulasuriya and Ajwad, 2007; Kurukulasuriya and Mendelsohn, 2008; Molua, 2008; Fleischer et al., 2008; Ajetomobi et al., 2011). In addition, a dummy variable was used to represent the season cultivation period on farmers' profitability. It is worth to note that currently more planted areas are devoted to the main season due to the fact that the off-season period requires more water supplies for irrigation. However, farmers in some areas were unable to cultivate paddy during the off-season period due to lack of irrigation infrastructures. For instance, in Kemasin Semarak, the irrigation infrastructures only cover about 50%* of its area. However, in major granary areas such as MADA in Kedah, the size of planted area during the main season and off-season is about the same. The development of the irrigation infrastructure in some areas being implemented and in some is expected to complete in the Tenth Malaysia Plan period. These include the construction of flood mitigation, drainage and irrigation. The season's dummy variable on the other hand seeks to explain whether profitability could be explained by cultivation season.

Methods

To attain the above objectives, this study applies the Ricardian method (Mendelsohn et al., 1994). The Ricardian method is a cross-sectional approach to study agricultural production. Cross-sectional observations across different farm can reveal the climate sensitivity for each farm. The method assesses economic impacts of climatic changes as well as capturing farmers' adaptation response to climate changes. Derived from David Ricardo's (1772 - 1823), original observation made was that land value would reflect its net productivity;the principle is shown in Equation 2a:

$$LV = \Sigma\, P_i,\, Q_i\, (X\,,\, F\,,\, H\,,\, Z\,,\, G)\, -\, \Sigma\, P_x X \qquad \text{(2a)}$$

where LV is the value of land, P_i is the market price of crop i, X is a vector of purchased inputs (except land), F is a vector of climate variables, H is water flow, Z is a vector of soil variables, G is a vector of socio-economic variables and P_x is a vector of input prices. The original Ricardian studies [for example, in US by Mendelsohn et al. (1994) and Canada by Reinsborough (2003)] estimate the impact of climate change on land value. However, in Malaysia, data pertaining land value is unavailable due to the absence of well functioning land markets. In this study, the granary level net revenue was used to estimate the climate response

---

* From personal interview with Mr. Mohd Suhaimi Mohd Rashid, Ministry of Agriculture's Officer.

Table 4. Determinants of net granary revenue per hectare.

Net revenue per hectare	
Variable	Coefficient
Rain	-0.6809**
Squared rain	0.000132*
Temperature	11759.30*
Squared temperature	-221.2114*
Planted area	-0.023389
Squared planted area	1.59E-06
Dummy cultivation season	-368.9814**
Constant	-155182.5*
N	160
R^2	0.61

**, *, Significant at 5 and 10%, respectively.

function and reflect the impact of climate change on the eight granary areas namely MADA, KADA, KETARA, IADP Pulau Pinang, Krian Sg. Manik, Barat Laut Selangor, Seberang Perak and Kemasin Semarak. Theoretically, the Ricardian model relies on a quadratic formulation of climate:

$$NR/ha_i = \beta_0 + \Sigma (\beta_1 T_{it} + \beta_2 T_{it}^2 + \beta_3 R_{it} + \beta_4 R_{it}^2) + \Sigma \beta_s L_{it} \mu_{it} \quad (2b)$$

where, NR/ha_{it}, Net revenue per hectare in period t for granary I; β_0 is intercept term; $\beta_1, \ldots \beta_s$, regression coefficients that measure the responsiveness of the net revenue per hectare in period t to a particular variable; T_{it} and R_{it}, vector of climate variables including temperature and rainfall in period t for granary i; $T_{it}^2 \& R_{it}^2$, vector of climate variables squared including temperature and rainfall in period t for granary I; L_{it}, other relevant explanatory (*non-climatic*) variables in period t for granary I; μ_{it}, error term.

When the quadratic term is positive, the net revenue function is u-shaped and when negative, the function is inverted u-shaped. It is expected that granary level net revenue per hectare hold a u-shaped relationship with temperature. The same goes to the size of the granary, while the main season is expected to have a positive relationship with net revenue per hectare. To estimate the above model, pooled time series cross-section analysis is used based on seasonal data over the period of 1999 to 2008, which cover all the eight granary areas in Malaysia. The model attempts to discover one principal hypothesis, that is, whether paddy farmers' net revenue per hectare is sensitive to climate. To test the hypothesis, the net revenue per hectare is regressed with climate and two controlled variables, which involve planted area and planting season. To analyze the hypothesis, a non-linear (quadratic) model has been constructed for ease of interpretation. Figures from Table 2 implied that the main season might have a negative relationship with farmers' net revenue. Hence, by employing a dummy parameter for 'cultivation season' (value of 1 for the main season and 0 if otherwise), this model attempts to discover the impact of cultivation season on farmers' net revenue.

In the initial runs, different fixed effect regressions are tested. The variables that fitted the model best are the one with cross-section fixed effects. Therefore, the results presented in this paper have eliminated the variable bias arising from omitted factors that vary across granary but were constant over time within a granary. Such variables may include geographical characteristics such as soil type, elevation and many more that are different across granaries but constant over time. Since the net revenue was

defined in season wise, thus, only one definition for climate can be made. This definition is based on data for planting and harvesting time period of granaries for each season in each year.

The climate variables coefficient from Equation 2b were used to assess the projected marginal impact of changes in one climate variable on granary level net revenue, which is evaluated based on following equations:

$$E [\Delta NR/ha /\Delta T_i] = \beta_{1,it} + 2 * \beta_{2,it} * E [T_i] \quad (3a)$$

$$E [\Delta NR/ha /\Delta R_i] = \beta_{3,it} + 2 * \beta_{4,it} * E [R_i] \quad (3b)$$

Finally, the above equations were used to simulate the impact of future climate change scenarios (Table 3) on net revenue function. The only variables that are subject to change were the climate variables, while other factors will remain unchanged. Obviously, this will not be the case over time as scale of economy, price and cost are subject to change over time which has tremendous impacts on future granary net revenue. Therefore, the purpose of this simulation is only to examine the impacts of climate in future, rather than to predict the future as such.

RESULTS AND DISCUSSION

Table 4 presents the results of the model. This model was estimated using generalized least square. Similar to Fleischer et al. (2008), the small value of R^2 is in this study as compared to the results by Mendelsohn et al. (1994) are due to three reasons: (1) land values are more stable compared to seasonal granary net revenue, which tends to fluctuate; (2) this study data comprises only eight granaries as observation, whereas Mendelsohn et al. (1994) observation are based on country averages and (3) Malaysia is small country relative to the US where the range of average daily temperature is only marginal ranging from 26 to 28°C compared to US which is from -12 to +34°C.

More planted area in the main season contributes towards a negative relationship towards farmers' net revenue. However, the positive sign for 'planted area' quadratic term implies a u-shaped relationship of granary size with net revenue, which suggests that the eight granaries are operating at increasing returns to scale. A similar relationship is observed by Kurukulasuriya and Ajwad (2007) and Kurukulasuriya and Mendelsohn (2008). Nonetheless, the 'planted area', which is considered in the regression to account for economies of scale, has an insignificant repercussion on farmer's net revenue. Hence, the other reason for the higher mean net revenue during the off-season period could be due to lesser cost of cultivation in this period. In the off-season period, some of the input cost could probably be absorbed or acquired during the main season period. For instance, cost of additional fertilizers, and costs for controlling weeds, insects and disease.

All the climate and climate squared variables are significant, thus, implying that climate has a non-linear effect on net revenues. The positive sign of squared rain and the negative sign of squared temperature indicate

Table 5. Marginal effects of temperature and precipitation on Net revenue per hectare

Marginal changes in climate variable	Net revenue per hectare
1 °C Increase in temperature	RM-442.42
1% Drop in rainfall	RM -0.01

Source: Author's simulation.

that the net revenue function is u-shaped (convex) in rainfall and inverted u-shaped (concave) in temperature. This means that temperature or rainfall affects the net revenue positively or negatively up to a certain level, above or below it causes damage to the crops. This is supported by the work carried out by Ajetomobi et al. (2011) that asserts second order temperature coefficient for irrigated rice farms in Nigeria is negative. The work also reveals that most of the agronomic research found that crops are consistently exhibited by inverted u-shaped relationship with annual temperature, although the maximum point varies with the crop. Such findings also correspond to finding by Baharuddin (2007) that claims rice grain yield might decline by 9 to 10% for each 1 °C rise. Kurukulasuriya and Ajwad (2007) on the other hand found that rainfall has a concave effect towards net revenue per hectare.

Drop in rainfall will affect crops that need wet conditions such as paddy. Given the small coefficient's value for rain, thus the marginal impact of rain is expected to be far smaller than temperature. The presence of irrigation water may partly affect the intense relationship of rainfall towards paddy. To some extent, irrigation water could also buffer the crop from being intensely influenced by temperature (Ajetomobi et al., 2011). Thus, the exclusion of irrigation water in this analysis could actually underrate the impact of rainfall and overrate the impact of temperature on paddy. Study by Fleischer et al. (2008) implied the relevancy to include the farm water inflow to overcome overestimation problems particularly for a crop that depends heavily on irrigation water.

Table 5 shows the marginal impact analysis result, which assesses the effect of an infinitesimal change in temperature and rainfall. As expected, the marginal effects of precipitation on farmers' net revenue were much smaller and not even close to 1% from temperature's effect. Drop in total rainfall in each season by 1% is expected to incur a loss at an average of RM 0.01% per hectare. On the other hand, 1 °C increases in average temperature for each season could result in a huge loss. The net revenue could fall at an average of RM 442.42 per hectare as a result of 1 °C increases in average temperature.

Forecasting

The simulation results for net revenue model are shown

in Table 6. It was estimated using Equations 3a and 3b, based on the climate scenario in Table 5. The expected losses per hectare and average net revenue per hectare are estimated based on total effects of changes in temperature and rainfall. The simulation effects presented marked variation in the granaries' net revenues during the main season and off-season. Increase in temperature during the main season period is less harmful as compared to the off-season period. The average losses associated as a result of increase in temperature in the off-season period compared to main season periods are 17.37% more. A 13.4% reduction in rain during the 2020 to 2029 also affects the off-season more. However, in both seasons, increase in rain for the period of 2050 to 2059 and 2090 to 2099 are expected to lessen the losses experienced earlier in 2020 to 2029. Nevertheless, the reductions are slightly higher during the main season period. Since the mean net revenue during the off-season period is slightly higher than the main season period, thus, the average net revenue per hectare in the off-season period is expected to be greater. In the period of 2020 to 2029 and 2050 to 2059, average net revenue per hectare for off-season period are 22.16 and 67.61% more respectively, as compared to main season. In the period of 2090 to 2099, both seasons recorded negative net revenue per hectare.

In contradiction with the study by Al-Amin et al. (2010) which found that, a variation in temperature of 0.3 to 1.4 °C and rainfall ±32% reduces incomes up to RM1280.60 per hectare per year, this study predicts that an increase of 1.3 °C in temperature and drop in rainfall by 13.4% leads to a total loss of RM 1841.73, which is 44% higher. However, both empirical results provide conclusive evidence that the paddy sector in Malaysia is highly responsive to climate change.

Indeed the results of this study suggest that paddy farmer's net revenue per hectare is sensitive to marginal change in climate variables in both seasons and the impact could be more during the off-season period, which is more hot and dry.

CONCLUSIONS AND RECOMMENDATIONS

This study attempts to analyze the impact of climate change on Malaysian paddy sector, focusing from the eight designated granary areas in Malaysia, namely MADA, KADA, KETARA, IADP Pulau Pinang, Krian Sg.

Table 6. Forecasts of average net profits per hectare based on future climate scenario predicted by MMD.

Climate variable	Net revenue per ha			
	Main season		Off- season	
	(RM)	(%)	(RM)	(%)
2020 to 2029				
Temperature	-823.20	-67.57	-1017.87	-67.84
Rainfall	-0.30	-0.02	-0.36	-0.02
Expected loss/ha	-823.50	-67.59	-1018.23	-67.86
Expected average NR/ha	394.71	32.41	482.16	32.13
2050 to 2059				
Temperature	-1088.66	-89.37	-1283.32	85.53
Rainfall	-0.22	-0.01	-0.30	-0.02
Expected loss/ha	-1088.88	-89.38	-1283.62	-85.55
Expected average NR/ha	129.33	10.62	216.77	14.45
2090 to 2099				
Temperature	-1619.57	-132.94	-1812.23	-120.78
Rainfall	-0.19	-0.02	-0.27	-0.02
Expected loss/ha	-1619.76	-132.96	-1812.50	120.80
Expected average NR/ha	-401.56	-32.96	-312.11	-20.80

Source: Author's simulation.

Manik, Barat Laut Selangor, Seberang Perak and Kemasin Semarak. It uses secondary climate, production and cost data to conduct the Ricardian cross-sectional approach to measure the relationship between the paddy farmers' net revenue with climate variables, planted area and cultivation season. The result of the study shows that climate affects paddy farmers' net revenue in Malaysia. The marginal impact of temperature on net revenue shows that if the temperature increases by 1°C, the net paddy farmers' revenue falls by RM442.42 per ha, whereas if the rainfall reduces by 1%, the net revenue drops by RM0.01 per ha.

Using simulated scenarios from MMD for the period of 2020 to 2029, 2050 to 2059 and 2090 to 2099; this study forecasts the impact of changes in climate towards paddy farmers' net revenue. The climate change scenarios' analysis indicates that scenarios of increasing temperature with decreasing rainfall will cause more damage to paddy sector in Malaysia. However, the scenarios of increasing temperature and increasing rainfall will not be preferable because the positive impact of rainfall will not be able to compensate for the adverse effect of the warming. Warming is expected to be harmful to Malaysian paddy sector but increase in rainfall would only be slightly beneficial. Therefore, it can be concluded that rainfall gains could not outweigh temperature losses.

In general, the anticipated effects would indeed impede government's attempt to improve rice self-sufficiency level and reduce reliance on imported rice. This study however managed to discover empirically two possible adaptation strategies. Expanding the irrigation infrastructure is vital as the off-season period is found to be more profitable as compared to main season. Currently, more planted areas are devoted to main season due to the fact that off-season period requires more water supplies from irrigation but some of the granary areas are not well equipped with irrigation infrastructure. If government decides to rely on the existing eight granaries, thus, the development of irrigation infrastructure in those areas is the best alternative available to increase total planted area. Perhaps, farmers are able to reduce cost of planting if they could plant more paddy during the off-season period, as the total cost of production will be spread over as output increases.

Finally, by expanding the eight granaries planted area, cost of planting in both seasons could be reduced due to an opportunity of enjoying an increasing return to scale. Although the result of 'planted area' is found insignificant, however, the u-shaped relationship provides another option that can be considered by the government. Expanding may not be in the form of adding the total acreage of the current granary areas, but perhaps could be in the form of finding a means to allow farmers to cultivate paddy three times a year rather than twice a year. Paddy farmers in Myanmar for instance, are able to cultivate paddy three times per year.

Generally, the agriculture sector in many countries is vulnerable to experience a negative impact from changes in climate in the future. However, magnitude of the impacts depends on the location, level of development, technological advancement and the institutional setting in the countries. Farmers or government could mitigate such impact through channel of adaptation strategies that could range from crops switching to farmland converting. Paddy farmers in Malaysia are also expected to experience the deleterious global effect of climate change. In this study, the adaptation strategies that can be undertaken are believed to come from an improvement in irrigation efficiency, expanding the irrigation infrastructure and area under cultivation.

LIMITATION OF THE STUDY

This study only represents the eight major granary areas in PM, excluding paddy farms outside the granary areas, as the panel data for these farms are not sufficient. The panel data observation regarding the farmers' social information is also unavailable, hence, deterring this study to observe such factor. Since the farm-level panel data is not available, thus this research is done based on the secondary data. This might cause biases in the analysis, as: (1) yield variability at the farm-level is usually bigger and (2) yield and weather relationships may not be the same from aggregated to farm level analysis. Even though, the farm-level data might be more reliable, however the collection of primary data from eight different granary areas will need huge funding and human resources. The primary data could also help future researcher to incorporate other non-climatic varia-bles such as socio-economic factors. Future research could also further expand this research model by considering paddy fields outside the eight granary areas.

In addition to the Ricardian method, there are also other alternative approaches that can be applied such as Bio-Economic model (Finger et al., 2011) or regression model using repeated cross-sectional method (Deschenes and Greenstone, 2007). An econometric model to analyze the impacts of climate change on the mean, variance and covariance of crop yields can also be deployed especially for research that intends to focus its analysis on the implications towards allocation of multiple crops (Isik and Devadoss, 2006). Apart from changes in mean weather characteristics, the construction of climate change scenarios can also be represented by the changes in climate variability using the combination of both global climate model (GCM) and the stochastic weather generator. These tools can be used to simulate farm-specific daily weather data and have been employed in studies applying the Bio-Economic model.

ACKNOWLEDGEMENTS

The authors would like to thank MMD Malaysia and Department of Agriculture Malaysia for providing the data.

REFERENCES

Ajetomobi A, Abiodun A, Hassan R (2011). Impacts of climate change on rice agriculture in Nigeria. Trop. Subtrop. Agroecosyst. 14:1-10.

Al-Amin AQ, Azam MN, Yeasmin M, Fatimah K (2010). Policy challenges towards potential climate change impacts: In search of agro-climate stability. Sci. Res. Essays 5(18):2681-2685.

Amien I, Rejekiningrum P, Pramudia A, Susanti E (1996). Effects of interannual climate variability and climate change on rice yield in Java, Indonesia. Water, Air. Soil Pollut. 92:29-39.

Baharuddin MK (2007). Climate change: Its effects on the agricultural sector in Malaysia, in National Seminar on Socio-Economic Impacts of Extreme Weather and Climate Change. June, Malaysia. pp. 21-22.

Behnin JKA (2008). South African crop farming and climate change: An economic assessment of impacts. Glob. Environ. Change 18:666-678.

Chamhuri S, Alam M, Murad W, Al-Amin AQ (2009). Climate change, agricultural sustainability, food security and poverty in Malaysia. Int. Rev. Bus. Res. 5(6):309-321.

Department of Agriculture (2009a).Paddy production survey reportMalaysia: Main Season 2009.Department of Agriculture Peninsular, Putrajaya, Malaysia.

Department of Agriculture (2009b).Paddy statistics of Malaysia 2007.Department of Agriculture Peninsular, Putrajaya, Malaysia.

Deschenes O, Greenstone M (2007). The economics impacts of climate change: Evidence from agricultural output and random fluctuations in weather. Am.Econ. Rev. 97:354-385.

Deutsch A, Hidayat T (2011). Indonesia battles rice shortfall. [http://www.ft.com/intl/cms/s/0/7c2ef448-2efd-11e0-88ec-00144feabdc0.html] site visited on 14/01/2011.

Dinar A, Mendelsohn R, Evenson R, Parikh J, Sanghi A, Kumar K, McKinsey J, Lonergan S (1998). Measuring the impact of climate change on Indian agriculture. World Bank Tech. Pap. 402:Washington.

Elenita CD, Ema DS (2005).Public sector intervention in the rice industry in Malaysia. In state intervention in the rice sector in selected countries: Implications for the Philippines. SEARICE and Rice Watch Action Network, Quezon City.

FAO (2011). Despites Asian floods FAO forecasts record rice harvests in 2011. [http://www.fao.org/world/regional/rap/home/news/detail/en/?news_uid=94541] site visited on 23/10/2011.

Finger R, Hediger W, Schmid S (2011). Irrigation as adaptation strategy to climate change - a biophysical and economic appraisal for Swiss maize prod. Climatic Change 105:509-528

Fleischer A, Lichtman I, Mendelsohn R (2008). Climate change, irrigation, and Israeli agriculture: Will warming be harmful. Ecol. Econ. 65:508-515.

Gbetibouo GA, Hassan RM (2005). Economic impact of climate change on major South African field crops:A Ricardian approach. Glob. Planet Change 47:143-152.

Isik M, Devadoss S (2006). An analysis of the impact of climate change on crop yields and yield variability. Appl. Econ. 38:835-844

IFPRI (2009). Impact of climate change on agriculture: Factsheet on Asia. [http://www.ifpri.org/publication/impact-climate-change-agriculture-factsheet-asia] site visited on 20/01/2011.

Kurukulasuriya P, Ajwad I (2007). Estimating the impact of climate change on small holders: A case study on the agriculture sector in Sri Lanka. Climatic Change 81:39-59.

Kurukulasuriya P, Mendelsohn R (2008). A Ricardian analysis of the impact of climate change on African cropland. Afr. J. Agric. Resour. Econo. 2(1):1-23.

Kumar K, Parikh J (2001). Indian agriculture and climate sensitivity. Glob. Planet Change 11:147-152.

Liu H, Li X, Fisher G, Sun L (2004). Study on the impacts of climate change on China's Agriculture. Climate Change 65:125-148.

McCarl BA, Adams RM, Hurd BH (2001).Global climate change and its impact on agriculture. [http://agecon2.tamu.edu/people/faculty/mccarl-bruce/papers/879.pdf] site visited on 19/01/2011.

Mendelsohn R (2007). Measuring climate impacts with cross sectional analysis. Climatic Change 45:1-7

Mendelsohn R, Nordhaus W, Shaw D (1994). The impact of global warming on agriculture: A Ricardian analysis.Am. Econ. Rev. 84(4):753-771.

Mendelsohn R, Nordhaus W, Shaw D (1996). Climate impacts on aggregate farm values: Accounting for adaptation.Agric. For. Meteorol. 80:55-67.

MMD (2009).Climate change scenarios for Malaysia 2001–2099.Malaysian Meteorological Department Scientific Report, Petaling Jaya.

Molua EL (2008). Turning up the heat on African agriculture: The impact of climate change on Cameroon's agriculture. Afr. J. Agric. Resour. Econ. 2(1):45-64.

Nguyen NV (2005). Global climate changes and rice foods security.FAO, Rome. pp.24-30.

Reinsborough MJ (2003). A Ricardian model of climate change in Canada.Canadian J.Econ. 36:21-40.

Sanghi A, Mendelsohn R (2008). The impacts of global on farmers in Brazil and India.Global Environ. Change 18:655-665.

Schelenker W, Hanemann WM, Fisher AC (2004). Will U.S. agriculture really benefit from global warming? Accounting for irrigation in the hedonic approach. Am. Econ. Rev. 95(1):395-406.

Seo SN, Mendelsohn R (2005). Climate change and agriculture in Sri Lanka: A Ricardian valuation. Environ. Develop. Econ. 10:581-596.

Thanyarat D (2011). Thai flooding kills hundreds. [http://news.theage.com.au/breaking-news-world/thai-flooding-kills-hunderds-20111012-lljkwl.hrml]site visited on 18/12/2011.

Determinants of access to agricultural credit among crop farmers in a farming community of Nasarawa State, Nigeria

K. I. Etonihu[1], S. A. Rahman[1] and S. Usman[2]

[1]Faculty of Agriculture, Nasarawa State University, Keffi, Nigeria.
[2]National Agricultural Extension and Research Liaison Services, Ahmadu Bello University, Zaria, Nigeria.

Farmers' limited access to agricultural credit facilities is one of the major factors responsible for the declining agricultural productivity in Nigeria. Hence, this study aims to identify determinants of access to agricultural credit among smallholder farmers in Doma Local Government Area (LGA) of Nasarawa State, Nigeria. The data were obtained from 125 farmers by administered structured questionnaire in 2008 production season through a two stage random sampling technique. Descriptive statistics and stepwise linear regression model were used to analyze the data. The study observed that education, distance to source of credit and types of credit source were significant factors affecting farmers' accessibility to agricultural credit in the study area. Hence, government policy that intends to improve the accessibility to agricultural credit facilities should create enabling environment to ease farmers' access to education and credit facilities.

Key words: Agricultural credit, farming community, Nasarawa State.

INTRODUCTION

Agricultural growth in Nigeria is increasingly recognized to be central to sustainable economic development. The sector plays a very significant role in addressing food insecurity, poverty alleviation and human development challenges. However, in more recent years, there has been a marked deterioration in the productivity of Nigeria's agriculture (Amaza and Maurice, 2005). Many reasons have been advanced for the declining agricultural productivity in Nigeria. One of the factors attributed to the declining productivity of the sector is farmers' limited access to credit facilities (Nwaru, 2004; Manyong et al., 2005). According to Alfred (2005), acquisition and utilization of credit for agricultural purposes promote productivity and consequently improved food security status of a community. Increase

productivity depends on adoption and technical efficiency of improved farming technologies (Obwona, 2002). In an effort to increase adoption rate among farmer, their purchasing power to acquire modern agricultural technologies should be improved. Most of the Nigerian farmers are smallholder trapped in vicious cycle of poverty. It has been argued that when agricultural credits are made accessible to farmers it will go a long way in breaking this cycle of poverty and liberating the farmers to improve their adoption of modern farm technologies which could enhance productivity and farmers' income. Adebayo and Adeola (2008) observed that agricultural credit enhances productivity and promotes standard of living by breaking vicious cycle of poverty of the resource poor farmers. Similarly, Nwaru et al. (2006) observed that credit facilitates adoption of innovations leading to increased farm productivity and income, encourages capital formation and improves marketing efficiency. There are two major sources of agricultural credit (that is, formal and informal sources). In the formal credit, institutions provide intermediation between depositors and lenders charge relatively low rates of interest that usually are government subsidized. In informal credit markets money is lent by private individuals. The informal sources of credit to smallholder farmers as identified in the study area were family/friends, money lenders, produce buyers and farmers' cooperatives, while the formal sources of credit were Nigerian Agricultural Co-operative and Rural Development Bank (NACRDB) and commercial bank. The Nigeria Agricultural and Co-operative Bank Limited was established in 1972. The NACRDB evolved recently from the merger of the Nigerian Agricultural and Co-operative Bank with the People's Bank. The bank's broad mandate encompasses savings mobilization and the timely delivery of affordable credit to meet the funding requests of the teeming Nigerian population in the agricultural sector of the National economy. But this has not been the case with smallholder farmers as the problem of accessibility hinders them from reaching formal financial institutions for production loans (Etonihu, 2010). This study was, therefore, conducted primarily to identify determinants of access to agricultural credit to provide information for effective policy intervention that will improve farmers' purchasing power to acquire modern agricultural technologies.

MATERIALS AND METHODS

The study was conducted in Doma Local Government Area (LGA), located in the southern zone of Nasarawa State in Nigeria. The LGA lies between latitude 0.9°33' north of the equator and longitude 0.9°32' east with distinct wet (March to October) and dry (November to February) seasons (Akwe, 2008). The population of the LGA is 138,991 with annual population growth rate of 3.2% (NPC, 2006). The projected population figure for 2010 of the LGA is 156,783. The average annual rainfall is approximately 1500 mm with the mean daily maximum and minimum temperature of 36.8 and 22.7°C, respectively (Akwe, 2008). The economic activity in the area is largely agrarian with majority of the people being engaged in cultivating crops such as yam, sesame, rice, cassava, sorghum, millet, cowpea and groundnut.

A two stage random sampling technique was used to obtain a sample of 125 crop farmers for the study. In the first stage, six (6) wards were selected randomly out of 10 wards in Doma LGA. In the second stage, 10% of the sampling frames of the total smallholder crop farmers were randomly selected in each of the selected wards. Thus, a total of 125 crop farmers constitute the working population for the survey. Primary data were collected based on 2008/2009 cropping season from the sampled farmers using structured questionnaire and interview schedule. The information collected from the crop farmers include: socio-economic characteristics of the farmers, types of agricultural credit available to them, credit needed and credit obtained by the farmers.

Descriptive statistics (frequency and percentage) and stepwise linear regression were used to analyze the data. In deciding the best set of explanatory variables for a regression model, researchers often follow the method of stepwise regression (Gujarati, 2007). In this method, one proceeds by introducing the explanatory variables one at a time (Stepwise forward regression).

The stepwise regression model in this study involved nine linear regression equations based on the number of independent variables in the model. The change in coefficient of multiple determinations (R^2-change) as a result of stepwise inclusion of factors was used to measure the proportion of variation in the accessibility to credit as induced by each factor included in the model.

The regression model in its general form is expressed as:

$$Y = f(X_1, X_2, X_3, X_4, X_5, X_6, X_7, X_8, X_9, U) \tag{1}$$

Where, Y = Rate of accessibility to credit, which is measured in percentage as:

$$Y = \frac{Amount\ of\ credit\ obtained}{Amount\ of\ credit\ applied} \times 100 \tag{2}$$

X_1 = Education (years); X_2 = Marital status measured as dummy (married= 1, single = 0); X_3 = Farming experience (years); X_4 = Farm size (hectare); X_5 = Household size (actual number); X_6 = Income per annum from crop production (₦); X_7 = types of credit source measured as dummy formal = 1 and informal = 0; X_8 = Distance to source of credit (km); X_9 = Experience in cooperatives (number of years spent in cooperative); U = Error term

RESULTS AND DISCUSSION

Descriptive statistics of the socio-economic characteristics of the farmers

The summary statistics of the variables obtained from the farmers are reported in Table 1. The study observed that 36% of the farmers fell within the age range of 41 to 50 years as an indication of majority being in the active and productive age group with a mean age is 46 years. Only 14% of the farmers were above 60 years of age. Only 10% of them were not married. The study also indicated that 67% of the farmers had over 10 years of farming experience and had farm size within the range of 2 to 4 ha with a mean of 2.13 ha implying that the farmers were smallholders with a mean farming experience of 14 years. The household size of the farmers for over 88% of them ranged from 6 to 20. The average household size was 13 persons implying that there was large source of family labour and high food demand in most of the farm households. The active mean age and years of experience can influence adoption of improved production practices, which invariably requires credit.

Determinants of access to agricultural credit among the farmers

Stepwise linear regression analysis was done to determine the relationship between socio-economic characteristics of the farmers and their rate of

Table 1. Distribution of the farmers according to their socio-economic characteristics.

Variable	Frequency	Percentage
Age (years)		
21 - 30	7	6
31 - 40	25	20
41 - 50	45	36
51- 60	30	24
61 - 70	18	14
Total	125	100
Education		
Primary/Qur'anic	64	51
Junior secondary school	13	10
Senior secondary school	27	22
Tertiary education	21	17
Total	125	100
Marital status		
Married	113	90
Single	12	10
Total	125	100
Farming experience (years)		
1 - 10	41	33
11 - 20	44	35
21 - 30	24	19
31 - 40	16	13
Total	125	100
Farm size (ha)		
<2	13	10
2 - 4	84	67
4 - 7	21	17
8 - 10	7	6
Total	125	100
Number of years in cooperative		
≤5	33	43
6 - 10	25	32
11 - 15	16	21
16 - 20	3	4
Total	125	100
Household size		
1 - 5	10	8
6 - 10	61	49
11 - 20	49	39
21 - 30	5	4
Total	125	100
Annual income (N)		
<100,000	48	38.4
100,000 - 150,000	31	24.8
151,000 - 200,000	34	27.2
>200,000	12	9.6
Total	125	100

Source: Field survey (2010).

Table 2. Estimated stepwise linear regression equations for factors determining accessibility of farmers to agricultural credit.

			Regression coefficient						Constant	F- value	R^2	R^2 change
X_1	X_2	X_3	X_4	X_5	X_6	X_7	X_8	X_9				
1.118									45.013	4.461	0.076	-
(2.112)**									(8.540)*			
1.128	3.682								41.506	2.318	0.080	0.004
(2.114)**	(0.488)								(4.645)*			
1.193	3.582	7.874E-02							39.506	1.573	0.083	0.003
(2.082)**	(0.471)	(0.394)							(3.823)*			
1.223	3.625	0.108	0.235						39.535	1.168	0.084	0.003
(2.082)**	(0.472)	(0.433)	(0.199)						(8.790)*			
1.236	4.032	0.354	0.292	-0.922					42.967	1.135	0.102	0.018
(2.104)**	(0.524)	(1.010)	(0.247)	(-1.001)					(3.913)*			
1.257	4.503	0.434	0.506	-1.030	-2.76E-05				42.759	0.993	0.108	0.006
(2.122)**	(0.578)	(1.151)	(0.283)	(-1.090)	(-0.599)				(3.867)*			
1.185	3.635	0.399	0.307	-0.754	-2.14E-05	-28.369			70.120	1.404	0.170	0.062
(2.048)**	(0.478)	(1.034)	(0.176)	(-0.809)	(-0.476)	(-1.887)***			(3.881)			
1.150	4.451	0.389	0.832	-0.829	-2.44E-05	-30.569	-9.410E-02		68.048	1.937	0.248	0.078
(2.065)**	(0.608)	(1.097)	(0.491)	(-0.924)	(-0.562)	(-2.109)**	(-2.208)**		(3.910)*			
1.204	4.007	0.418	0.628	-1.134	-2.87E-05	-30.777	-9.155E-02	0.407	69.850	1.793	0.260	0.012
(2.142)**	(0.544)	(1.172)	(0.366)	(-1.171)	(-0.656)	(-2.117)**	(-2.136)**	(0.854)	(3.973)*			

Source: Field survey (2010). Figures in parenthesis are t-values. *, ** and ***significant at 1, 5 and 10%, respectively.

accessibility to agricultural credit. The result of the stepwise regression analysis is presented in Table 2 and reveals that about 26% of the variation in rate of farmers' accessibility to agricultural credit was explained by all the nine independent variables included in the regression model. Of all the explanatory variables included in the model, only education, type of credit sources and distance to credit source were found as significant factors affecting individual rate of accessibility to agricultural credit in the study area.

Education was positively and significantly related to the rate of credit accessibility at 5% level. Ideally, educated farmers are likely to understand the benefits of credit in modern production and comprehend extension information on sources and utilization of credit. This is in line with the findings of Ozowa (1995) who reported that the literacy level promote the understanding of extension activities among farmers in the rural areas. As expected, distance to type of credit source shows a negative relationship with the rate of credit accessibility and significant at 5% level. This implies that the farther the source of credit from the farmer's home, the more it becomes difficult for the farmer to access agricultural credit. The coefficient for the types of credit source was also negative and significant at 5% level. This implies that there was significant difference in amount of credit available to the farmers between the formal and informal sources. The credit services from the informal sources targeted to the smallholder farmers in the rural areas are readily and closely available and required less paper work without collaterals needed for loans advancement.

Conclusion

The role of agricultural credit in the development of agricultural sector cannot be overemphasized. Credit enhances farmers' purchasing power to enable them acquires modern technologies for their farm production. Access to the credit seems to be limited among smallholder farmers due to certain constraints. This study has identified education, distance to credit source and types of credit source as major factors that influenced farmers' access to agricultural credit. This study therefore, recommends that:

a) Enabling environment should be created to improve farmers' accessibility to educational facilities. This can be achieved through mass education for rural dwellers and functional extension activities.
b) Most of the formal sources of credit (for example, commercial banks) should be encouraged to open branches in rural areas and promote rural micro finances to make credit easy for farmers to obtain.

REFERENCES

Adebayo OO, Adeola RG (2008). Sources and Uses of Agric Credit by Small Scale Farmers in Surulere LGA of Oyo State. Anthropologies. 10(4):313-314.

Akwe A (2008). Prevalence of Embryonic Losses due to Slaughtering of Cow, Ewes and Does at Doma and Lafia Abattoirs, Unpublished B. Agric Project, Faculty of Agriculture, Nasarawa State University, Keffi, Nigeria.

Alfred SDY (2005). Effect of extension information on credit utilization in a democratic and deregulated economy by farmers in Ondo State of Nigeria. J. Agric. Exten. 8:135-140.

Amaza PS, Maurice DC (2005). Identification of Factors that Influence Technical Efficiency in Rice-Based Production Systems in Nigeria. Paper presented at Workshop on Policies and Strategies for Promoting Rice Production and Food Security in Sub-Saharan Africa:- November 2005, Cotonou (Benin), pp. 7-9.

Etonihu IK (2010). Farmers' Accessibility to Agricultural Credit for Crop Production in Doma Local Government Area of Nasarawa State, Nigeria. Unpublished B. Agric Project, Faculty of Agriculture, Nasarawa State University, Keffi, Nigeria.

Gujarati DN (2007). Basic econometrics fourth edition. Mc Graw-Hill, New Delhi p. 354.

Manyong VM, Ikpi A, Olayemi JK, Yusuf SA, Omonona BT, Okoruwa V, Idachaba FS (2005). Agriculture in Nigeria: Identifying Opportunities for Increased Commercialization and Investment. IITA, Ibadan, Nigeria. p. 159.

National Population Commission (NPC) (2006). Human population figures of 2006 census in Nigeria.

Nwaru JC (2004). Rural Credit Market and Resource Use in Arable Crop Production in Imo State of Nigeria, PhD Dissertation Micheal Okpara University of Agriculture, Umudike, Abia State, Nigeria.

Nwaru JC, Onyenweaku C E,Nwosu AC (2006). Relative Technical Efficiency of credit and Non-credit User Crop Farmers. Afr. Crop Sci. J. 14(3):241-251.

Obwona M (2002). Determinants of Technical Efficiency amongst small and medium scale farmers in Uganda: A case of tobacco growers. Economic Policy Research Centre (EPRC) Uganda. Occasional, pp. 19-24.

Ozowa VN (1995). Information Needs of Small-scale Farmers in Africa. The Nigerian Example. Q. Bull. Assoc. Agric. Inf. Spec. 40(1):1-3.

3

Risk of failure to meet regional and national goals relevant to agricultural development and poverty reduction in South Africa

Simbarashe Ndhleve and Ajuruchukwu Obi

Department of Agricultural Economics and Extension, University of Fort Hare, South Africa.

This study examined whether the Eastern Cape province's district municipalities are on course to achieve local and regional goals of improving agricultural production and reducing poverty. Results show significant strides towards locally set targets and high uncertainty in meeting the regional targets. Out of seven district municipalities, five municipalities are still about 5% points below the designated 10% of the annual budget to agriculture, and only three district municipalities scored more that the set 6% agricultural growth rate. All the seven district municipalities are not in a position to meet the target of 7% economic growth rate. Econometric simulations using the Hodrick-Prescott filter on data dating from 1995 to 2010 shows that all the seven district municipalities are off-track the set millennium development Goal 1 target with one even retrogressing from that goal. Failure to significantly reduce poverty is largely attributed to the province's failure to boost agricultural production which is an outcome of low and inefficient public expenditure management, inconsistent and misaligned policies. Regional policies should be built on local policies to bring in the desired impact on local development. This study made a strong case for frequent independent evaluation of set goals.

Key words: Agricultural productivity, growth, poverty.

INTRODUCTION

Countries are struggling to improve agriculture and reduce poverty. The millennium development Goal 1, the comprehensive Africa agriculture development programme (CAADP), the Southern Africa development community (SADC) regional indicative strategic development plan (RISDP) and South Africa's provincial development goal (PGDP) highlights the challenges faced by countries in improving the state of agriculture and reducing poverty. Local, regional and global development initiatives have always been recognized as important milestones for development but their overall economic impact has been always mixed (Benin et al.,

2010). In most cases, these initiatives represent a set of time-bound targets.

South Africa is a signatory to all the above mentioned initiatives (African Union, 2003; Somma, 2008; Mwape, 2009; Republic of South Africa, 2010). These initiatives have important implications on improving agricultural production and reducing poverty. Besides these initiatives, South Africa is failing to improve the state of agriculture and reduce poverty significantly (Machethe, 2004; Manona, 2004; Hall and Aliber, 2010; May, 2010; Ndhleve and Obi, 2011). In the Eastern Cape and Limpopo provinces, agricultural production has failed to improve in approximately more than 15 years whereas agriculture remains the backbone of their economies (Hall and Aliber, 2010). Such a growth pattern makes the sector's impact on poverty highly dormant and makes farming households susceptible to increased poverty (NEPAD, 2003).

The targeted times for most of these initiatives are either the 2014 or 2015. Since less than five years are remaining, governments are under increasing pressure to meet set targets. In recent years, South Africa has witnessed country wide strikes and demonstrations on service delivery issues and many development related issues. Failures of similar programmes like the rural development plan (RDP) and the growth, employment and redistribution (GEAR) macro-economic policy, everyone would like to know whether the current policies are on track to meet the intended goals. With the target years on the horizon, it is important to put these new policy dimensions under scrutiny. This study quantifies a set of scenarios for agricultural development and poverty reduction in the study area. It examines whether growth in agricultural expenditure is in accordance with the CAADP goal and SADC RISDP goals, whether progress towards meeting the 6% agricultural growth target set under the CAADP initiative through accelerated agriculture production is reachable and whether it is possible to meet the provincial PGDP goal and the MDG 1 of halving the 1994 level of poverty by 2014 and 2015, respectively.

Description of initiatives

Following independence in 1994, the first democratic government of South Africa adopted the rural development programme (RDP), the growth employment and redistribution development programmes in order to address past injustices in the distribution of productive resources. Following failure of these two policies, the government made available the accelerated and shared growth initiative for South Africa (ASGISA), PGDP, land reform policy, the agricultural black economic empowerment (AgriBEE) and adopted various regional initiatives in order to develop the economy. All these latter initiatives are assumed to have been informed by both the failure of previous policies and important development theories, therefore positive results are highly expected. The following description provides information on the PGDP, CAADP, SADC RISDP and the MDG1.

Eastern Cape's provincial growth and development plan (PGDP) 2004 to 2014

The Eastern Cape province adopted the PGDP in 2004. This programme signifies the province's compliance to the MDGs (Premier of Eastern Cape, 2009). It contains similar goals as that of the MDGs. It aims to maintain an economic growth rate of between 5 and 8% per annum and target to the 2004 level of poverty by 50% (Premier of Eastern Cape, 2009). The PGDP was designed to deal with the continuing spread and increase in the incidence of poverty and unemployment, as well as spatial inequality between different regions. This programme recognises the importance of agricultural transformation, infrastructural development, secondary sector, tourism sector, service delivery and the impacts of HIV/AIDS in solving the problem of poverty. In 2009, the Eastern Cape's office of the premier reported that the PGDP is yet to achieve its desired impact of halving poverty by 2014 in the Eastern Cape as almost half of the population of the Eastern Cape has no income and a further 22% live on less than R800/month.

Regional initiatives

The problem of poverty and under development in many developing countries has brought about a number of regional initiatives to harness governments and regional partnership roles in reducing poverty and enhancing economic growth. In the following sections regional initiatives aimed at reducing poverty and improving economic growth in South Africa namely; the Africa union new partnership for Africa's development (AU/NePAD)'s CAADP of Southern Africa, the RISDP, and the first millennium development goal (MDG1), are reviewed. South Africa is a signatory to these initiatives (African Union, 2003).

AU/NEPAD's comprehensive Africa agriculture development programme (CAADP)

In 2001, the new partnership for Africa's development (NEPAD) was formed by the Assembly of Heads of State in Africa to foster growth and development and addresses the challenges of poverty and under development facing the African continent. The comprehensive Africa agriculture development programme (CAADP) under the Africa union NEPAD recognised the importance of agriculture as the cornerstone of sustained growth and poverty reduction. Its main goal is enhance agriculture-led economic growth, eliminate hunger, reduce poverty, eliminate food and nutrition insecurities, and enable the expansion of exports. As targets for a successful implementation, the CAADP employs the millennium development goal (MDG) of reducing poverty and hunger by half by 2015, through the pursuit of a 6% average annual growth in the agriculture sector and allocating an average of 10% of national budgets to the sector (Benin et al., 2010). In partnerships with AU/NEPAD, South Africa is mandated to achieve these two goals (African Union, 2003). A report on progress towards the CAADP goals by Somma (2008) concluded that after five years, only a handful of Africa's 53 nations have reached the designated 10% target. According to NEPAD's 2007 tally, thirteen countries managed to spend from 5 to less than 10% on agriculture, and 15 more invested less than 5%. The remaining 18 countries, South Africa included, did not report on this. According to Mwape (2009), the number of countries spending more than 10% increased from 11% in 2003 to 22% in 2006.

SADC's regional indicative strategic development plan (RISDP)

SADC countries adopted the concept of the poverty reduction strategy papers (PRSPs) in order to formulate strategic plans and earmark financial resources for achieving their poverty reduction goals. Among other goals, the SADC RISDP Target 1 for food security is to achieve a gross domestic product GDP growth rate of at least 7% per year and halve the proportion of the population that lives below the poverty line between 1990 and 2015 (SADC, 2008).

Millennium development goal (MDG1)

The millennium declaration adopted by all 191 member states of the United Nations commits these countries to implement eight target-oriented millennium development goals (MDGs) that were formulated to significantly improve human lives by 2015 (UNDP, 2010). The MDG1 targets to halve the proportion of the population living on less than US$1 per day between 1990 and 2015. South Africa is committed to achieve the MDGs within the stipulated time, that is, by 2015. South Africa's MDG's Country Report (RSA, 2010) clearly indicated that the country is well on course to meet all the MDGs. Although doubts have been expressed about the feasibility of halving poverty by 2015 (Alemu, 2010).

MATERIALS AND METHODS

Study area

Eastern Cape's six district municipalities and one Metropolitan constituted the sample frame. The Eastern Cape Province is highly rural and essentially agrarian in nature and has a wide range of diversity in terms of both geographical and socio-economic characteristics across its municipalities that can add value to the study. It is richly endowed with farming land. Households share some common village resources and using communal land (Hebinck and Lent, 2007). Two-third of the population in the province lives below the poverty line and largely dependents on social grants and surviving through subsistence agriculture (Hall and Aliber, 2007).

For years, the government of Eastern Cape has relentlessly invested in develop the agricultural sector but the sector remains severely underdeveloped.

Data used

This assessment is based on the data secured from statistics South Africa and Eastern Cape socio-economic consultative council (ECSECC) (2010), the repository of data on all the socio-economic indicators of Eastern Cape. Indicators from 1995 to 2010 were used to tell a credible story of how processes and investments associated with CAADP initiative, SADC PRSP, MDG1 and PGDP are influencing economic growth and poverty in Eastern Cape. Each indicator was evaluated against a purposively selected goal. The selected set goals and indicators which are of interest in this evaluation process are presented in Table 1.

Choosing the base year is an important aspect of trend analysis (Benin et al., 2010). The baseline for all the indicators in this study is 1995, that's the time South Africa began assembling data for the Eastern Cape Province. The baseline year for the CAADP is 2003, the year the Maputo declaration was made (Benin et al., 2010).

Data analysis

Simple comparative analyses that provide graphical snapshots of municipal current economic indicators and compare it with the set targets of the provincial and key regional set targets were employed. Simulations were made in order to assess whether MDG1 is achievable by 2015 or not. Simulations provide a logical and consistence framework that provides analysts and policymakers with a valuable presentation a sector's performance (Seventer, 2002). Exponential smoothing following the Hodrick-Prescott Filter was used to make the projections. This method allows the projections of future incidence of poverty using past trends. This method is widely used among macro-economists to obtain a smooth estimate of the long-term trend component of a series. It was used by Cogley and Nason (1995) and Bardsen et al. (1995) in a working paper analysing business cycles. Technically, the Hodrick-Prescott filter (HP) is a two-sided filter that computes the smoothed series of τ of y by minimizing the variance of y around τ, subject to a penalty that constrains the second difference of τ. That is, the HP filter chooses τ to minimise:

$$\sum_{t=1}^{T}(y_t - \tau_t)^2 + \lambda \sum_{t=2}^{T-1}((\tau_{t+1} - \tau_t) - (\tau_t - \tau_{t=1}))^2$$

(1)

Let y_t for t = 1, 2, T denotes the logarithms of a time series variable. The series y_t is made up of a trend component, denoted by τ and a cyclical component, denoted by c such that $y_t = \tau_t + c_t$. The penalty parameter λ controls the smoothness of the series σ. The larger λ is, the smoother the σ. As λ approaches infinity, a linear trend emerges from the τ.

In reporting progress towards MDG1, this study used the same system as the one used by UNDP (2010). See Table 2 for detailed information. The simulated performance of each district municipality with regard to MDG1 follows the above categorizations after comparing it with the target values set. This set of information enables quantification of the deviation from the target and an evaluation to check if the deviation is acceptable. These analyses are narrowed only to assess progress towards the MDG1 only because all the initiatives are assumed to feed into the MDG1 as the mother goal. Understanding progress towards MDG1 will, therefore, also help us understand all the other targets.

Table 1. Selected Indicators to monitor changes in Agricultural Expenditure, agricultural GDP, total GDP and the incidence of poverty.

Development initiative, goals and targets	Indicators
PGDP	
Target 1: Halve, between 2004 and 2015, the proportion of people whose income is less thanUS$1 a day	Proportion of the population below US$ 1 a day
CAADP initiative	
Target 5: To invest at least 10% of budget to agriculture	Share of agriculture expenditure in the budget
Target 1: To achieve at least 6% agricultural GDP growth rate	Agricultural GDP growth rate
SADC RISDP	
SADC RISDP Target 1: To achieve 7% GDP growth per year	GDP growth rate
MDG1	
Target 1: Halve, between 1990 and 2015, the proportion of people whose income is less thanUS$1 a day	Proportion of the population below US$ 1 a day

Table 2. Progress towards MDG1.

Category	Explanation
Early achiever	It has already reached the target.
On track	It is likely to reach the target by 2015 or any set date.
Off track/slow	It has been making progress, but only slowly, so may not reach the target before 2015 or any set date.
Off track/regressing/no progress	It has made no progress and may even have regressed, moving further away from the target.

RESULTS AND DISCUSSION

The following section presents the results of the evaluations. To contextualize this discussion, a comparative analysis with relevant provincial-level data will be presented.

Eastern Cape's public expenditure evaluation outcome for 2010

The gaps between the set target expenditure and actual expenditure are shown in Figure 1. The results of the agricultural expenditure tracking survey for Eastern Cape presented in Figure 1 indicates that all the district municipalities have not reached the 10% target using budget figures for 2010, and the bulk of the district municipalities, five out seven district municipalities are almost within 5% points of the set threshold.

Some important implications emerge from the above finding. The emphasised role of public investment in agriculture in literature calls for increased spending by the province. It is expected that all the district municipality

programs should continue to foster an increase in the size of public spending. Countries that sustained agricultural growth during the Green Revolution were committing 11% of their budget expenditure to the agricultural sector (Mwape, 2009). This, therefore, implies that there is a need to increase the scale and size of public expenditure across the Eastern Cape Province district municipalities since it is acknowledged to be essential in increasing agricultural productivity.

Agricultural growth rate

Using the most recent data for 2009 to 2010, Figure 2 presents both the growth rates in provincial GDP and GDP from the agricultural sector for South Africa and the Eastern Cape province and all its district municipalities. Total GDP growth rates are compared with the SADC RISDP Target 1 of reaching a 7% GDP growth rate. The agricultural GDP is evaluated against the CAADP target of 6% agricultural growth rate.

The dashed line shows the 7% GDP growth rate target set under the SADC RISDP target and the bold line

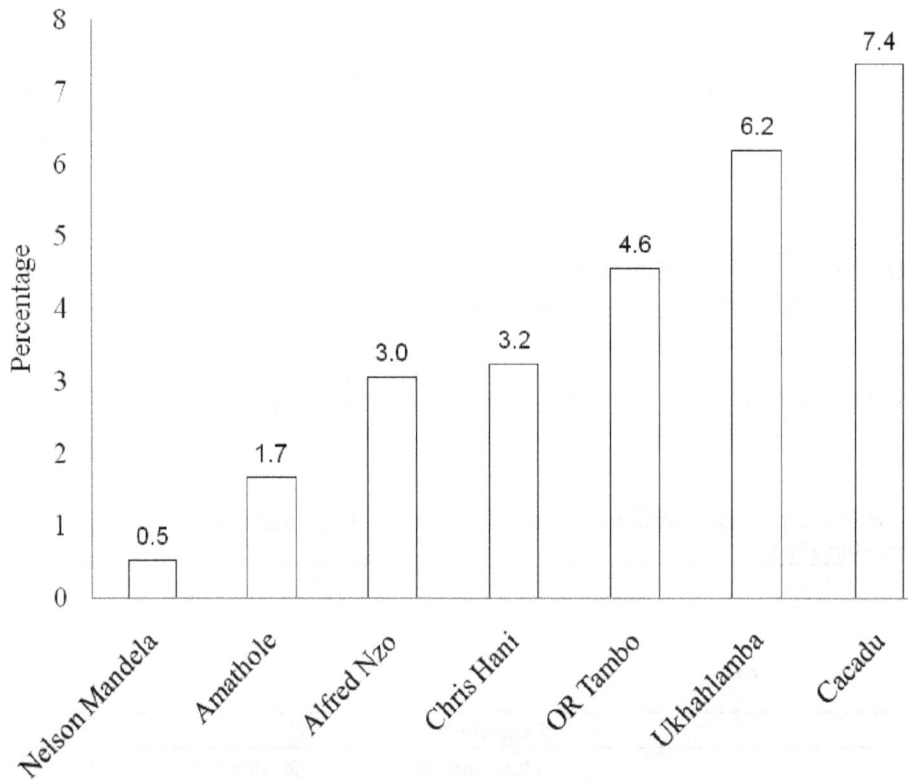

Figure 1. Agricultural Expenditure as a share of Total Expenditure (2009 to 2010), Source: Authors' calculations based on data from Government publications.

shows the 6% agricultural GDP growth rate set under the CAADP. The presented results shows that the district municipalities are doing relatively well in progressing towards the CAADP target and less well in progressing towards the SADC RISDP Target 1. Using 2009 to 2010 figures, OR Tambo, Alfred Nzo and Amatole district municipalities scored more than the set 6% growth in the agricultural sector. This is a reflection of significant progress towards the 6% agricultural growth rate set by the CAADP. When comparing figures for 2009 to 2010 for Cacadu and Chris Hani district municipalities, the chances of reaching the CAADP target are in great doubt. These two districts scored a growth rate of less than 3%.

Relative to the SADC RISDP target, the situation across district municipalities is depressing. When comparing progress towards the SADC RISDP target with the CAADP target discussed above, progress towards this goal is less impressive. Using figures for 2009 to 2010, Figure 2 shows that prospects towards the achievement of the 7% growth rate in GDP remains gloomy as all the district municipalities scored a growth rate less than the set 7% growth rate. Four district municipalities are far from reaching both the CAADP target of 6% growth in agricultural GDP and the SADC RISDP target of 7% growth in total GDP. It is virtually impossible to meet the MDG1 in the province in the

absence of agricultural growth and GDP growth. Meeting MDG1 requires support to the agricultural sector and all the other sectors that make up the total GDP. An effort should be made to duplicate the development activities implemented in the three prospering municipalities that have achieved the set targets or Tambo district municipality should be used as model of growth.

Progress towards MDG1 and SADC RISDP

Figure 3 presents a snapshot of the margin between the current level of poverty and the targeted value for the incidence of poverty under these two initiatives. The results in Figure 3 show that all the district municipalities have not halved poverty levels as the graphs present the variations between the set target for both the MDG1 and the SADC RISDP and the current level of poverty in 2009 to 2010.

With five years left, the prospects of reducing the level of poverty across the seven district municipalities are bleak with the most doubtful prospect being in respect to five district municipalities namely Amatole, Chris Hani, UKhahlamba, O. R. Tambo and Alfred Nzo. Poverty levels in all these five districts are well above the set SADC RISDP and the MDG1 target, with some even recording more than 60%.

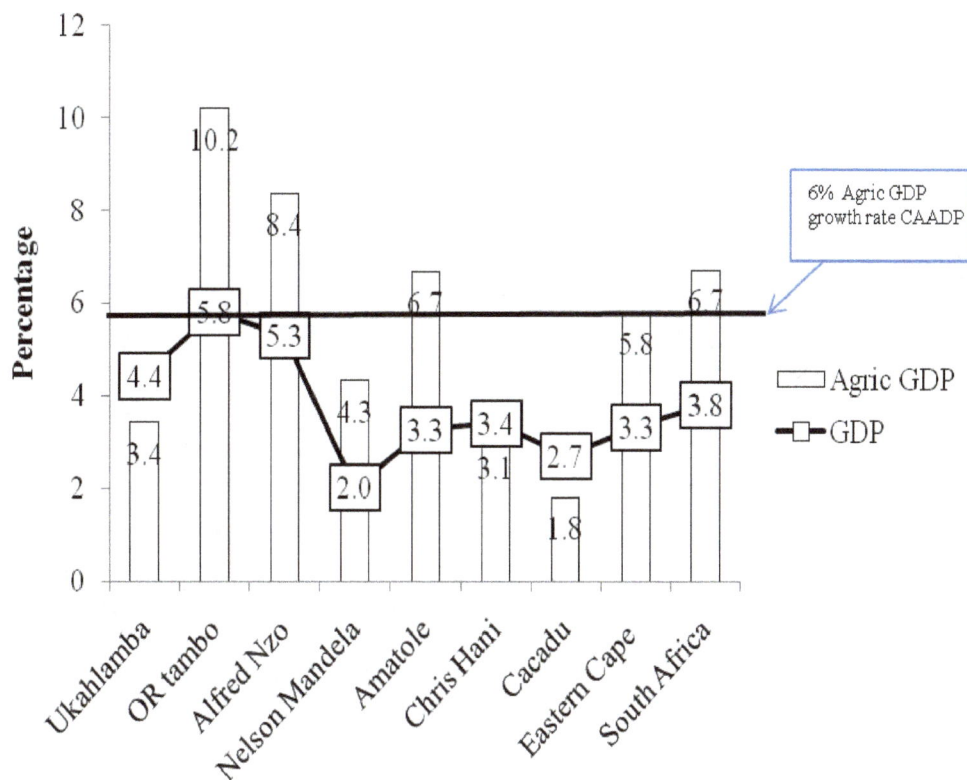

Figure 2. Total GDP and Agricultural GDP Growth Rates; Source: Authors' calculations based on data from ECSECC (2010) database.

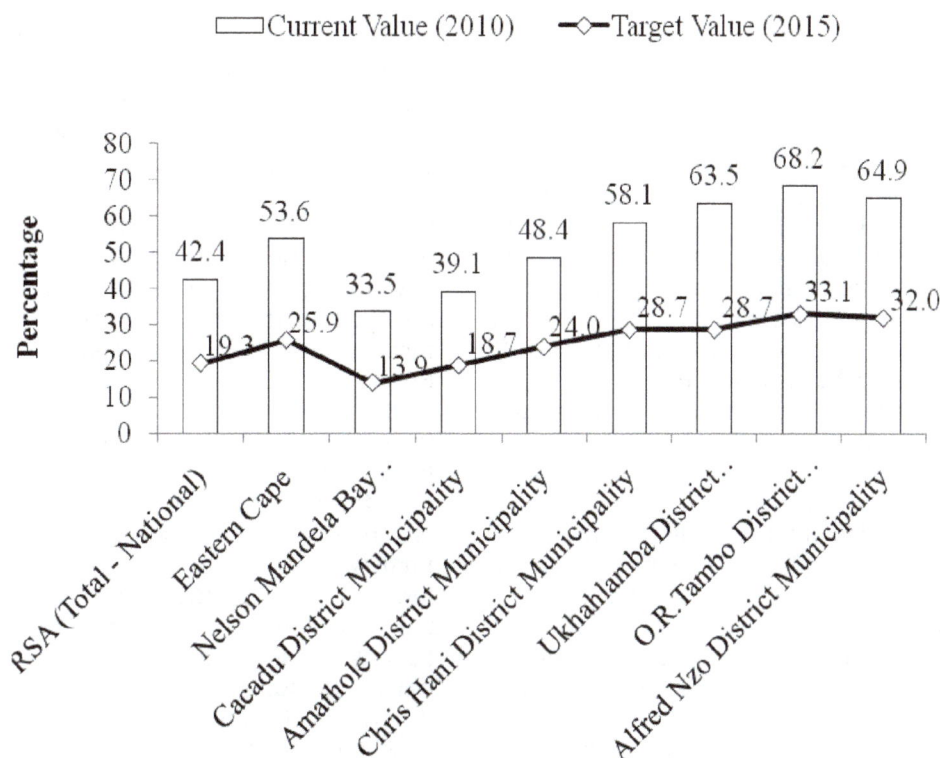

Figure 3. Changes in Poverty and progress towards MDG1 and SADC RISDP; Source: Authors' calculations based on data from ECSECC.

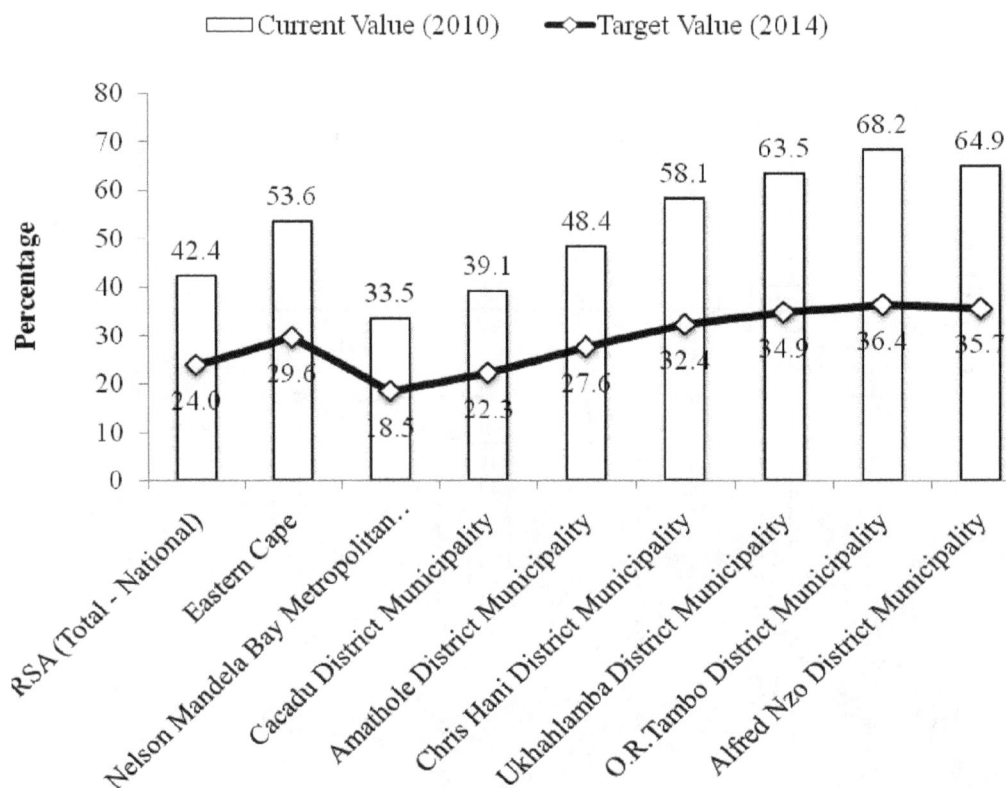

Figure 4. Changes in Poverty and progress towards PGDP, Source: Authors' calculations based on data from ECSECC (2010) database.

Level of progress towards PGDP

Using 2004 as the base year, Figure 4 compares the seven district municipalities' 2009 to 2010 poverty levels with the target set under the PGDP with respect to poverty. Huge variations still exist between the set target and the figures for 2009 to 2010. However, considering that the PGDP initiative was only implemented less than six years ago, in 2004, the district municipalities seem to have reduced poverty by noticeable percentage points. With five years left, the prospects of achieving the PGDP target are good across all the districts municipalities since the reductions in poverty shown in Figure 4 are noticeably linked to the initial level of poverty for 2004.

The MDG1 and the PGDP share a similar goal of halving the proportion of the poor although the target incidence of poverty is different due to altered base year. It is therefore deemed important to compare progress towards these targets in the province under study. Figures 3 and 4 suggest that all the district municipalities are pprogressing relatively well towards the PDGP compared to the MDG1. Both figures show a large variation between the target value and the achieved value for 2010. The average margin between the target percentage and the percentage for 2010 is relatively lower under the PGDP than what appears in Figure 3

showing progress towards the MDG1. The PGDP seems to be presenting a slightly impressive picture of progress towards poverty reduction relative to the MDG1. This might imply that the set of planning under the PGDP is better than that of the MDG.

The adoption of both the PDGP and the MDG1 was a landmark decision for the Eastern Cape Province's district municipalities. From the above results, it is obvious that both the 2014 deadline for PGDP and the 2015 deadline for MDG1 were optimistic. The above results show that the prospects of achieving both the PGDP and MDG1 by 2014 and 2015 are highly unlikely as the differences between the current value and the set targets are still high with less than five years left. Even though there is limited progress, the observed changes from the base year provides enough justification for the province to commit resources in line with the two policy frameworks. Further inferential statistics on progress towards the set poverty level that takes into consideration the past trend in annual changes in the incidence of poverty and the forecasted trend over the next five years will be presented in the following section, succeeding discussion of the above results.

Although, it is difficult to isolate the impacts of PDGP and MDGs, the above results show slight variation in progress towards the PGDP and MDG1. The outcome that the Eastern Cape province is progressing better towards the PGDP target than the MDG1 target reflects importantly on the variations in the impact of regional policy as compared to local policies. PGDP is a local policy framework designed for the Eastern Cape province and the MDGs are global programmes. It is important to note that a good set of policy frameworks and set targets may not necessarily produce commensurate gains in terms of poverty reduction. In fact, for many countries the effects of regional or global policies have been noticeably smaller than that of local policies. This suggests that the province had a potential for reducing poverty faster using local development policy frameworks rather than regional policies. This situation is not surprising given the variation between the impact of local policy and that of regional policies observed in across Eastern Cape district municipalities. The fact that the PGDP is an initiative specifically meant for the Eastern Cape Province cannot be taken lightly. Progress noted under the PGDP is probably attributed to the suitability and appropriateness of the set of policies drawn for the programme relative to the MDG which is a set of umbrella policies for all countries. It is probable that local policy could have a magnified impact on poverty relative to a regional policy because of stronger involvement of either country or provincial leadership and stakeholders in the implementation of the local programmes unlike global programmes.

Another factor that must also be taken into account is the process by which the policies have been derived, under either the MDG1 or the PGDP. Since the Eastern Cape Province guided the PGDP, the provincial government officials and consultants have been quite influential in facilitating the process, thus the better results. The MDGs have been driven by world leaders without sufficient prior national or grassroots consultations with inhabitants of the Eastern Cape province or other entities for that matter. This is mostly the case with all countries. This might be the probable reason why the PGDP has a higher marginal impact on poverty than the MDG1. New agricultural development programmes should be tailor made to support the current programme in place, not necessarily replacing them. The advent of new growth frameworks needs to accommodate the previous capacities upon which the sector is presently working on. The success of these development arrangements is highly possible when implementing.

The use of provincial policy in conjunction with regional policy or global policy can improve the efficiency of resource use in the Eastern Cape province. Policy making at provincial level can improve efficiency in the use of public resources, especially the allocation of public funds. The theoretical premise of decentralization of policy making for greater efficiency in the provision of public goods and services to meet local demand seems to be materialising in South Africa when comparing progress towards MDGs with that of the PGDP. The sub-national governments are always assumed to be efficient in both policy and use of public funds as they are

relatively better in aligning either expenditures or polices with local priorities (Momoniat, 1998).

Eastern Cape's district municipalities' progress towards MDG1

Inferential statistics showing trends in the incidence of poverty up to 2010 and the consequence of continuing with the observed trends in poverty reduction over the next five years are presented in Table 3 where figures for the base year and current status are compared to the estimates for 2015 and 2025 assuming a Business as Usual Scenario[1].

Using 1995 to 2010 estimates, various district municipalities will not meet the poverty target, by neither 2015 nor 2025. Except for Nelson Mandela Metropolitan, all the district municipalities have been making progress albeit, slowly. In Nelson Mandela Metropolitan poverty is even increasing implying retrogression, moving further away from the set target. The observed retrogression in Nelson Mandela Metropolitan might be the effect of urbanization. In the Eastern Cape province, amongst all the seven district municipalities, none is considered on track to achieve the MDG1. The MDG1 should be effectively be integrated into all the district municipalities development planning process and municipalities should be increasingly used as a vehicle through which governments seek to operationalize their agriculture and poverty reduction strategies.

Conclusion

Progress towards both local and regionally set targets is not uniform across district municipalities. Progress towards the CAADP target, the SADC RISDP and the PGDP presented mixed results across municipalities. However, all the municipalities are yet to register notable declines in poverty as the most challenging goal is the MDG1. This goal is seemingly unreachable both in 2015 and 2025. This implies that the province would require more robust pro-poor growth well above historical rates. This result calls for immediate action in improving the sectors that contributes to high marginal reductions in poverty in every municipality. In most municipalities, failure to meet the PGDP goal and the MDG1 is largely attributed to the province's failure to boost agricultural production. Therefore, a case is made for articulation of strengthened provincial comprehensive agricultural public expenditure programs that build a consensus for increased levels and efficiency of public expenditure for agriculture development to levels above historical rates.

---

[1]Baseline scenario that examines the consequences of continuingcurrent trends in the population, economy, technology and human behaviour.

Table 3. Achievement of poverty targets in Eastern Cape.

District municipality	Base year 1994 to 1995	Current status 2010	Estimate 2015	Progress towards target by 2015	Progress towards Target by 2025
Amatole	50.9	45.6	47.1	Off track/slow	Off track/slow
Alfred Nzo	68.2	61.2	62.9	Off track/slow	Off track/slow
Cacadu	40.4	36.7	38.5	Off track/slow	Off track/slow
Chris Hani	62	54	56.1	Off track/slow	Off track/slow
Nelson Mandela Metropolitan	30.8	30.9	34.2	Off track/retrogressing/ no progress	Off track/retrogressing/ no progress
O R Tambo	69.8	64.1	66.4	Off track/slow	Off track/slow
UKhahlamba	62.9	60.0	63.3	Off track/slow	Off track/slow

Source: Author's simulations based on data from ECSECC (2010) using Hodrick-Prescott Filter (HP), (see Equation 5.6).

Abbreviations: CAADP, Comprehensive Africa agriculture development programme; **GDP,** Gross Domestic Product; **MDG,** millennium development goals; **SADC RISDP,** Southern African development community regional indicative; strategic development plan; **PGDP,** Provincial growth and development plan; **ECSECC,** Eastern Cape socio-economic consultative council; **NEPAD,** new partnership for Africa's development; **AgriBEE,** agricultural black economic empowerment; **ASGISA,** accelerated and shared growth initiative for South Africa.

REFERENCES

African Union (2003). Assembly of the African Union: Maputo Declaration. 10-12 July. Available online: www.africaunion.org/root/au/Documents/Decisions/hog/12HoGAssembly2003.pdf.Accessed on the 28th March 2009.

Alemu ZG (2010). Measuring poverty, deprivation, and progress in service delivery: Application to the Eastern Cape Province, South Africa. Seminar presented at National Agricultural Marketing Council, Pretoria August 18, 2010. Development Bank of South Africa.

Bardsen G, Fisher P, Nymoen R (1995). "Business Cycles: Real facts or fallacies? Working Paper presented in the Centennial of Ragnar Frisch, Oslo.

Benin S, Johnson M, Omilola B, Beintema N, Bekele H, Chilonda P, Davis K, Edeme J, Elmekass A, Govereh J, Kakuba T, Karugia J, Makunike R, Massawe S, Mpyisi E, Nwafor M, Olubode-Awosola F, Sanyang S, Taye B, Wanzala M, Yade M, Zewdie Y (2010). Monitoring and Evaluation (M&E) System for the Comprehensive Africa Agriculture Development Programme (CAADP).ReSAKSS Working Paper No. 6. Washington, D.C.: International Food Policy Research Institute (IFPRI).

Cogley T, Nason J (1995). "Effects of the Hodrick-Prescott Filter on Trend and Difference Stationary Time Series: Implications for Business Cycle Research". J. Econ. Dyn. Control Vol. 19.

ECSECC (2010). Statistical Database. Available online: www.ecsecc.org/statistics-database. Accessed on the 20 October 2011.

Hall R, Aliber M (2010). The Case for Re-Strategising Spending Priorities to Support Small-Scale Farmers in South Africa. Institute for Poverty, Land and Agrarian Studies (PLAAS), Working paper 17. University of the Western Cape.

Hebinck P, Lent PC (2007). Livelihoods and landscapes: The people of Guquka and Koloni and their resources. Leiden, Boston.

Machethe LC (2004). Agriculture and poverty in South Africa: Can agriculture reduce poverty? Department of Agricultural Economics,

Extension and Rural Development and Postgraduate School of Agriculture and Rural Development, University of Pretoria.

Manona WW (2004). Impact of health, water and sanitation services on improving the quality of life of poor communities. Unpublished Doctoral dissertation, Stellenbosch, Stellenbosch University.

May J (2010). Poverty Eradication: The South African Experience. United Nations Conference Centre, Addis Ababa, Ethiopia. Available online: http://www.un.org/esa/socdev/social/meetings/egm10/documents/May%20paper.pdf Accessed on the 11th January, 2011.

Momoniat I (1998). Fiscal decentralisation in South Africa: A practitioner's perspective. Available online:http://info.worldbank.org/etools/docs/library/128819/Momoniat%202001%20South%20Africa.pdf. Accessed on the 14 October 2011.

Mwape F (2009). How are countries measuring up to the Maputo declaration? CAADP Policy Brief. June 2009. Available online: http://www.sadc.int/index/browse/page/104. Accessed on 14 April 2011.

Ndhleve S, Obi A (2011). "Determinants of household activity choice, rural income strategies and diversification". Institutional Constraints to Smallholder Development in Southern Africa, Wageningen Academic Publishers, The Netherlands.

NEPAD (2003). Comprehensive Africa Agriculture Development Programme. Process and scope of the Agriculture Programme. Available online: www.nepad.org/system/files/caadp.pdf. Accessed on the 13 March, 2011.

Premier Eastern Cape (2009). Assessment of the Eastern Cape Provincial Growth and Development Plan. Final Report. Bisho, Eastern Cape.

Republic of South Africa (2010). Millennium Development Goals: Country Report 2010. StatSA: Pretoria.

SADC (2008). Regional Indicative Strategic Development Plan. Available online: http://www.sadc.int/index/browse/page/104. Accessed on 14 April 2011.

Seventer EV (2002). Evidence-based Employment Scenarios: Appropriate analytical tools for economy-wide analysis of employment creating policies in South Africa. Trade and Industrial Policy Strategies (TIPS). Human Sciences Research Council.

Somma A (2008). The 10% that could change Africa. IFPRI Forum: International Food Policy Research Institute, Washington, D.C.

UNDP (2010). Millennium Development Goals country report 2010. Available online: www.statssa.gov.za/news_archive/Docs/MDGR_2010.pdf. Accessed on the 13 May 2011.

4

Measuring diet quantity and quality dimensions of food security in rural Ethiopia

Degye Goshu, Belay Kassa and Mengistu Ketema

School of Agricultural Economics and Agribusiness, Haramaya University, P.O. Box: 05, Haramaya University, Ethiopia.

Food insecurity is an overriding problem of most developing countries like Ethiopia, which requires empirical evidence pertinent to food security policy formulation and implementation. This paper investigates food security situation of households by surveying 260 farm households randomly and proportionately sampled from the major farming systems in Ethiopia. Households' daily calorie availability and dietary diversity were measured to capture the diet quantity and quality dimensions of food security of households. A seemingly unrelated regression (SUR) model results of the two measures suggested that the mean daily calorie intake per adult equivalent and dietary diversity level of households were about 1871 kcal and 6.8, respectively, with significant differences in farming systems and household idiosyncratic characteristics. The univariate probit model results show that the likelihood of households to be food security was 42.3%, while their probability to have semi-diversified diet was 37.2%. However, food security status and dietary diversity status were weakly interdependent and their determinant factors were significantly different. The major contribution of this paper is that it employs econometric estimation of dietary diversity scores and status and measures their interactions with diet quantity scales and food security status at household level.

Key words: Food security, dietary diversity, seemingly unrelated regression (SUR) model, bivariate probit.

INTRODUCTION

The main development objective of the Ethiopian Government is poverty eradication and the country's development policies and strategies are geared towards this end (MoFED, 2006; FDRE, 2012). Smallholder farming is the dominant livelihood activity for the majority of Ethiopians, but it is also the major source of vulnerability to poverty and food insecurity (Brown and Teshome, 2007). To combat this problem, the Ethiopian Government has designed food security policy and strategy which was first issued in 1996 within the framework of Ethiopia's Poverty Reduction Strategy (FDRE, 1996; FDRE, 2004). In this regard, agriculture is assumed to be a strong option for spurring growth, over-

coming poverty, and enhancing food security. It is a vital development tool for achieving the Millennium Development Goals (MDG), one of which is to halve by 2015 the share of people suffering from extreme poverty and hunger (World Bank, 2008).

In the 1970s, definitions of food security emphasized a nation's aggregate food production but since then the focus is the ability of poor households to gain access to food in the necessary amounts. According to FAO (1996), food security is assumed to exist "when all people, at all times, have physical and economic access to sufficient, safe and nutritious food to meet their dietary needs and food preferences for an active and healthy life". The four dimensions of food security are food availability, stability of supply; accessibility of food, and quality and safety of food. Access, sufficiency, and quality are important aspects of the definition of food security which should be addressed by food security indicators.

In the literature, there are three categories of indicators of food security each with limited capacity to capture the extent of food security and hunger: Process, outcome, and trend indicators (Hoddinott, 1999). Household calorie acquisition and dietary diversity are the two basic outcome indicators of diet quantity. Diet quantity measures of this kind include daily food energy consumption per capita or per adult equivalent and percentage of households or people that are food energy–deficient (Radimer et al., 1990; FAO, 1996; Hoddinott, 1999; Swindale and Ohri-Vachaspati, 2005; Smith and Subandoro, 2007). If the estimated total energy in the food that the household acquires daily is lower than the sum of its members' daily requirements, the household is classified as food energy-deficient or commonly known as 'food insecure'.

On the other hand, the three basic indicators of diet quality are diet diversity, percentage of food energy from staples, and quantities of foods consumed daily per adult equivalent (Hoddinott, 1999; Smith and Subandoro, 2007; FAO, 2007; Kennedy et al., 2011). It might be quite possible to a household to meet its energy requirement but to be prevented from leading an active, healthy life of the household members due to deficiencies of other nutrients. Improved diet quality is associated with improved birth weight and child nutritional status and with reduced mortality (Ruel, 2002, 2003). It is, therefore, critically important that indicators of the nutritional quality of the food people eat are included in any analysis of food security. Diet diversity reflects how varied the foods typically consumed by a household are (Smith and Subandoro, 2007). Dietary diversity indicators based on food groups predict nutrient adequacy better than those based on individual foods (Ruel, 2002). The percentage of food energy acquired from staples at household level is measured as the percentage of dietary energy available from food staples in the total dietary energy available. A higher value indicates lower diet quality, because energy-dense starchy staples have small amounts of bio-available protein and micronutrients, leaving those consuming large amounts of them compared to other foods vulnerable to protein and micronutrient deficiencies (Smith and Subandoro, 2007).

In the latest theoretical and empirical literature of food security, there are about five methods of analyzing food security, each measuring different aspects of food security situation at different levels. The method of individual food intake data is used to measure diet quantity (calorie intake) of an individual in a household, while the method of household calorie acquisition is used to measure the same variable at household level

(Radimer et al., 1990; FAO, 1996; Swindale and Ohri-Vachaspati, 2005; Smith and Subandoro, 2007). Dietary intake method, on the other hand, is used to measure the diet quality dimensions (nutritional content) of households or individuals (FAO, 2007; Smith and Subandoro, 2007; Kennedy et al., 2011). The United States Department of Agriculture (USDA) 18-item core module is being used recently in some countries of the world to measure diet quantity. This method categorizes households into three groups by their food security situation as 'food secure', 'food insecure without hunger', and 'food insecure with hunger' (Bickel et al., 2000). The method of index of household coping strategies is a measure of coping mechanisms based on how households adapt to the presence or threat of food shortages (Radimer et al., 1990; Maxwell and Frankenberger, 1992; Maxwell, 1996). This paper employs household calorie acquisition and dietary intake methods of food security analysis whereby moderate accuracy is maintained and misreporting is kept low (Hoddinott; 1999).

Empirical evidences on food security in Ethiopia indicate the prevalence of high level of food insecurity with significant idiosyncratic and spatial characteristics. There is much variation in household consumption patterns in Ethiopia, depending on specific geographical and sociocultural characteristics where calorie consumption is low, a high percentage of this consumption coming from cereals, and per capita intake of calories is relatively higher in rural than in urban areas (Guush et al., 2011). According to Samuel (2004), grain production and food security are not affected by the same factors but household size and age of the household head significantly determine household food security in Ethiopia. To achieve national food security in Ethiopia, it is necessary to improve market functioning, invest in infrastructure which reduces food transaction costs and provide incentives for increased production (Berhanu, 2004). For the period between 1989 and 1994, Block and Webb (2001) identified that households in Ethiopia which had initially more diversified income subsequently experienced a relatively greater increase in income and calorie intake. On the other hand, the analysis on the effects of food crisis in 2008 in Ethiopia shows broad deterioration of household food security (Hadleya et al., 2011). Rural income transfer programs (food-for-work and productive safety nets) in Ethiopia serve as temporary safety nets for food availability, but they are limited in boosting the dietary diversity of households and their coping strategies (Uraguchi, 2012). In addition to income transfer projects as determinants of household food security, socio-demographic variables of education and family size as well as agricultural input of land size are found to be significant in accounting for changes in households' food security due to these programs.

With regards to calorie adequacy, significant proportion (45%) of rural households in the Amhara regional state of

Ethiopia are food insecure (Freihiwot, 2007). Nearly 61% of the sample household characterized by poor access to oxen, livestock and farm land in central Ethiopia (Eastern Shewa) are food insecure (Hailu, 2012). On the other hand, about 80% of households in Eastern Ethiopia are food insecure with food insecurity gap of 30% (Zegeye and Hussien, 2011). But, Abebaw et al. (2011) estimates that 66% of the sampled households in this region of the country are food insecure, with food insecurity gap of 27%. Food insecurity status of household in this region (Dire Dawa area) is significantly determined by family size, annual income, amount of credit received, access to irrigation, age of household head, farm size, and livestock owned (Bogale and Shimelis, 2009).

These specific food security studies generally suggest that depth and intensity of food insecurity is high, influenced by poor functioning of marketing systems and other household and socioeconomic factors. However, all the studies focused on one aspect of food security situation, specifically the percentage of households facing calorie shortage. To account for such shortcomings, the objectives of this study are to measure food security situations of households and identify their determinants and interactions in rural Ethiopia through two major outcome indicators of food security: Diet quantity and quality.

RESEARCH METHODOLOGY

Sampling technique and the dataset

This study used primary data collected from four districts selected from two major sedentary farming systems in Ethiopia, Central and Hararghe highlands. These two sedentary farming systems cover about 40% of the total sedentary farming systems in Ethiopia (Getahun, 1980; Dercon and Hoddinott, 2009). Because the study areas are heterogeneous in terms of their food security situation, two-stage stratified random sampling technique was employed. In the first stage, districts were stratified into two as highland and non-highland. Two districts from highland areas were randomly selected from each farming system. In the second stage, kebeles – the lowest administrative levels – were stratified by their food security situation as better-off and worse-off. One kebele was randomly selected from each stratum.

Finally, a total of 260 rural households were randomly selected from eight kebeles proportionate to the number of households in each district and kebele. Because weights were given to farming systems, the samples were also proportionate at the two farming systems level. Stratification procedures at each stage were carefully employed to increase homogeneity within a food security stratum and heterogeneity between strata so that precision and sampling efficiency would be maximized. To overcome the problem of bias in collecting data on food security indicators, the survey was conducted in two phases depending on the harvesting periods and fasting months in the two farming systems. This timing of survey periods was expected to minimize the variability of household consumption prevalently observed to be skewed left and right during harvest and before harvest periods, respectively.

Methodologically, it is simple if continuous measures of food security are estimated. However, it is also often useful, both for policy and research purposes, to simplify the food security scale into a small set of categories, each one representing a meaningful range of severity on the underlying scale, and to discuss the percentage of the population in each of these categories. In this study, the measures of food security were treated as both continuous and categorical variables. A household was treated as either food secure or insecure based on its amount of calorie availability per adult equivalent, or may fall at a lower or higher level on the food insecurity continuum. Dietary diversity levels were measured from the counts of food items consumed by households. Households were also treated as having medium and low dietary diversity status depending on the number of food groups they consumed.

The endogenous variables used as food security indicators were daily calorie availability per adult equivalent, food security status, dietary diversity scores/levels, and dietary diversity status. These food security indicators were hypothesized to be primarily determined by a household's resource endowment. As such, the expected determinants of food security measures were categorized into five based on their resource endowment as humane capital (family size, farming experience, dependency ratio, literacy status, sex of the household head), social capital (degree of civic engagement and/or responsibility as a proxy for social networks and social class), physical capital (pattern of cultivated land allocation, livestock holding, number of oxen owned, total assets owned), financial capital (income earned, access to credit or amount of credit received), natural capital (irrigation water use, proportion of land under irrigation), and a dummy variable for the farming systems as a proxy to capture omitted location-specific characteristics (Table 1).

Analytical methods

Two groups of outcome measures of food security situation were estimated in this paper: the first is household calorie acquisition including daily calorie intake per adult equivalent to determine food availability and status of food security situation. The second outcome measure was dietary diversity to capture diet quality. Daily calorie availability can be measured in two ways, based on consumption per equivalent male adult and consumption based on age and sex without converting equivalent male adult. Food balance sheet and aggregate household calorie consumption were constructed for this purpose. Food security condition was estimated based on calorie requirement, according to sex, age, and activity level of household members, as recommended by FAO and WHO (1985).

Household calorie availability was computed from each food item consumed and grouped into seven food groups[1], adjusted for food processing to obtain the net weekly calorie availability. The net weekly calorie availability was divided by seven to obtain the household daily calorie intake. The family size of each household was converted into adult equivalent family size which considers age, sex, and activity level of each family member in the household. The daily net calorie consumption of the household was divided by the adult equivalent family size to obtain the daily calorie availability per adult equivalent of the household. Households with daily calorie consumption greater than or equal to 2200 kcal per day were categorized as 'food secure', and those households whose calorie intake fallen below this food security threshold grouped as 'food insecure'.

The association between household dietary diversity scores and dietary energy availability indicates that increasing household

---

[1] These food groups are (1) cereal, roots and tubers, (2) pulses and legumes, (3) dairy products (4) meats, fish and eggs (5) oils and fats, (6) fruits, and (7) vegetables.

Table 1. Definition and notation of variables.

Variable name	Notation	Measurement	Expected effect on food security indicators	
			Calorie intake level/status	Dietary diversity level/status
Daily calorie availability (log)	lncalav	Continuous		
Food security status	secur	Binary (1 if calav >= 2200, 0 otherwise)		
Household dietary diversity score	hdds1	Counts (food groups)		
Household dietary diversity score	hdds2	Counts (food items)		
Household dietary diversity status	hds	Binary (1 if hdds1 >= 4, 0 otherwise)		
Female heads	femal	Binary (1 if female, 0 otherwise)	±	±
Family size	famsz	Continuous (head count)	-	-
Child dependency ratio	depc	Continuous (%)	-	-
Literacy status	literat	Binary (1 if literate, 0 otherwise)	+	+
Land cultivated	land	Continuous (ha)	+	+
Land allocated to staples	landst	Continuous (ha)	+	+
Irrigation water use	irrig	Binary (1 if user, 0 otherwise)	+	+
Proportion of irrigated land	irrigp	Continuous (%)	+	+
Quantity of fertilizer	frtqt	Continuous (qt)	+	+
Livestock holding (TLU)	tlu	Continuous (tropical livestock unit)	+	+
Annual gross income (log)	lninom	Continuous (ETB)	+	+
Social capital	social	Binary (1 if socially networked, 0 otherwise)	+	+
Off-farm activity	ofarm	Binary (1 if participant, 0 otherwise)	+	+
Distance to nearest road	road	Continuous (km)	-	-
Access to credit	credit	Binary (1 if participant, 0 otherwise)	+	+
Production of major cash crop	mcash	Binary (1 if *khat* producer, 0 otherwise)	+	+
Farming system	farmsy	Binary (1 if Hararghe highlands, 0 otherwise)	±	±

dietary diversity improves energy availability; but the reverse may not hold true (Hoddinott and Yohannes, 2002). Dietary diversity scores have potential for monitoring changes in dietary energy availability, particularly when resources are lacking for quantitative measurements. The association between dietary diversity and mean micronutrient density adequacy of complementary foods indicate the positive correlations in age groups (FANTA, 2006). Generally, dietary diversity scores have been shown to be valid proxy indicators for dietary energy availability at household level and micronutrient adequacy of diets of young children and women of reproductive age.

The seemingly unrelated regression (SUR) model representation of daily calorie intake per adult equivalent and household dietary diversity scores[2] was employed to simultaneously estimate the two linear outcome measure of food security situation of households (Zellner, 1962; Greene, 2012):

$$\ln calav_i = \mathbf{x}'\boldsymbol{\beta}_1 + \varepsilon_{1i}$$
$$hdds_i = \mathbf{x}'\boldsymbol{\beta}_2 + \varepsilon_{2i_i} \qquad (1)$$

[2] The number of food items is a count data. If there is overdispersion in the data, linear estimation of count data is assumed to result in inefficient, inconsistent, and biased parameter estimates. However, the count data model (Poisson and negative binomial regression) results of dietary diversity scorers of food items were not largely different from the linear SUR model results because overdispersion was not detected.

where $\ln calav_i$ is the log of daily per adult equivalent calorie intake of household i; $hdds_i$ is household dietary diversity scores measured by the number of food items consumed per week; and $\mathbf{X}$ is a vector of factors determining daily calorie intake and dietary diversity level of households; $\boldsymbol{\beta}_1$ and $\boldsymbol{\beta}_2$ are the respective vectors of coefficients, ε_{1i} and ε_{2i} are their random terms.

Household and socioeconomic determinants of food security status and their likely effects were estimated by a univariate probit representation. The latent variable regression model for binary regression models was specified by the structural equation as (Maddala, 1983; Long, 1997; Cameron and Trivedi, 1998; Long and Freese, 2005; Cameron and Trivedi, 2009):

$$\sec ur_i^* = \mathbf{x'}\boldsymbol{\beta} + \varepsilon_i \qquad (2)$$

where $\sec ur_i^*$ is binary latent variable for food security status (observed if $\sec ur_i^* > 0$, 0 otherwise); $\mathbf{X}$ is a vector of household specific and other socioeconomic factors determining food security status; $\boldsymbol{\beta}$ is a vector of parameters of interest, and ε_i random error.

The link between the observed binary y and the latent $\sec ur_i^*$ is made with a simple measurement equation:

$$secur_i = \begin{cases} 1 & if \ \ secur_i^* = \mathbf{x'}\boldsymbol{\beta} + \varepsilon_i > 0; \\ 0 & if \ secur_i^* \le 0. \end{cases} \qquad (3)$$

In addition to estimating the dietary diversity level of households, their dietary diversity status was determined from the frequency distribution of the number of food groups consumed and categorized into two as medium and poor dietary diversity status. This binary data was represented by a univariate probit model which is analogous to the model specified earlier to estimate the food security status of households in Equation (2).

The presence of bivariate interdependence between food security and dietary diversity status was tested by employing the bivariate probit model (Hardin, 1996; De Luca, 2008; Greene, 2012):

$$\sec ur_i^* = \mathbf{x}_1' \boldsymbol{\beta}_1 + v_{1i}$$
$$hds_i^* = \mathbf{x}_2' \boldsymbol{\beta}_2 + v_{2i} \qquad (4)$$

Where $\sec ur_i$, and hds_i are the food security status and the dietary diversity status, respectively; and v_{1i} and v_{2i} are their respective error terms in the seemingly unrelated bivariate probit and assumed to be normally distributed with $N\left[\begin{pmatrix}0\\0\end{pmatrix}\begin{pmatrix}1 & \rho\\\rho & 1\end{pmatrix}\right]$, where ρ is the tetrachoric correlation between the latent variables.

Accordingly, the latent variables, observed and unobserved, were specified as:

$$secur_i = \begin{cases} secur_i^* = \mathbf{x}_1'\boldsymbol{\beta}_1 + v_{1i} & if \ \sec ur_i^* > 0; \\ 0 & if \ secur_i^* \le 0. \end{cases} \qquad (5)$$

$$hds_i = \begin{cases} hds_i^* = \mathbf{x}_2'\boldsymbol{\beta}_2 + v_{3i} & if \ hds_i^* > 0 \\ 0 & if \ hds_i^* \le 0 \end{cases}$$

RESULTS AND DISCUSSION

Data description

Two important dimensions of food security were analyzed through two basic indicators of food security: diet quantity and diet quality. The dichotomous classification of households based on their daily calorie availability suggested that 42.7% of the households were food secure, while the rest majorities (57.3%) were food insecure or calorie-deficient. The frequency distribution of counts of food items suggested that most households consumed five kinds of food items grouped under three food categories. The binomial classification of households by their level of diet diversity into two status as 'medium diversity' and 'low diversity' showed that only 40% of the households consumed more than three food groups, suggesting that the rest majority faced low diet quality.

The estimated income and consumption inequalities among households were decomposed into their constituent income and calorie sources as reported in Table 2. The consumption inequality measured by the Gini coefficient was 0.21 and 0.22, respectively, in Central and Hararghe highlands which are nearly similar to the 0.27 national rural consumption inequality estimated in the year 2010/11 (FDRE, 2012). The four food groups used as major sources of household calorie consumption were cereals, roots, and tubers, pulses and legumes, livestock products, and fats and oils. Consumption of cereals, roots and tubers and oils and fats were household calorie sources enabling to reduce consumption inequality in rural Ethiopia. As expected in developing countries, the estimated share in total inequality suggested that consumption inequality was predominantly contributed by consumption of cereal, roots and tubers. The marginal effects indicated that a unit percentage increase in consumption of cereals, roots and tubers reduced consumption inequality by about 0.06%. However, consumption on pulses and legumes and livestock products was source of consumption inequality. On the other hand, the sample households had total income inequality of about 0.45, higher than the national rural income inequality estimated in the same period (0.30), while it was lower in Central highlands (0.38) but higher in Hararghe highlands (0.52). This income inequality was twofold higher than the consumption inequality. Production of crops seems the sole determinant of income inequality in both farming systems. A unit percentage increase in crop income reduced total income inequality by about 1.4%.

Diet quantity and diversity levels

Assuming that daily calorie availability is correlated with

Table 2. Decomposition of income and consumption inequalities by farming systems (marginal effects).

Source	Marginal effects (% change)		
	Central highlands	Hararghe highlands	All
Total calorie (Gini)	0.208	0.219	0.222
Cereal, roots and tubers	-0.049	-0.077	-0.061
Pulses and legumes	-0.007	0.031	0.016
Livestock products	0.004	0.013	0.012
Oils and fats	-0.006	0.033	-0.001
Total income (Gini)	0.380	0.522	0.446
Crop income (Gini)	-2.160	-0.799	-1.393
Livestock income	0.000	0.000	0.000
Off-farm income	-0.129	-0.788	-0.501

Source: Authors' computation (2012).

Table 3. Simultaneous estimation results of daily calorie availability and dietary diversity.

Equations and variable	Coefficients	
	Daily calorie availability	Dietary diversity scores
Female heads	0.13*	0.90***
Family size	-0.05***	-0.01
Literacy status	0.10**	1.21***
Land cultivated	0.16**	-0.08
Land allocated to staples	-0.08	0.16
Irrigation water use	-0.18***	-0.46*
Quantity of fertilizer	0.08***	0.24
Livestock holding (TLU)	0.003	0.12**
Annual gross income (log)	0.06	0.74***
Off-farm activity	-0.14***	0.02
Distance to nearest road	-0.02***	-0.01
Farming system	0.05	3.72***
Constant	7.22***	-2.94*
R^2	0.2746	0.4948
Predicted value $\left(e^{7.54}\right)$	1878.71	6.77
Predicted value (Hararghe highlands)	1828.13	8.92
Predicted value (Central highlands)	1916.69	5.19
Predicted value (literate)	1990.57	7.46
Predicted value (illiterate)	1798.26	6.25
Cross-equation correlation of residuals	0.20	
Breusch-Pagan test of independence	0.001	

***, ** and *, respectively, signify significance levels of 1, 5 and 10%. Source: Authors' computation (2012).

dietary diversity scores of food items, underlying common determinants of food security measures were identified by estimation of the linear SUR model of the two equations. The cross-equation correlation of residuals was strongly significant at 1% level and the null that the two equations are independent was rejected, suggesting that their simultaneous estimation was appropriate (Table 3). About 27.5 and 49.5% of the variation, respectively, in daily calorie availability and dietary diversity scores of food items were explained by the SUR model. The SUR model results demonstrated that the determinants

enhancing daily calorie availability were female heads, literacy status, land cultivated, quantity of chemical fertilizer used, and other shocks captured by the constant term, all of which were in line with the expectations depicted in Table 1 (Hoddinott, 1999; Kennedy et al., 2011). Factors adversely affecting households' daily calorie availability were family size, irrigation water use, participation in off-farm activity, and road distance. The negative effect of off-farm activity on food security reflects the Ethiopian context in the farming systems for the fact that most households engaged in commercial and other

Table 4. Univariate probit estimation of determinants of household food security status.

Determinant	Coefficients (secure=111)	Marginal effects		
		All	Hararghe highlands	Central highlands
Female heads	0.45*	0.18*	0.18*	0.18*
Family size	-0.14***	-0.06***	-0.06***	-0.06***
Land cultivated	0.73***	0.28***	0.28***	0.29***
Land allocated to staples	-0.66**	-0.26**	-0.25**	-0.26**
Irrigation water use	-0.56***	0.21***	-0.20***	-0.21***
Quantity of fertilizer	0.30*	0.12*	0.12*	0.12*
Livestock holding (TLU)	0.03	0.01	0.01	0.01
Annual gross income	0.24*	0.09*	0.09*	0.09*
Off-farm activity	-0.26	-0.10	-0.10	-0.10
Access to credit	0.41*	0.16*	0.16*	0.16*
Farming system	-0.16	-0.06	-0.06	-0.06
Constant	-1.95			
Predicted probability (all)		0.423	0.388	0.449
Predicted probability (with irrigation)		0.270	0.241	0.292
Predicted probability (without irrigation)		0.480	0.444	0.506
Predicted probability (with credit)		0.559	0.523	0.585
Predicted probability (without credit)		0.397	0.362	0.423
Log likelihood		145.06		
Pseudo R^2		0.18		

***, **, and *, respectively, signify significance levels of 1, 5 and 10%. Source: Authors' computation (2012).

off-farm activities are those which are less food secure and resource-deficient. They use off-farm activities as a coping strategy to overcome their food insecurity situation. On the other hand, factors enhancing diet diversity were female heads, literacy status, livestock holding, annual gross income, and the farming system in which the households operate. Factors adversely affecting diet diversity were irrigation water use and other exogenous shocks. In terms of magnitude, the most important determinants were the farming system followed by literacy status and female-headed households. With the exception of irrigation water use, the factors of daily calorie intake and dietary diversity scores were generally in line with the empirical evidences in Ethiopia by Birhanu and Moti (2010), Moti and Gardebroek (2008), Adane (2009), and Mamo et al. (2009).

The predicted values of daily calorie availability per adult equivalent and number of food items consumed were 1878.7 and 6.8, respectively. However, households' daily calorie availability and dietary diversity scorers were significantly different between households in the two farming systems and between their literacy statuses. Households in Central highlands, on average, obtained relatively higher daily calorie (1916.7 kcal) per adult equivalent and lower number of food items (5.2) a week as compared to their counterparts in Hararghe highlands, which obtained some 1828.1 kcal and 8.9 number of food items. The results generally indicated that households in

Central highlands were relatively better-off in their daily calorie availability and worse-off in their dietary diversity. The role of literacy status in enhancing the likelihood of food security and dietary diversity was relatively higher (1990.6 kcal and 7.5 food items) than that of illiterate households (1798.3 kcal and 6.3 food items).

Diet quantity and diversity status

Households' food security status was estimated by a univariate probit model and the results reported in Table 4. Food security status was significantly enhanced by female heads, cultivated land, quantity of chemical fertilizer used, annual gross income, and access to credit. However, their food security status was adversely affected by family size, land allocated to staples, and irrigation water use. The negative effect of irrigation water use can be explained by the prevalent situation most parts of Ethiopia that households with irrigation water access were promoted to produce crops more demanded in the market (cash crops) and the risk involved in the marketing of these crops was eventually harming their food security status.

The marginal effects and the associated probabilities were computed at different combinations of the farming systems, irrigation water use, and credit access. Food security status of households was predominantly

Table 5. Univariate probit estimation of determinants of household dietary diversity status.

Determinant	Coefficients	Marginal effects		
		All	Hararghe highlands	Central highlands
Female heads	0.12	0.05	0.04	0.03
Family size	0.02	0.01	0.01	0.01
Literacy status	0.47***	0.18***	0.16***	0.12**
Land cultivated	0.09	0.03	0.03	0.02
Irrigation water use	-0.22	-0.08	-0.08	-0.05
Quantity of fertilizer	-0.11	-0.04	-0.04	-0.03
Livestock holding (TLU)	0.11***	0.04***	0.04***	0.03***
Annual gross income (log)	0.41***	0.15***	0.14***	0.10***
Off-farm activity	-0.10	-0.04	-0.03	-0.02
Access to credit	-0.13	-0.05	-0.04	-0.03
Distance to nearest road	-0.04	-0.01	-0.01	-0.01
Farming system	1.57***	0.54***	0.54***	0.54
Constant	-5.40***			
Predicted probability (all)		0.372	0.708	0.167
Predicted probability (literate)		0.477	0.792	0.243
Predicted probability (illiterate)		0.299	0.635	0.122
Log likelihood		-135.04		
LR $\chi^2(12)$		79.89		
Pseudo R^2		0.23		

***, **, and *, respectively, signify significance levels of 1, 5 and 10%. Source: Authors' computation (2012).

increased by total cultivated land (28%), female heads (18%), access to credit (16%), quantity of fertilizer used (12%), and annual income (9%). Their food security status was mainly decreased by land allocated to staples (26%), irrigation water use (21%), and family size (6%). The huge and negative marginal effect of land allocated to production of staples suggests that households land allocation to staples could not lead to better food security situation unless optimum land allocation decisions are made. However, the marginal probabilities of households to be food secure in the two farming systems were not largely different.

The likelihood of households to be food secure was about 42.3%, but lower for households in Hararghe highlands (38.8%) as compared to the likelihood of their counterparts in Central highlands (44.9%). These results are consistent with the SUR model results since households' food security status in Central highlands was relatively better-off, even if they were relatively worse-off in their dietary diversity. Households using irrigation water had lower probability to be food secure (27%) compared to those households without irrigation (48%). This negative effect was relatively lessened in Central highlands. But, credit access was relatively more consequential in enhancing food security of households (55.9%) compared to those households without credit (39.7%). This scenario was verified in both farming systems where households with credit access in Central

highlands were more probably foods secure (58.5%) and those without credit less probably food secure (42.3%). Credit access was relatively more effective in improving household food security in Central highlands.

Households' diet diversity status was estimated by a univariate probit model as reported in Table 5. The factors determining households' diet diversity were literacy statuses, livestock holding, annual income, farming system, and the constant term (negatively). The probability of households to diversify their diet quality was largely increased by the farming system (54%) followed by literacy status (18%) and annual income (15%). The likelihood of households to have semi-diversified diet was 37.2%, but higher for those in Hararghe highlands (70.8%) and lower in Central highlands (16.7%). Literate households had higher probability of having better diet diversity (47.7%) compared to the illiterate ones (29.9%). The role of literacy status was also different in the two farming systems as indicated by the predicted probabilities in Hararghe (79.2%) and Central highlands (24.3%), suggesting that food security effects of literacy was more pronounced in Hararghe highlands.

The expected interdependence of food security status and dietary diversity status was disproved by employing the bivariate probit estimation of the two equations (Table 6). The null that food security status and dietary diversity status are independent was accepted since their tetrachoric correlation was weak (13%). Unlike the SUR

Table 6. Bivariate probit estimation of food security and dietary diversity status.

Variable	Coefficients (Equations)	
	Food security equation	Dietary diversity equation
Female heads	0.56**	0.12
Family size	-0.14***	0.02
Literacy status	0.25	0.46***
Land cultivated	0.67***	0.09
Land allocated to staples	-0.57**	-
Irrigation water use	-0.69***	-0.21
Quantity of fertilizer	0.29**	-0.11
Livestock holding (TLU)	0.03	0.11***
Annual gross income (log)	0.23	0.42***
Off-farm activity	-0.32*	-0.10
Access to credit	0.42**	-0.12
Production of major cash crop	-0.37	-
Distance to nearest road	-0.04*	-0.04
Farming system	-	1.51***
Constant	-1.82	-5.45***
Athrho, $\hat{\rho}$	0.19	
Rho, ρ	0.19	
Log pseudo likelihood	-275.541	
Wald $\chi^2 (25)$	133.76	
Wald test of rho=0: $P > \chi^2 (1)$	0.13	

Food security status and dietary diversity status are not significantly interdependent, which suggests that reporting of joint marginal effects is less important. ***, **, and *, respectively, signify significance levels of 1, 5 and 10%. Source: Author's computation (2012).

model results, the bivariate probit representation of the two equations suggested that food security status and dietary diversity status were not significantly interdependent and their univariate estimation was correct. Food security status and dietary diversity status had no common underlying determinants in rural Ethiopia. Though there is strong and positive interdependence between the two linear measures of food security scales, the interdependence between nonlinear measures of diet quantity and quality is very weak. This result suggests that nonlinear measures of diet quantity and quality aspects of household food security in rural Ethiopia were determined by different factors.

To verify the reliability of the negative effects of irrigation water access on food security situation, households' crop choice and other relevant indicators were further investigated with households' status of irrigation water access. The mean comparison test of crop choices as measured by their land allocation suggested that the mean values of most crops chosen by households were significantly different between irrigation water users and nonusers. Nonusers allocated relatively more land to staples (sorghum, wheat, barley, and horse

bean) and less risky cash crops like *khat,* while irrigation water users allocated relatively more land to more risky perishable cash crops (carrot and onions) and less productive cash crops like fenugreek. There was no significant difference between households with and without irrigation water in their land allocation to other crops. This evidence verifies that own production of food was an integral component of smallholders' food security in Ethiopia. There should be innovation and introduction of productive, high value and less-risky cash crops from which households can choose. Moreover, households with irrigation water access should be promoted to engage in the production of such crops. In the existing situation in Ethiopia, the debate on ability of a smallholder-dominated subsistence farm economy to diversify into riskier, high-value crops is intuitive (Birthal et al., 2007; Birhanu et al., 2007; Samuel and Sharp, 2007; Hendriks and Msaki, 2009; Birhanu and Moti, 2010; Langat et al., 2011).

The mean daily calorie intake of irrigation water users was only 1870 kcal, which is relatively lower compared to the daily calorie intake by nonusers (2107 kcal). About 70% of the food insecure households were users of irrigation water. About 65% of the households who participated in crop output markets were irrigation water users, which verify their engagement in the production and marketing of relatively more marketable crops. However, this does not mean that irrigation water use, *ceteris paribus*, adversely affects food security situation. Rather, irrigation water users were relatively less food secure because their crop mix and diversification and commercialization patterns were changed when they used irrigation water. Their increased market participation was not supported by better scale of crop diversification to cope up with the additional risk they most likely faced when they participated in the output markets. Irrigation users were less likely to produce staples (food crops) and more likely to face market risk and to be calorie-deficient. In addition, the negative effect of irrigation water use on food security situation was verified by many empirical evidences specifically in areas where market risk is pronounced (Birhanu et al., 2007; Birhanu and Moti, 2010).

CONCLUSIONS AND POLICY IMPLICATIONS

Food insecurity in Ethiopia, like most developing countries, is an overriding problem of development policy agenda. A number of empirical studies conducted on the subject have proven that food security policies and intervention mechanisms require relevant and inclusive empirical evidence on factors related to poverty reduction and enhancement of food security. This study profoundly examines the food security situation of farm households in Hararghe and Central highlands of Ethiopia and estimates the link between food security measures. Using the major indicators, food security situation of rural households was very low or poor, 57.3% of them suffering from food insecurity problems, primarily dependent on staples for their food energy source, and consuming on a few number of food groups. Food security problems were significantly different across farming systems and household idiosyncrasy. Households in Central highlands were relatively better-off in their daily calorie intake and food security status, while they were worse-off in their dietary diversity level and status.

The daily calorie intake and dietary diversity scores of households were positively interdependent, verifying the expectation that households with better dietary diversity were able to have better diet quantity. The simultaneous estimation of household daily calorie intake and dietary diversity scores suggested that the most important determinants of daily calorie intake were female heads, family size, literacy status, total cultivated land, irrigation water use, quantity of fertilizer used, participation in off-farm activity, road distance, and other exogenous shocks. On the other hand, dietary diversity was determined by female heads, literacy status, irrigation water use, live-

stock holding, annual income, farming system, and other exogenous factors. On average, households obtained about 1878.7 kcal per day per adult equivalent and consumed only 6.8 number of food items a week, with significant difference by farming systems in which the households operate.

Households' food security status was determined by female heads, cultivated land, quantity of chemical fertilizer used, annual gross income, access to credit, family size, land allocated to staples, and irrigation water use. The likelihood of households to be food secure was 42.3% with largely different probabilities and marginal effects among farming systems, irrigation water use, and credit access. Households' dietary diversity status, on the other hand, was determined by literacy statuses, livestock holding, annual income, farming system, and the constant term. The probability of households to have semi-diversified diet was 37.2% with significant difference between farming systems and literacy status of household heads. The bivariate probit estimation of food security status and dietary diversity status of households suggested that these two nonlinear measures were not strongly and significantly interdependent. Accordingly, it is imperative to design appropriate food security intervention strategies since households are more likely to fail in achieving these two-pronged objectives because of limited resource endowments, marketing problems, and other exogenous shocks.

Very important policy implications are derived from this study. One of the basic problems of developing countries like Ethiopia is lack of budge to assess food security situations of households. Estimation of diet quantity available to households is costly, cumbersome and more susceptible to misreporting. The positive linear interdependence between diet quantity and diet diversity is an important evidence to employ cost-effective method of assessing food security situations in Ethiopia. Because dietary diversity of households is associated with their calorie intake, the government can initiate extensive and rapid food security assessment schemes with limited budget in order to formulate and implement relevant food security policies, strategies, and programs. Monitoring of effectiveness of such food security programs will also be cost-effective if dietary intake method of food security analysis is employed.

Idiosyncratic features were significantly influencing food security status of households. Family size was strongly and adversely affecting food security status which necessitates accelerated policy interventions in family planning in order to lessen its negative effects. Literacy status, on the other hand, was an important characteristic feature of household heads in improving daily calorie intake and scale and status of dietary diversity. The current effort in Ethiopia to have educated farmers will have to improve households' calorie supply and dietary diversity in the long run. However, it is also vital to promote adult education in order to improve positive nutritional effects of literacy of farm household in the short run. Moreover, households in the two farming systems have strongly and significantly differentiated socioeconomic characteristic features and other unobserved heterogeneities which would influence their food security situations differently. This is a strong empirical evidence to suggest the need to formulate policies and strategies which should take into account these differentials. Policies related to household food security should incorporate these idiosyncratic features and spatial covariate changes in order to achieve food security objectives in different farming systems of the country.

Land allocation pattern of households was adversely and strongly affecting their food security condition. But quantity of chemical fertilizer used was significantly enhancing food security of households. Promoting and supporting smallholders to make optimal land allocation decision and to use production inputs like fertilizer will further improve household food availability through increased production and productivity. Asset holdings were also important factors influencing diet quality dimensions of household food security, calling the need to improve physical and financial asset holdings through income diversification interventions.

ACKNOWLEDGMENTS

The authors wish to express their sincere acknowledgements to the Swedish International Development Cooperation Agency (SIDA) and Ethiopian Strategy Support Program (ESSP-II) of the Ethiopian Development Research Institute (EDRI) and International Food Policy Research Institute (IFPRI) for their financial support.

REFERENCES

Abebaw S, Janekarnkij P, Wangwacharakul V (2011). Dimensions of food insecurity and adoption of soil conservation technology in rural areas of Gursum district, eastern Ethiopia. Kasetsart J. (Soc. Sci) 32:308-318.

Adane T (2009). Impact of Perennial Cash Cropping on Food Crop Production and Productivity. Eth. J. Econ. 18(1):1-34.

Bogale A, Shimelis A (2009). Household level determinants of food insecurity in rural areas of Dire Dawa, Eastern Ethiopia. AJFAND 9(9):1914-1926.

Berhanu A (2004). The food security role of agriculture in Ethiopia. eJADE 1(1):138-153.

Birhanu G, Adane H, Kahsay B (2007). Feed marketing in Ethiopia: results of rapid market appraisal. Addis Ababa, Ethiopia.

Birhanu G, Moti J (2010). Commercialization of smallholders: Does market orientation translate into market participation? ILRI (International Livestock Research Institute), Addis Ababa, Ethiopia.

Birthal PS, Joshi Pk, Thorat A (2007). Diversification in Indian agriculture towards high-value crops: The role of smallholders. Available at http://www.ifpri.org/sites/default/files/publications/ifpridp00727.pdf, accessed October 20, 2012.

Bickel G, Nord M, Price C, Hamilton W, Cook J (2000). Guide to measuring household food security. United States Department of Agriculture (USDA), USA.

Block S, Webb, P (2001). The dynamics of livelihood diversification in post-famine Ethiopia. Food Policy 26(4):333-350.

Brown T, Teshome A (2007). Implementing policies for chronic poverty in Ethiopia. Addis Ababa, Ethiopia.

Cameron AC, Trivedi TK (1998). Regression Analysis of Count Data. Cambridge: Cambridge University Press.

Cameron AC, Trivedi TK (2009). Microeconometrics Using Stata. StataCorp Ld, USA.

De Luca G (2008). SNP and SML estimation of univariate and bivariate binary-choice models. Stata J. 8:190–220.

Dercon S, Hoddinott J (2009). The Ethiopian rural household surveys 1989-2004: Introduction. FPRI, Addis Ababa, Ethiopia.

FANTA (Food and Nutrition Technical Assistance) (2006). Developing and validating simple indicators of dietary quality and energy intake of infants and young children in developing countries: Summary of findings from analysis of 10 data sets (available at www.fantaproject).

FAO and WHO (1985). Energy and protein requirements. Technical Report Series 724. Geneva.

FAO (2007). Guidelines for measuring household and individual dietary diversity. Rome, Italy.

FAO (1996). World food summit plan of action (available at www.fao.org).

FDRE (Federal Democratic Republic of Ethiopia) (1996). Food Security Strategy. Addis Ababa, Ethiopia.

FDRE (2004). Food Security Program 2004-2009. Addis Ababa, Ethiopia.

FDRE (2012). Ethiopia's Progress towards eradicating poverty: An interim report on poverty analysis study (2010/11). MoFED. Addis Ababa, Ethiopia.

Freihiwot F (2007). Food security and its determinants in rural households in Amhara Region. Ethiopian Development Research Institute (EDRI), Addis Ababa, Ethiopia.

Getahun A (1980). Agro-climates and agricultural systems in Ethiopia. Agricultural Systems, 5(1): 39-50.

Greene WH (2012). Econometric Analysis (7th Edition). New Jersey: Pearson Hall. pp. 292-293.

Guush B, Zelekawork P, Kibrom T, Seneshaw T (2011). Foodgrain consumption and calorie intake patterns in Ethiopia. Available at http://img.static.reliefweb.int/sites/reliefweb.int/files/resources, accessed on July 9, 2012.

Hadleya C, Linzerb DA, Tefera B, Abebe G, Fasil T, Lindstrome D (2011). Household capacities, vulnerabilities and food insecurity: Shifts in food insecurity in urban and rural Ethiopia during the 2008 food crisis. Soc. Sci. Med. 73:1534-1542.

Hailu M (2012). Causes of Household Food Insecurity in Rural Boset Woreda: Causes, Extent and Coping Mechanisms to Food Insecurity. LAP LAMBERT Academic Publishing.

Hardin JW (1996). Bivariate probit models. Stata Technical Bulletin Reprints, College Station, TX: Stata Press. 6:152-158.

Hendriks SL, Msaki M (2009). The impact of smallholder commercialization of organic crops on food consumption patterns, dietary diversity and consumption elasticities. Agrekon, 48(2):184-199.

Hoddinott J, Yohannes Y (2002). Dietary diversity as a food security indicator. FANTA 2002, Washington, DC. (available at http://www.aed.org/Health/upload/dietarydiversity.pdf, accessed 12/07/2012: 4:25 am).

Hoddinott J (1999). Choosing outcome indicators of household hood security. IFPRI, Washnigton, D.C, USA.

Kennedy G, Ballard T, Dop M (2011). Guidelines for measuring household and individual dietary diversity. FAO, Rome.

Langat BK, Ngéno VK, Sulo TK, Nyangweso PM, Korir MK, Kipsat MJ, JS, Kebenei JS (2011). Household food security in a commercialized subsistence economy: A case of smallholder tea famers in Nandi south district, Kenya. J. Dev. Agric. Econ. 3(5):201-209.

Long JS, Freese J (2005). Regression Models for Categorical Dependent Variables Using Stata. 2nd edition, Stata Press, USA.

Long JS (1997). Regression Models for Categorical and Limited

Dependent Variables (RMCLDV), Thousand Oaks, CA: Sage Press.

Maddala GS (1983). Limited-Dependent and Qualitative Variables in Econometrics. Cambridge, Cambridge University Press.

Mamo G, Assefa A, Degnet A (2009). Determinants of smallholder crop farmers' decision to sell and for whom to sell: Micro-level data evidence from Ethiopia. In: Getnet Alemu and Worku Gebeyehu (eds), Proceedings of the Ninth International Conference on the Ethiopian Economy, Addis Ababa, Ethiopia. pp. 47-76,

Maxwell S, Frankenberger T (1992). Household Food Security: Concepts, Indicators, and Measurements. A Technical Review. UNICEF/IFAD.

Maxwell D (1996). Measuring food insecurity: The frequency and severity of coping strategies. *Food Policy 21:291–303.*

MoFED (Ministry of Finance and Economic Development) (2006). Ethiopia: Building on Progress: A Plan for Accelerated and Sustained Development to End Poverty (PASDEP), Addis Ababa, Ethiopia.

Moti J Gardebroek C (2008). Crop and market outlet choice interactions at household level. Eth. J. Econ. 7(1):29-47.

Radimer K, Olson C, Campbell C (1990). Development of indicators to assess hunger. J. Nutr. 120:1544–1548.

Ruel MT (2002). Is dietary diversity an indicator of food security or dietary quality? A review of measurement issues and research needs. Food Consumption and Nutrition Division Discussion Paper 140, IFPRI, Washington, D.C.

Ruel MT (2003). Operationalizing diet diversity: A review of measurement issues and research priorities. J. Nutr. 133(11):3911S–3926S.

Samuel G (2004). Food insecurity and poverty in Ethiopia: Evidence and lessons from Wollo. Ethiopian Economics Association (EEA) Working Paper No. 3/2004, Addis Ababa, Ethiopia.

Samuel G, Sharp K (2007). Agricultural commercialization in coffee growing areas of *Ethiopia. Eth.J. Eco.,16(1): 89-118.*

Smith LC, Subandoro A (2007). Measuring Food Security in Practice. IFPRI, Washington, D.C.

Swindale A, Ohri-Vachaspati P (2005). Measuring household food consumption: A technical guide. FANTA Project, Academy for Educational Development (AED), Washington, D.C.

Uraguchi ZB (2012). Rural income transfer programs and rural household food security in Ethiopia. J. Asian Afri.Stud., 47(1):33-51.

World Bank (2008). Agriculture for Development. World Development Report 2008, The World Bank. Washington, DC.

Zegeye T, Hussien H (2011). Farm households' food insecurity, determinants and coping strategies: The case of Fadis district, eastern Oromia, Ethiopia. Eth. J. Agric. Econ. 8(1):1-35.

Zellner A (1962). Further properties of efficient estimators in seemingly unrelated regression equations. Int. Econ. Rev. 3:300-313.

5

Livelihood diversification and welfare of rural households in Ondo State, Nigeria

Adepoju Abimbola O. and Obayelu Oluwakemi A.

Department of Agricultural Economics, University of Ibadan, Oyo State, Nigeria.

Agriculture, the main source of livelihood in Nigeria, especially in the rural areas, is plagued with various problems. As a result, most of the rural households are poor and are beginning to diversify their livelihoods into off and non-farm activities as a relevant source of income. This study examined the effect of livelihood diversification on the welfare of rural households in Ondo State. Primary data used in the study were obtained from 143 respondents selected employing a multistage sampling technique. Data were analyzed using descriptive statistics, multinomial logit and the logit regression models. The distribution of respondents by the type of livelihood strategy adopted revealed that almost three-quarters of the respondents adopted the combination of farm and nonfarm strategy. Econometric analysis showed that household size, total household income and primary education of the household head were the dominant factors influencing the choice of livelihood strategies adopted. Income from non-farm activities, as well as income from a combination of non-farm and farming activities, impacted welfare positively relative to income from farming activities. The study recommends the promotion of non-farm employment as a good strategy for supplementing the income of farmers as well as sustaining equitable rural growth.

Key words: Ondo State, livelihood diversification, welfare, rural households, Nigeria.

INTRODUCTION

In Africa, various studies have shown that while most rural households are involved in agricultural activities such as livestock, crop or fish production as their main source of livelihood, they also engage in other income generating activities to augment their main source of income. A majority of rural producers have historically diversified their productive activities to encompass a range of other productive areas. In other words, very few of them collect all their income from only one source, hold all their wealth in the form of any single asset, or use their resources in just one activity (Barrett et al., 2001). In Nigeria, the agricultural sector is plagued with problems which include soil infertility, infrastructural inadequacy, risk and uncertainty and seasonality among others. Thus,

rural households are forced to develop strategies to cope with increasing vulnerability associated with agricultural production through diversification, intensification and migration or moving out of farming (Ellis, 2000). In other words, the situation in the rural areas has negative welfare implications and predisposes the rural populace to various risks which threaten their livelihoods and their existence. As a result of this struggle to survive and in order to improve their welfare, off-farm and non-farm activities have become an important component of livelihood strategies among rural households in Nigeria. Further, the growing interest in research on rural off-farm and non-farm income in rural economies is increasingly showing that rural peoples' livelihoods are derived from diverse sources and are not as overwhelmingly dependent on agriculture as previously assumed (Gordon and Craig, 2001). This could be owing to the fact that a diversified livelihood, which is an important feature of rural survival and closely allied to flexibility, resilience and stability is less vulnerable than an undiversified one, this is due to the likelihood of it being more sustainable over time and its ability to adapt to changing circumstances. In addition, several studies have reported a substantial and increasing share of off-farm income in total household income (Ruben and van den Berg, 2001; de Janvry and Sadoulet, 2001; Haggblade et al., 2007). Reasons for this observed income diversification include declining farm incomes and the desire to insure against agricultural production and market risks (Matsumoto et al., 2006). In other words, while some households are forced into off-farm and non-farm activities, owing to less gains and increased uncertainties associated with farming (crop and market failures), others would take up off-farm employment when returns to off-farm employment are higher or less risky than in agriculture. Mainly, households diversify into non-farm and off-farm activities in their struggle for survival and in order to improve their welfare in terms of health care, housing, sustenance, covering, etc. Thus, the importance and impact of non-agricultural activities on the welfare of rural farm households can no longer be ignored.

An understanding of the significance and nature of non-farm and off-farm activities (especially its contribution to rural household income or resilience) is of utmost importance for policy makers in the design of potent agricultural and rural development policies. Further, the rising incidence of low level of welfare of rural households in Nigeria, that remains unabated despite various policy reforms undertaken in the country, requires a deeper understanding of the problem and the need to proffer solutions to the problem through approaches that place priority on the poor and ways on which rural households through diversification can maintain their livelihood.

LITERATURE REVIEW

In Africa, the average share of rural non-farm incomes as proportion of total rural incomes, at 42%, is higher than in Latin America and higher still than in Asia (Reardon et al., 2000). Most evidence shows that rural non-farm activity in Africa is fairly evenly divided across commerce, manufacturing and services, linked directly or indirectly to local agriculture or small towns, and is largely informal rather than formal. Also, while households earn much more from rural nonfarm activity than farm wage labour, non-farm wage labour is still more important than self-employment in the non-farm sector (Reardon, 1999; Haggblade et al., 2007). Hussein and Nelson (1998) in their study on sustainable livelihood and livelihood diversification concluded that while livelihood diversification is normal for most people in rural areas of

developing countries in Africa, non-agricultural activities are critical components of the diversification process. Further, livelihood diversification is pursued for a mixture of motivations and these vary according to context: from a desire to accumulate, invest and the need to spread risk or maintain incomes, to a requirement to adapt to survive in eroding circumstances or some combination of these. In addition, the character of livelihood diversification is dependent primarily upon the context within which it is occurring (the differential access to diversification activities and the distribution of the benefits of diversification). However, the poorest rural groups probably have the fewest opportunities to diversify in a way that will lead to accumulation for investment purposes.

According to Babatunde and Quaim (2009), the pattern of income diversification among rural households in Nigeria, showed that majority of the households have fairly diversified income sources. On the average, while only 50% of the total household income is generated from farming, the rest comes from different off-farm sources. However, there are notable differences across income strata. While farming remains the dominant income source for the poorest, off-farm occupation especially self-employed activities are the main sources of income for relatively richer households. Also, Ellis (2000) using regression models, showed that households have unequal abilities to diversify their income sources and that education, asset, endowment, access to credit, and good infrastructure conditions, increase the levels of household diversification. These factors improve the opportunity to start own business and find employment in the higher paying non-farm sector.

In other words, resource-poor households in remote areas are constrained in diversifying their income sources. Ibekwe et al. (2010) using double log regression, noted that a distress diversification hypothesis is supported by the negative relationship between nonfarm income and the farm output per hectare of land in South Eastern Nigeria. They accounted for household's involvement in nonfarm activities by reference to their demographic features and to other household specific characteristics such as occupation, education level, number of spouse(s), family size and land holding as well as farm output. It could be inferred from the result that land holding size, years of workers education, per hectare value of agricultural output, occupation and age of household head are the major determinants of nonfarm income at the household level in South Eastern Nigeria . The study suggested that economic and social factors should matter in nonfarm sector policy in Southeast Nigeria if diversification is to be encouraged.

MATERIALS AND METHODS

This study was carried out in Odigbo Local Government Area of Ondo State, Nigeria. Odigbo Local Government is headquartered in the town of Ore town. It has an area of 1,818 km², a population of 230,351 persons and 11 wards (NPC, 2006). The major vegetation type in the area is rainforest with a slopy topography. The area is predominantly agrarian and notable food and cash crops grown in the area include: plantain, banana, cassava, maize, yam, cocoa, oil palm and kola. The region has averagely high temperature which ranges between 21 to 29°C and high relative humidity with two distinct seasons namely: the rainy season which lasts from March/April to October/November and the dry season which lasts from the rest of the year October/November till March/April. Primary data used in this study were obtained in a cross-sectional survey of rural households in the study area. The collection of data involved the use of structured questionnaires to obtain information on socio-economic and demographic characteristics such as household size, level of education, age of household heads, land holdings etc. as well as consumption expenditure, other indicators of well-being of the rural households and diversification activities of the respondents.

A multistage sampling technique was employed in selecting the representative households used for this study. The first stage was the purposive selection of Odigbo Local Government Area out of the eighteen Local Government Areas in Ondo State owing to the predominantly rural nature of the area. In the second stage, three wards out of the eleven wards in the Local Government were randomly selected while the third stage involved the selection of households based on probability proportionate to size of the households in the wards. Consequently, a random sample of 54 respondents were sampled in Oniparaga ward, 45 respondents from Ago-Alaye ward and 51 respondents from Araromi-Obu ward making a total of 150 households. However, due to incomplete questionnaire information by seven of the respondents, only information from 143 households was used for the study. These 143 households constituted the sample size used for the study. The analytical techniques employed in this study include: descriptive statistics, multinomial logistic regression and the logit regression models.

Multinomial logistic regression

When there is a dependent categorical variable, the multinomial logistic regression model is commonly used. The regressors are the same across all choices for each observation. The model is specified as:

$$P_{r(Y_i=j)}= \frac{\exp\left(X_i\beta_j\right)}{\sum_{j=0}^{J}\exp\left(X_i\beta_j\right)} \quad j=0----2$$

Where Y_i = 3 unordered categories of livelihood strategies adopted by the respondents: Y1 = those who adopt non-farm strategy alone; Y2 = those who adopt a combination of farm and nonfarm strategies; Y0 = those who adopt farm strategy alone; Y0 is the reference case.

Welfare measurement

Following the adoption of Foster, Greer and Thorbecke- FGT (1984) class of poverty measures, households' total monthly expenditure was used to determine households' poverty status. The poverty line was constructed as two-thirds of the mean monthly per-capita expenditure of all households. This approach has been used by several researchers and institutions (NBS, 2005; Oni and Yusuf, 2008) as a measure of welfare. Households were then classified into their poverty status based on the poverty line:

$$FGT_\alpha = \frac{1}{N}\sum_{i=1}^{H}\left(\frac{z-y_i}{z}\right)^\alpha$$

Where Z is the poverty line; N is the total number of people; H is the number of poor who are below Z; y_i is the expenditure of the ith individual; α is a "sensitivity" parameter which can take values between 0 and 2.

Hence, non-poor households were those whose monthly expenditure was above or was equal to two-thirds of the mean per capita expenditure of all households while those whose per capita expenditure was below two-thirds of the mean monthly per capita expenditure were classified as poor.

Logit regression model

Logit regression analysis examines the influence of various factors on a dichotomous outcome by estimating the probability of the event's occurrence. It does this by examining the relationship between one or more independent variables and the log odds of the dichotomous outcome by calculating changes in the log odds of the dependent as opposed to the dependent variable itself. The log odds ratio is the ratio of two odds and it is a summary measure of the relationship between two variables (Olayemi et al., 1995). The Logit model is presented as:

$$P=\frac{\exp(z)}{1+\exp(z)} \tag{1}$$

Where P is the proportion of occurrence.

$$Z = \beta_0 + \beta_1X_1+ \ldots\ldots +\beta_nX_n \tag{2}$$

Where $X_1 \ldots X_n$ are the explanatory variables. The inverse relation of Equation 1 is:

$$Z = \ln\left(\frac{P}{1-P}\right) \tag{3}$$

That is, the natural logarithm of the odds ratio, known as the logit. It transforms P which is restricted to the range [0, 1] to a range [−∞, ∞].

Y = Poverty status of households (Poor = 1, 0 otherwise).

The independent variables include:

X_1 = Age of the respondents (in years);
X_2 = Gender of household head (male = 1, 0 if otherwise);
X_3 = Primary education (yes = 1, 0 if otherwise);
X_4 = Secondary education (yes = 1, 0 if otherwise);
X_5 = Tertiary education (yes = 1, 0 if otherwise);
X_6 = Primary occupation of household head (farming = 1, 0 if otherwise);
X_7 = Own house (yes = 1, 0 if otherwise);
X_8 = Household size;
X_9 = Marital status (married = 1, 0 if otherwise);
X_{10} = Total household income (₦).

RESULTS AND DISCUSSION

Table 1 presents the socio-economic characteristics of

Table 1. Socio- economic characteristics of the respondents.

Variable	Frequency	Percentage
Age		
20-39	36	25.2
40-59	79	55.2
≥60	28	19.6
Sex		
Male	121	84.6
Female	22	15.4
Marital status		
Single	6	4.2
Married	117	81.8
Seperated/divorced	7	4.9
Widowed	13	9.1
Household size		
1-6	86	60.1
7-12	55	38.5
>13	2	1.4
Educational status		
No formal education (NFE)	23	16.1
Primary	55	38.5
Secondary	42	29.4
Tertiary	23	16.0
Primary occupation		
Farming	79	55.2
Artisan	16	11.2
Trading	31	21.7
Govt. salaried Job	17	11.9
Type of livelihood strategy		
Strategy adopted		
Farm only	10	7.0
Non farm only	28	19.6
Farm and non farm	105	73.4
Monthly income (N)		
11,000 - 30,000	50	35.0
31,000 - 50,000	56	39.2
51,000 - 70,000	18	12.6
>70, 000	19	13.2
Total	**143**	**100.0**

Source: Field survey (2012).

the respondents. Results revealed that more than four-fifths (84.6%) of the households were headed by males while more than half (55.3%) of the respondents were in their economic active age. The average age of the respondents stood at 47.5 ± 11.9 years in the study area. While married household heads were in the majority (81.8%) in the study area, about three-fifths of the respondents had household sizes of between 1 and 6

Table 2. Reasons for diversification.

Reason for diversification	Frequency	Percentage
Limited agricultural income	7	4.9
Large family	2	1.4
Availability of nonfarm opportunities	3	2.1
Seasonal nature of agric produce	3	2.1
Favourable demand for goods and services	7	4.9
To live well	11	7.7
Limited agricultural income and large family	67	46.9
Limited agricultural income, large family and availability of non farm opportunities	20	14.0
Seasonal nature of agric produce and availability of non-farm opportunities	23	16.0

Source: Field survey (2012).

Table 3. Factors influencing the choice of livelihood strategy adopted.

Variable	Nonfarm		Combination of farm and nonfarm	
	dy/dx	Z	dy/dx	Z
Gender	14.10	0.01	12.63	0.01
Age	-0.057	-1.07	-0.001	-0.02
Household size	0.89	2.92*	0.65	2.53*
Total income	0.07	2.11**	0.001	1.86***
Own house	0.26	0.18	0.13	0.09
Married	-14.20	-0.01	-14.04	-0.01
Primary education	-2.55	-1.79***	-1.74	-1.70***
Secondary education	14.97	0.01	14.81	0.01
Tertiary education	-1.66	-0.84	-2.16	-1.26

Source: Field survey (2012). * significant at 1%, **significant at 5%, ***significant at 10%. Number of observation = 143. LR chi^2 (18) = 59.58. Prob> chi^2 = 0.0000. Log likelihood = -73.056464., Pseudo R^2 = 0.2896.

members. The average household size stood at 6.1 ± 2.6 in the study area. With respect to the educational status of the respondents, almost two-fifths of the respondents had primary education while only 16.1 had no formal education. This implies that most of the respondents have one form of formal education or the other. Highlights of the occupational analysis of the respondents revealed that more than half of the respondents were engaged in farming as their primary occupation, indicating that farming is the predominant occupation in the study area. This is expected as most households in the rural areas depend mainly on agriculture as their primary source of livelihood. However, literature has shown that diverse income portfolio, creates more income and distributes income more evenly. Thus, it is easier to adopt the combined livelihood strategies than switching full time between either of them (Ellis, 2000). In line with this, as shown in the table, very few of the respondents obtained income from only one source as almost three-quarters of the household heads engaged in a combination of farm and nonfarm activities. With respect to the monthly income distribution of the respondents, more than half of

the respondents earn between ₦31,000 and ₦70,000 monthly while a little above one-tenth of the sampled respondents earn over ₦70,000 per month. The average monthly income of the respondents in the study area stood at ₦46,533 ± ₦24,315.

As presented in Table 2, most of the respondents had various reasons for diversifying into other activities. Some of these reasons include limited agricultural income, large family size, availability of non-farm opportunities, seasonal nature of agricultural produce, favourable demand for goods and services or a combination of these, among others. However, the main reason for diversification reported by almost half of the respondents in the study area was a combination of limited agricultural income and large family size.

The result of the multinomial regression analysis of the factors influencing the choice of livelihood strategies adopted by the respondents in Odigbo Local Government of Ondo state is presented in Table 3. The chi-square value of 59.58 which was significant at 1% level shows that the model has a good fit for the data. The marginal effects result of the regression is reported as follows.

Table 4. Poverty status of households.

Poverty status	Frequency	Percentage
Non-poor	82	57.3
Poor	61	42.7
Total	143	100.00

Source: Field survey (2012).

The coefficient of household size of 0.89 was significant at 1%, implying that an increase in the household size by one member increased the likelihood of adopting the only non-farm strategy by 0.89 relative to the adoption of the only farm strategy. That is, the larger the household size, the higher the likelihood of opting for the only non-farm livelihood strategy. This result is inconsonance with the findings of Harjes (2007), in which increase in household size increased the likelihood of adopting nonfarm activities. Similarly, the coefficient of total income of household of 0.07 was positive and significant, implying that a naira increase in total household income increased the likelihood of adopting the only non-farm strategy relative to the only farm strategy.

This may be owing to the fact that nonfarm activities give higher returns in terms of income. This finding corroborates the findings of Babatunde and Qaim (2009). On the other hand, the coefficient of primary education was negative (-2.55) and significant indicating that household heads with primary education are less likely to adopt the only non-farm strategy relative to the only farm strategy where they are likely to have better prospects. This result is supported by the findings of Norsida and Sadiya (2009) that individuals who have more years of schooling have a higher likelihood of participating in non-farm work. In other words, the higher the level of education, the higher the likelihood of participation in non-farm activities.

With respect to the choice of the combination of farm and non-farm strategy as a livelihood option, the coefficients of household size and total household income were positive and significant suggesting that a member increase in the household size and a naira increase in total household income increased the likelihood of adopting a combination of farm and nonfarm strategy. This could be owing to the fact that in large sized households, limited resources are spread thinly on maintaining a large number of people in terms of meeting their basic and other needs and the fact that increased household size is also synonymous with more depen-dants who do not contribute to household income. Thus, households in order to augment household income for meeting the basic needs of the family will engage in a combination of farm and non-farm strategy relative to the choice of the farm strategy only. This result corroborates

the findings of Babatunde and Qaim (2009) and Ellis (2000).

On the other hand, the coefficient of primary education was negative and significant implying that household heads with primary education are less likely to adopt a combination of farm and nonfarm strategy. From these findings, it is evident that the major factors influencing the choice of livelihood strategy adopted in Odigbo Local Government area of Ondo State are household size, total household income and primary education of the household head. Per-capita expenditure was used as a proxy for welfare in this study. Based on this, the poverty line constructed as two-thirds of the mean per-capita expenditure of all the households stood at ₦2,752.03. This implies that households whose per capita expenditure fall below ₦2,752.03 were classified as poor while households whose per capita expenditure equaled or was above the poverty line were classified as non-poor.

Based on the poverty line, households were classified into their poverty status as either non-poor or poor as presented in Table 4. The table shows that 42.7% of households in Odigbo local government area of Ondo state are poor while 57.3% are non-poor. Table 5 presents the effect of livelihood diversification as well as other socio-economic factors that influence rural households' welfare in Odigbo Local Government area of Ondo State. The 'chi square' value of 107.35 which was significant at 1% indicates that the model has a good fit. The results of the marginal effects after Logit are reported as follows:

The coefficient of gender was negative and significant implying that households headed by males have a lower level of welfare than their female counterparts. Specifically, being a male headed household increased the likelihood of being poor by 0.313. Similarly, the coefficient of secondary education was negative implying that household heads with secondary education have a lower likelihood of being poor relative to those with no formal education. On the other hand, the coefficient of household size was positive indicating that a member increase in household size increased the likelihood of being poor by 0.132. This could be as a result of greater burden on the actively working members of the household.

While the coefficient of the use of firewood as a source of energy for cooking was positive, the coefficient of living in a flat/apartment was negative. This implies that households using firewood as a source of energy for cooking have a higher likelihood of being poor, while households living in a flat/apartment have a lower likelihood of being poor. These are reflections of the level of welfare of the households as these variables are usually determined by the level of income of such households. Income from non-farm activities as well as income from a combination of non-farm and farming activities, impacted welfare positively relative to income

Table 5. Effect of livelihood diversification on household welfare.

Variable	dy/dx	Coefficient	Z
Gender	-0.313	-3.546	-2.34**
Age	-0.086	-0.050	-1.54
Household size	0.132	0.772	3.56*
Married	-0.295	-1.424	-0.57
Primary education of HH	-0.209	-1.393	-1.30
Secondary education of HH	-0.287	-1.997	-1.74***
Tertiary education of HH	-0.149	-1.087	-0.86
Firewood	0.355	2.577	3.23*
Own house	0.127	0.836	1.24
Protective well	-0.100	-0.563	-0.86
Flat/apartment	-0.215	-1.351	-2.01**
Non farm income	-0.036	-3.299	- 4.52*
Farm + non-farm income	-0.411	-2.501	-3.09*

Source: Field survey (2012). *Significant at 1%, **significant at 5%, ***significant at 10%. Number of observation = 143. LR chi^2 (14) = 107.35. Prob> chi^2 = 0.0000. Log likelihood = -43.897308. Pseudo R^2 = 0.5501.

from farming activities only. This is expected as agriculture in the rural areas of Nigeria is largely characterized by low capital involvement, use of crude implements, poor infrastructural and storage facilities and human drudgery which ultimately leads to lower average earnings. Hence, nonfarm activities and a combination of farm and non-farm activities were pursued as strategies to increase household welfare in the study area.

CONCLUSION AND RECOMMENDATIONS

This study has shown that non-farm income plays a very important role in augmenting farm-income as almost three-quarters of the respondents adopted a combination of farm and nonfarm strategy. This is an indication that farming alone is not an adequate source of revenue for the rural households. Therefore, promoting non-farm employment may be a good strategy for supplementing the income of farmers as well as sustaining equitable rural growth. This could be achieved through training programmes directed towards training farmers in skills that can be used in non-farm jobs in their vicinity as well as improvements in infrastructure, education and financial markets.

Specifically, engagement in non-farm activities, apart from reducing income uncertainties and providing a source of liquidity in areas where credit is constrained, could increase agricultural productivity as it provides the resources necessary for investment in advanced agricultural technologies. The adoption of better technology is expected to be highly profitable and will encourage the transition from traditional to modern agriculture. Therefore, there is a need for the government to formulate policies to increase the availability of non-

farm jobs in the rural areas. Further, the private sector should be encouraged to create income-generating activities in the rural areas to enhance their livelihood diversification activities and ultimately their living standard.

REFERENCES

Babatunde RO, Qaim M (2009). The role of off-farm income diversification in rural nigeria: driving forces and household access. Conference paper presented on 23 mar 2009 at the Centre for the Study of African Economies (CSAE), Economics Department, Oxford. http /conferences/2009-EDiA/papers/051-Babatunde.pdf-[28/02/10].
Barrett CB, Reardon T, Webb P (2001). "Nonfarm income diversification and household livelihood strategies in rural Africa: concepts, dynamics and policy implications, Food Pol. 26:315-331.
De Janvry A, Sadoulet E (2001). Income strategies among rural households in mexico. The role of off- farm activities. W. Dev. 29(3):467-480.
Ellis F (2000). The determinants of rural livelihood diversification in developing countries. J. Agric. Econ. 51(2):289-302.
Foster J, Greer J, Thorbecke E (1984). A Class of Decomposible Poverty Measures. Econometrica 2(81):761-766.
Gordon A, Craig C (2001). Rural non-farm activities and poverty alleviation in sub-Saharan Africa." Social and economic development department. Natural Resources Institute. Policy Series. P. 14.
Haggblade S, Hazell P, Reardon T (2007). Transforming the rural non-farm economy. John Hopkins University Press Baltimore.
Harjes T (2007). Globalization and income inequality: An European perspective," IMF Working Paper 07/169 (Washington: International Monetary Fund).
Hussein K, Nelson J (1998). Sustainable livelihoods and livelihood diversification, IDS working, Brighton: Institute of development studies. P. 69.
Ibekwe UC, Eze CC, Ohajianya DO, Orebiyi JS, Onyemauwa CS, Korie OC (2010). Determinants of nonfarm income among farm households in South East Nigeria. Acad. Ari. 2(8):29-33.
Matsumoto T, Kijima Y Yamano T (2006). The role of local nonfarm activities and migration in reducing poverty: evidence from Ethiopia, Kenya, and Uganda. Agric. Econ. 35:449-458.
National Population Commission (2006). National Census Report.

NBS (2005). Annual Needs and Livelihood Assessment. Millenium Development Goals Brochure, South Sudan.

Norsida M, Sadiya SI (2009). Off-farm employment participation among paddy farmers in the Muda Agricultural Development Authority and Kemasin Semerak granary areas of Malaysia" Asia-Pacific Dev. J. 16(2):141-153.

Olayemi JK (1995). A Survey of Approaches to Poverty Alleviation. A Paper presented at the NCEMA National Workshop on Integration of Poverty Alleviation Strategies into Plans and Programmes in Nigeria Ibadan, 27 November- 1 December.

Oni OA, Yusuf SA (2008). Determinants of Expected Poverty among Rural Households in Nigeria. AERC Research Report 183, September.

Reardon T (1999). Rural non-farm income in developing countries. Rome: Food and Agriculture Organization.

Reardon T, Taylor JE, Stamoulis K, Lanjouw P, Balisacan A (2000). Effects of nonfarm employment on rural income inequality in developing countries: An investment perspective. J. Agric. Econ. 51(2):266-288.

Ruben R, Van den Berg M (2001). Non farm employment and poverty alleviation of rural households in Honduras. World Develop. 29(3):549-560.

Yusuf SA, Oni OA (2008). Expected poverty profile among rural households in Nigeria. Afr. J. Econ. Pol. 15(1):139-163

6

Assessment of effect of climate change on the livelihood of pastoralists in Kwara State, Nigeria

I. F. Ayanda

Department of Agricultural Economics and Extension Services, Kwara State University, Malete, Nigeria.

The study examined socio-economic characteristics of pastoralists, investigated perceived effects of climate change on grazing land, herd's performance and changes in livelihood of the pastoralists. Through a multi-stage sampling technique, 140 pastoralists were randomly selected. Data were collected using interview scheduled and analyzed by percentages, frequency, tables and Chi square statistical tools. The result of the study showed that respondents were with an average age of 49.7 years. 10.8 and 5% of the pastoralists had primary and secondary education, respectively. Furthermore, 67.5% of the pastoralists strongly agreed that pattern of rainfall in recent time affects pasture availability while 47.5 and 52.5% reported a decline in milk production and an increase in herd mortality respectively. Pastoralists advanced diminishing land for cattle grazing, poor quality pasture, inadequate income and a decline in cattle productivity as reasons for diversifying into crop production and other enterprises. A significance relationship was established between herd's milk production and factor of climate change (calculated x^2 = 52.00, tabulated x^2 = 7.8147, $p \leq 0.00$). It was concluded that climate change adversely affected livestock performance. Pastoralists should be encouraged through extension services to diversify production while livestock rearing is not compromised. This in turn will fast track Nigeria's strive for self-sufficiency in food production and employment generation.

Key words: Irregular rainfall pattern, declining grazing land, low herd production, income, crop, other enterprises.

INTRODUCTION

Transhumance pastoralism was originally a way of life among communities whose lives and livelihood are inseparably intertwined with cattle, goats, sheep and other ruminant species that depend on natural rangeland for grazing resources. In spite of the advent of monetized economy, pastoralism has remained a veritable source of livelihood and food security as cattle, goats and sheep perform economic, as well as traditional, social and exchange functions. However, the world is witnessing the adverse effects of climate change which include frequency and intensity of storm, thunder, flood, drought,

hurricanes, increased frequency of fire, poverty, reduced agriculture productivities, adverse effects on grazing land and pasture quality. It had a cumulative effect on natural resources and disruption of eco-system. The impact of climate change can be vast. In Nigeria, this means that some stable ecosystems such as the Sahel Savanna may become vulnerable because warming will reinforce existing patterns of water scarcity, increasing the risk of drought in Nigeria and most countries in West Africa. It is obvious from the definition that climate change is an inherent attribute of climate, which is caused by both human activities (anthropogenic) and natural processes (bio-geographic) (IPCC, 1996). As a result of climate change, the pastoralists migrated from the northern parts of the country to southern parts in search of pasture and water. The migration increases pressure on land use. Climate change also influences the existing vegetation type which favours cattle production in many southern parts of Nigeria. Presently, some of the land-use practice of the pastoral Fulanis such as seasonal bush burning along the grazing orbits for regeneration of pasture, periodic movement of the huts or dwelling place within the settlement areas, intensification of land use, shifting cultivation with short fallow periods and lack of commitment to investment in long-term land improvement initiatives such as incorporation of leguminous species into pasture or grazing land cover, loss of bio-diversity is capable of compromising the integrity and resilience of the ecosystem (Ayoade, 2004) and are plausible reasons for the ubiquitous face off between crop farmers and pastoralists.

Climate change as suggested by some researchers could impact the economic viability of livestock production systems worldwide. Surrounding environmental conditions directly affect mechanisms and rates of heat gain or loss by all animals (NRC, 2002). Lack of prior conditioning of livestock to weather events often results to catastrophic losses in the domestic livestock industry. It also affects the feed intake of the animal because ingestion of food is directly related to heat production, any change in feed intake and /or energy density of the diet will change the amount of heat produced by the animal. The ambient temperature has the greatest influence on voluntary feed intake. The ever growing pressure on land in the past few years has been described by many experts as a clear manifestation of the impact of climate change across Nigeria with most states in the far North being the worst affected by these changes. This has put the pastoralists in a state of dilemma (Omotayo, 2010). The pastoral Fulani believes that animal reproduction does not depend on the fecundity of the breed but rather on proper nutrition. Current efforts to combat global warming focus on reducing the emission of heat-trapping gases, but do not fully address the substantial contribution of land use to climate change, since even small changes of 100^2 km in urban development or deforestation can change local rainfall patterns and trigger other climate disruptions (BNRCC, 2008).

Problem statement

The bulk of locally produced meat and milk in Nigeria are through transhumance pastoralists. The dwindling pastoral and water resources such as open rangelands, wetlands (Fadama land), watercourses and rivers present a new challenge to pastoralism (Adamu, 2008). This could be held responsible for the low productivity of their cattle over the years. The situation is aggravated by climate change which exposed the pastoralists and their herds to tougher weather situations especially drought, poor quality pasture, risk of contacting diseases, pests, conflict between the pastoralists and crop producers over land use. The ever growing pressure on land in the past few years arising from population explosion, industrialization and institutional development has been described by many experts as a clear manifestation of the impact of climate change across Nigeria (Heinrich Boll Foundation, 2000). The problem of the pastoralists is further compounded with various agricultural development programmes which made pumps available for agricultural production in fadama area. Increasingly, however, pastoralists discovered that the rivers where they grazed their animals are now blocked off by farms and gardens. The problematic issues of customary tenure surfaces once again.

In Kwara State, farmers tend to farm in the designated grazing reserves because the land is particularly fertile. This marked the beginning of conflict in Bankubu, Baruten Local Government Area of the state. Today, there is increasing number of conflicts in many parts of Kwara State and the country at large (Joseph, 2012; Ademola, 2012) which resulted in huge losses in lives and properties.

In recent times, drought and flood are unpredictable and are more frequently occurring. This had a cumulative effect on natural resources and disruption of eco-system. Climate change reduced available land for livestock production purposes because of desert encroachment currently moving at 600 m/annum (or 350,000 ha per annum) (IPCC, 2007b, Oyetade, 2007). Consequently, pastoralists migrated to the southern part of the country where pasture and water are better guaranteed, an action, which often results into conflict between crop farmers and pastoralists with attendant low productivity which engulfed the agricultural sector in Nigeria. This forced the government to rely mostly on food importation to the extent that in 2007 the Federal government expended a total of N78.026 on milk importation in 2008 (National Bureau of Statistics, 2009), a situation described as dangerous for the nation's economy (Olayemi, 2005).

Cattle, sheep and goat performed better (in terms of calving, growth, milk production, etc) within a temperature range between 10 and 20°C called Comfort Zone" (McDowell, 1980). The temperature range in Kwara State is between 30 and 35°C. This is above the comfort zone and is capable of predisposing the animals to thermal stress which in turn can undermine the productivity of the animals. Irregular rainfall pattern also drastically affected the availability of water and pasture. Conducive weather condition, water and food are important in the physiological processes of these animals. In Nigeria therefore, it is not out of place to assume that the

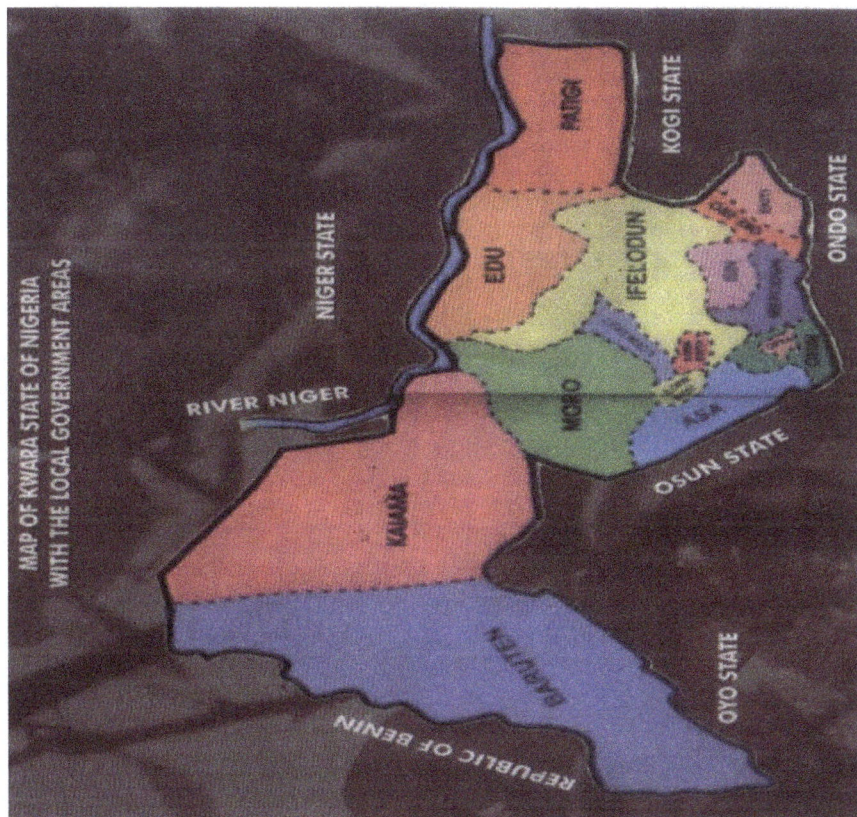

Figure 1. Map of Kwara State showing the sixteen local government areas.

prevailing environmental situations may be an eye opener for the pastoralists to look for alternative means of livelihood. These necessitate this study. The study was undertaken to provide answers to these research questions. What are the socio-economic characteristics of the pastoralists? To what extent has climate change affected grazing land? What are the perceived effects of climate change on the performance of the herds? And what are the changes in the means of livelihood of pastoralists due to climate change?

It is expected that the outcome of this study will assist the policy makers, planner, donors, public and private extension organizations to include the pastoralists in policy, programme planning and implementation relating to crop production and other enterprises aimed at increasing food production and employment generation.

Objectives of the study

The objectives of the study were to:

1) Describe the socio-economic characteristics of the pastoralists,
2) Determine perceived effects of climate change on grazing land in the study area,

3) Investigate perceived effects of climate change on herd's performance,
4) Identify changes in means of livelihood of the pastoralist due to climate change.

Hypothesis tested

The following hypotheses were tested:

1) There is no significant relationship between effect on grazing land and climate change,
2) There is no significant relationship between performances of the herd and climate change.

METHODOLOGY

The study area

The study was carried out in Kwara state, Nigeria which is located within the North Latitude 11° 2^1 and 11° 45^1. It falls between longitudes 2° 45^1 and 6° 40^1 East of Greenish meridian (Figure 1). The state is bounded in the south with Oyo, Ekiti and Osun State. It is bounded in the West by Benin Republic while in the North and the East, it is bounded by River Niger, and Kogi State, respectively. The state has a land area of 32,500^2 km (3,250,000 ha) with a temperature range between 30 and 35°C. The vegetation in the northern parts of the State is mainly savannah grass land while to the southern part is wooded Guinea Savannah. The rainfall pattern both in quantity (900 to 1500 mm) and distribution (6 to 7 months) and vegetation types favour production of cattle, goat, sheep and arable crops. The favourable climatic conditions are responsible for the exodus of Fulani from the northern parts of the country where adverse effects of climate change are mostly felt. The population of Kwara State is 2.3 million people (NPC, 2006). Kwara State is naturally endowed for livestock production. Crop production (rice, yam, cassava, guinea corn, maize, groundnut, sweet potato, cotton etc) is the major farming enterprise of the major tribes (Yoruba, Nupe and Baruba) in the State while livestock production is the major means of livelihood of the migrants Hausa/Fulani.

Target population

The target population for the study was the pastoralists in the sixteen local government areas (LGAs) of Kwara State. The local governments areas include Asa, Ilorin East, Ilorin West, Ilorin South (Kwara Central); Baruteen, Kaiama, Edu, Patigi and Moro (Kwara North); Irepodun, Ifelodun, Oyun, Offa, Ekiti, Oke-Ero and Isin (kwara South). There are preponderance of crop farmers and pastoralists in all the 16 LGAs in the state. The pastoralists constitute the sample frame from which the respondents were selected.

Sample size and sampling technique

The study used a multistage sampling technique. Stage one involved a random selection of seven (43.75% of the LGAs in the state) local government areas. These include Asa, Moro, Isin, Ifelodun, Kaiama, Edu and Baruteen LGAs. Stage 2 involved a random selection of five pastoralists' settlements (Gaa) in each LGA. The 'extension agents' in each LGA assisted in the compilation of the lists of the pastoralists, to the extent possible, within their areas of jurisdiction. Twenty (20) pastoralists were randomly selected from the five (5) Gaas in each LGA. Thus, a total of 140 pastoralists were selected from the seven (7) LGAs as respondents. Data were collected by means of structured interview schedule and analyzed with percentages, frequencies, tables and Chi-square statistics.

RESULTS AND DISCUSSION

Socio-economic characteristics of the pastoralists

The result of the study as shown in Table 1 indicated that 46.4% of the respondents were in the age range of 51 to 60 while the mean age of the respondents was 49.7 years. At this age, Ismaila et al. (2010) reported that farmers are incapable of handling tedious farming activities such as covering long distances to graze the animals. Unless the pastoralists are well-nourished, covering long distances may have implications on their health status. This can be subject of another research. However, low level (7.9%) of youth in the age bracket of 21 to 30 years was involved in transhumance pastoralism. It is possible that the youth diversified to other areas of the economy for their livelihood. Majority (90%) of the pastoralists were male. This implies that majority of the listed respondents were male; although,

the roles of female in pastoralism are also important especially in processing and marketing of livestock products. About half (50.8%) of the respondent acquired quranic education suggesting that pastoralists in the study area are mostly adherent of Islamic faith. However, 10.8 and 5% had primary and secondary education respectively, a reflection of the level of formal education among the pastoralists in the study area. Educational pursuit of the youth explains the low number of pastoralists (7.9%) that fell within the age bracket of 21 to 30 years of age.

It was also revealed that 32.1% of the respondents spent 31 to 40 years with an average of 29.57 years in transhumance pastoralism. It was 35.8% of pastoralists with herd size in the range of 21 to 30 heads of cattle while the average herd size was 21. Inability to maintain larger herd size could be linked with poor quality pasture, inadequate water resulting from increasing desertification in Nigeria. This supports the findings of Brenjo (2007) that the environment can no longer support all of its occupants when hectares of grazing land turn into desert in Sudan. This increases conflict and distrust and further separates the Arabs (pastoralists) and non-Arabs (farmers) in Sudan from reaching an agreement over land use (An-Naim, 2004). Many (62.8%) of the respondents fell within the age bracket of 51 years and above. Therefore, age factor might inform the basis for diversification into crop and other enterprises that require less of wandering and favour sedentary life. The results showed that 65% of pastoralist cultivated 1 to 3 ha with an average of 1.7 ha of land for crop production. This is greater than the national average of 0.57 ha per farmer (Ingawa, 2005). It follows that if the pastoralists were integrated into the national extension services delivery systems, they could be part of national progress to achieve self-sufficiency in food production. However, pastoralists should be exposed to modern animal husbandry practices to assist in coping with the adverse effects of climate change so that livestock rearing would not be compromised as this can affect local supply of animal protein.

Perceived effects of climate change on grazing land

The results (Table 2) revealed that 67.1% of pastoralists strongly agreed that irregular pattern of rainfall in recent time affects pasture availability implying that the pastoralists would have to wander a long distance in search of pasture and water. About half (52.5%) strongly disagreed that pasture and water is readily available throughout the year in their domain while 52.1% disagreed that prevailing temperature has no effect on the pasture. In addition, altogether, 65% disagreed that drought is not a common occurrence in their localities. These agreed with the findings of BNRCC (2008) that the impact of climate change can be vast. In Nigeria, this means that some stable ecosystems such as the Sahel

Table 1. Socio-economic characteristics of the pastoralists fulanis.

Characteristics	Frequency	Percentage = 100 N = 140
Age (years)		
21-30	11	7.9
31-40	14	11.0
41-50	27	19.3
51-60	65	46.4
> 61	23	16.4
Average	49.7	
Gender		
Male	126	90.0
Female	14	10.0
Marital status		
Single	19	13.6
Married	91	65.0
Widowed	20	14.3
Divorced	10	7.1
Household size		
≤ 5	97	69.3
6-10	40	28.6
11-15	3	2.1
Average	6	
Educational level		
No formal education	39	28.4
Adult education	8	6.0
Quaranic education	71	50.8
Primary education	15	10.8
Secondary education	7	5.0
Years spent in cattle rearing		
1-10	8	5.7
11-20	13	9.4
21-30	44	31.4
31-40	45	32.1
≥ 41	30	21.4
Average	29.57	
Size of herds		
≤ 10	24	20.0
11-20	37	30.8
21-30	43	35.8
31-40	16	13.4
Average	21	
Farm size (hectares)		
<1	44	31.4
1-3	91	65.0
-6	5	3.6
Average	1.7	

Source: Field survey (2012).

Table 2. Perceived effects of climate change on grazing land.

Perceived effects of climate change	SA	A	U	D	SD
Irregular pattern of rainfall in recent time affected pasture availability	94 (67.1)	33 (23.6)	4 (2.5)	5 (3.3)	4 (2.5)
Prevailing temperature has no effect on the pasture	7 (5)	29 (20.7)	0 (0)	73 (52.1)	31 (22.2)
Flood occurrence hinder pasture growth	35 (25.0)	28 (20.0)	2 (1.2)	71 (51.3)	4 (2.5)
Drought is not a common occurrence in your location	18 (12.5)	24 (17.5)	7 (5.0)	35 (25.0)	56 (40.0)
Pasture is readily available throughout the year	21 (15.0)	15 (11.2)	4 (2.5)	74 (52.5)	26 (18.8)
Water is readily available throughout the year	16 (11.2)	12 (8.8)	5 (3.8)	44 (31.2)	63 (45.0)
You cover long distance to grace your animals	46	65	4	18	17

Source: Field survey (2012). Figures in parenthesis represent percentages.

Table 3. Perceived effects of climate change on performances of the herds.

	SD (%)	A (%)	UD (%)	DA (%)	SD (%)
Milk production has reduced tremendously due to noticeable change	66 (47.5)	59 (42.5)	-	11 (7.5)	4 (2.5)
Herd mortality is on the increase	74 (52.5)	51 (36.7)	-	15 (10.8)	-
New type of disease are noticed	78 (56.2)	28 (20)	4 (2.5)	25 (17.5)	5 (3.8)
Pre-calving and post calving morality increases	50 (36.2)	47 (33.8)	-	31 (22.5)	10 (7.5)
Abortion in cattle increases	38 (27.5)	46 (32.5)	-	56 (40)	-
Abortion in cattle decreases	91 (65)	33 (23.7)	-	12 (8.8)	4 (2.5)

Source: Filed Survey (2012). Figures in parenthesis represent percentages.

Savanna may become vulnerable because warming will reinforce existing patterns of water scarcity and increasing the risk of drought. This explains the migration of pastoralists to southern parts of Nigeria and thus increases pressure on land use for cattle and crop production.

Perceived effects of climate change on performances of the herds

The result (Table 3) revealed that 47.5% of the respondents strongly agreed that the herd's milk production is reducing due to changes in climatic elements. More than half (52.5%) of the respondents strongly agreed that herd mortality is on the increase while 56.2% reported the emergence of new types of diseases. Furthermore, 60% of the respondents agreed that abortion in cattle increases while 40.8% reported incidence of pre- and post calving mortalities in their herds. This might not be unconnected with the quality of existing pasture and the need to cover long distances for grazing under harsh weather conditions. The findings agreed with NRC (2002) that climate change could impact the economic viability of livestock production systems worldwide.

Livelihood strategies of pastoralists due to climate change

Table 4 summarized the enterprises undertaken by the

respondents as their means of livelihood in view of prevailing adverse effects of climate change on livestock production in the study area. The result indicated that 75.5% of the pastoralists engaged in crop farming while 14.23, 15, 9.23, 7.86 and 13.57% engaged in trading, commercial transportation, farm labour, security guard and use of motor-cycle (Okada) for human transportation respectively to supplement the dwindling income from cattle production.

Pastoralists' reasons for venturing into other enterprises

Table 5 showed pastoralists reasons for venturing into other enterprises in descending order of importance. These include diminishing land for cattle grazing with a mean ranking of 5 using a five point likert- rating scale. Others include poor quality of existing pasture (4.6), inadequate income from cattle rearing to meet family requirements (4.39), land tenure system (3.57) and low cattle productivity (3.55) for diversifying into other enterprises.

These necessitated their venturing into crop production for food and to supplement their inadequate income from cattle production. It can be inferred that the planting of crops by the pastoralists might be suggestive to the farmers that the pastoralists intend to stay permanently on their land. This might be partly responsible for the constant hostilities between the two groups.

Table 4. Livelihood strategies of crop farmers and pastoralists.

Livelihood strategy	Pastoralists N=140	
	Frequency	Percentage
	106	75.5
	140	100
Trading	20	14.23
Commercial transportation	21	15
Farm labour	13	9.23
Security guard	11	7.86
	19	13.57

Source: field Survey, 2012.

Table 5. Pastoralists' reasons for venturing into other enterprises.

Reasons	SA	A	MD	D	SD	Mean
Income from cattle not adequate to meet family needs	91 (65)	33 (23.6)	2 (1.43)	8 (5.71)	6 (4.3)	4.39
Land for cattle grazing is diminishing	140 (100)	-	-	-	-	5.00
Pasture quality is becoming low	115 (82.1)	14 (10)	4 (2.9)	4 (2.9)	3 (2.1)	4.60
Low cattle productivity	97 (69.3)	29 (20.7)	2 (1.4)	7 (5)	5 (3.6)	4.14

Source: Field survey (2012). Figures in parenthesis represent percentages.

Table 6. Chi-square analysis of the relationship between the climatic factors and the performances of herds.

Variables	Degree of freedom	x^2 calculated	x^2 tabulated	Level of significance	Comments
Reduction in milk production	3	52.000	7.8147	0.000	Significant relationship exists
herd mortality is on the increase	2	20.725	5.9914	0.000	Significant relationship exists
Declining size of rangeland	3	53.500	7.8147	0.000	Significant relationship exists

Source: Field survey (2011).

Chi-square analysis of the relationship between the effects of climate change on grazing land and the performances of herds

The result of Chi square analyses (Table 6) established a significant relationship between climate change and declining size of grazing land; herd performances (milk production, x^2 = 52.00, tabulated = 7.8147, P≤0.05; herd mortality, x^2 = 20.725, tabulated = 5.9914, P≤0.05). The results confirm the findings of NRC (2002) that lack of prior conditioning of livestock to weather events such as temperature and drought often result to catastrophic losses in the domestic livestock industry. Ambient temperature has the greatest influence on voluntary feed intake. These explain the poor performances of local herds to cope with supply of animal protein required in Nigeria. This also agreed with Amogu (2009) that unfavourable environmental situations hinder livestock production in Nigeria.

CONCLUSION AND RECOMMENDATION

The study has shown that climate change has reduced grazing land, herds' milk production and increases mortality rate. The pastoralists have diversified into crop production and other enterprises to supplement their income from declining herd population. It was recommended that pastoralists should be encouraged, through extension services, to participate in crop and other enterprises as alternative ways of enhancing the dwindling income from livestock rearing. This in turn will fast track Nigeria's strive for attainment of self-sufficiency in local food production.

REFERENCES

Adamu B (2008). Keynotes address in Gefu JO, Alawa CBI, Maisamari B, (eds). The future of Transhumance Pastoralism in West and Central Africa. Strategies, dynamic, conflicts and interventions,

Proceeding of International conference on the future of transhumance pastoralism in West and Central Africa held in Abuja, Nigeria.

Ademola A (2012). When neighbours fight over cattle, farms. Nigerian Tribune, April 5, P. 27.

An-Naim AA (2004). Causes and solutions for Darfur. The San Diego Union Tribune. http//www.sigonsandiego.com/uniontrib/20040815/news_lz1e5darfur.html Retrieved 22/05/2012.

Amogu U (2009). Maximizing the animal production value addition chain in Nigeria. Invited presentation at Nigeria Institute of Animal science, Annual General Meeting, University of Calabar, July 28, 2009.

Ayoade JO (2004). Climate Change. Ibadan: Vantage Publishers. pp. 45-66.

Brenjo N (2007). Looking for water to find peace in Darfur. Alert Net blog. http://www.alertnet.org/db/t/b/06/30-100806-1htm Retrieved 22/05/2012.

Building Nigeria's Response to Climate Change (BNRCC, 2008). Vulnerability, impacts and adaptation to climate change in Nigeria. 1, Oluokun Street, Bodija, U, I. P. O. Box 22025, Ibadan, Nigeria. Email: info@nigeriaclimatechange.org.

Heinrich Boel Foundation (2000). Climate Change and Human Right. In New Perspetive quarterly, 17:14-26.

Ingawa S (2005). New Agricultural Technologies Adopted from China. Message Delivered at a Workshop for North –East and North –West. States of Nigeria. Nigerian Tribune, No 13, 714, September 1, 2005 P. 1.

Intergovernmental Panel on Climate Change (IPCC) (1996). Climate change synthesis report. www.ipcc.ch/pdf/assessment -report/a-4/pdf.

Intergovernmental Panel on Climate Change (IPCC) (2007b). Impacts, adaptation and vulnerability: The Working Group II Contribution to the Intergovernmental panel on climate change fourth assessment report. Cambridge University Press.

Ismaila U, Gana AS, Tswanya NM, Dogara D (2010). Cereals Production in Nigeria: Problems, constraints, and opportunities for betterment. Afric. J. Agric. Res. 5(12):1341-1350, 18 June 2010 Available online at http://www.academicjournals.org/AJAR. ISSN 1991-637X ©2010 Academic Journals.

Joseph O (2012). 11 killed, thousands flee Nasarawa State as Tiv, herdsmen clash. The Guardian, March 12, P. 8.

McDowell RE (1980).The influence of environment on physiological functions. In Animal Agriculture. The Biology, husbandry and use of domestic animals, Cole, H. H. and W. N. Garrett (eds.) Second Edition, Freeman and Company, San Francisco. P. 460.

National Bureau of Statistics (2009). Statistical News (http://www.nigerianstat.gov.ng). Retrieved 12/01/2013.

National Population Commission (NPC) (2006).The National Populaion Office, Ilorin Kwara State, Nigeria.

National Research Council (NRC), 2002: Abrupt climate change inevitable surprises. National Academy press, Washington, Dc. National Academy Press.

Olayemi JK (2005). Food Security in Nigeria". Research Report No 2. Development policy centre, Ibadan. 2(32):77-78.

Omotayo AM (2010). The Nigerian Farmer and Elusive Crown, 30th inaugural Lecture, Federal University of Agriculture, Abeokuta, Nigeria. P. 9.

Oyetade L (2007). Desertification in Nigeria (African Agriculture).Text of a lecture delivered at the 38th interdisciplinary research discourse, postgraduate school, University of Ibadan, 11th December, 2007.

7

Effects of property rights on agricultural production: The Nigerian experience

A. A. Akinola and R. Adeyemo

Department of Agricultural Economics, Obafemi Awolowo University, Ile- Ife, Osun State, Nigeria.

The study was conducted to examine the effects of property rights and other factors on the outputs of maize, yam and cassava in three zones of Osun State in Nigeria. This study employed a multi-stage sampling technique to select 105 farmers involving growers of maize, yam and cassava in the study area. Data were analyzed with the aid of descriptive statistics, budgetary techniques and a multiple linear regression model. The results of budgetary analysis showed that variable cost was highest in yam production. The average revenues per hectare for maize, yam and cassava were (N is Nigerian currency equivalent to about $0.0067) N104, 487.50, N583, 846.20 and N438, 208.50, respectively. However, the average net incomes were N19, 908.40, N432, 079.00 and N96, 543.90 for maize, yam and cassava, respectively. Based on the rates of returns, N1 invested in each of maize, yam and cassava production yielded N1.2, N3.4 and N3.1, respectively implying that yam was the most profitable crop in the study area. The result of the multiple regression model revealed that farm size significantly affect the outputs of the three crops. Land rights type (having either use right/use and transfer right) and security of land defined by duration of land use affected maize output while duration and ownership type affected yam output, whereas, duration only affect cassava output. There is therefore the need to review the land distribution and administration policies based on the identified significant factors affecting each crops.

Key words: Nigeria, Osun State, crops, maize, yam, cassava, land rights.

INTRODUCTION

Land is probably the most important factor of production. The unique feature of land is its fixed nature and this has generated a lot of policies administration in its use rights and transfer. The rights to land are an international issue with dynamisms depending on individual country's tenure arrangement. Property rights will determine land ownership related factors affecting the application of technologies for agricultural and natural resource management. Secured property rights give sufficient incentives to the farmers to increase their efficiencies in terms of productivity and ensure environmental sustainability. It is natural that without secured property rights, farmers do not feel emotional attachment to the land they cultivate, do not invest in land development and will not use inputs efficiently (Tenaw et al., 2009). There is broad agreement in the literature that secure individual land rights will increase incentives to undertake productivity enhancing land related investments. More secure property rights could affect productivity by improving household's security of tenure and thus their ability and readiness to make investments; providing better access to credit; and reducing the transaction costs associated with land transfers (Tenaw et al., 2009). Besley (1995) revealed that having more secure tenure to a plot increased the probability that individuals would plant trees, and undertake a wide range of other investments such as drainage, irrigation, mulching, etc. that would enhance better yield.

Fajemirokun (2000) indicated the need for secure ownership rights over a sufficiently long time horizon which needs not necessarily be a formal title to facilitate improvements emerges from most African countries.

Land policies and reforms in Nigeria

In Nigeria, reform has often sought to transform customary tenure land into state property or individualized private property. This was contained in the promulgation Land Use Act (LUA) of Nigeria, 1978. This brought about a fundamental change in land tenure systems through the abolition of private ownership of land (Fajemirokun, 2000). According to the Act, all land comprised in the territory of a State is vested in the State Governor who holds in trust for the use and common benefit of all Nigerians. Under this uniform system of land tenure, the highest interest in land is a right of occupancy. This can either be a statutory right of occupancy, which is granted by the State Governor in respect of land in both urban and non-urban areas and a customary right of occupancy, which is granted by a Local Government in respect of land in a non-urban area. Designation of urban and non-urban areas of a State is the exclusive responsibility of the State Government. During privatization, men (and particularly male heads of household) acquire complete and legal ownership of land (Davison, 1988). Individualized and private ownership transfers the few rights, such as cultivation rights, that women and minority groups may have land under customary rules to those men who are able to claim all rights to land (Lastarria-Cornhiel, 1997). More recently, there is the trend to recognize the previously existing customary tenure and land authorities which is still a problem to farmers especially the crop farmers. Rarely, however, has the effect of property rights on crops production been discussed in Nigeria and strong empirical evidence to test its effect and impact has been scarce and scattered due to paucity of literature. Several studies (Afeikhena, 2000; Besley, 1995; Feder et al., 1988) emphasized the effect of property rights on land conservation investment.

In Nigeria and other sub-Sahara countries, traditional land tenure system of ownership is still predominant. According to Deininger and Binswanger (1999), undefined property rights could affect economic growth in the following ways; Firstly, secure property rights will increase the incentives of households and individuals to invest, and often will provide them with better credit access, something that will not only help them make such

investments, but will also provide and assurance substitute in the event of shocks. Secondly, it has long been known that in traditional agriculture, the operational distribution of land affects output, implying that a highly unequal land distribution will reduce productivity. Even though the ability to make productive use of land will depend on policies in areas beyond land policy that may warrant separate attention, secure and well-defined land rights are key for household asset ownership, productive development and factor market functioning. Based on the afore-mentioned assertions, the situation of land tenure system and property rights prevalent in most of developing countries in sub-Saharan Africa, Nigeria inclusive is similar in many respects as long as the agricultural production output remains low (Tenaw et al., 2009). They stressed that the changing climatic conditions in many developing countries have impacts on agricultural production at local and country level. This is an important issue, which is worth paying attention to in order to prevent problems that may affect the population. Insecure land right or the lack of land ownership also restricts the farmers' access to credit that are necessary for improved agricultural land practices for better yield (Feder et al., 1988). This non-access to credit predisposed farmers to go for traditional land-use practices which will eventually generate poor yield (Bamire and Fabiyi, 2002). The traditional institutions in the country allow various land acquisition types such as rent, share cropping and lease hold systems. These rights are non-definite, non-directional and insecure. There is therefore the need to get empirical evidence of the effects of land rights on crop outputs. Our study, therefore, intends to examine the socio-economic characteristics of the respondents; the nature, ownership and distribution of property rights in land and how they are acquired; determine the costs and returns to maize, yam and cocoa and also examine the effect of property rights on the output of these crops.

The importance of this study lies in providing information on the effect of property rights on crops output to assist policymakers where such rights are practiced in promoting right accessibility that would enhance better farmers yield.

METHODOLOGY

Area of study

The study was conducted in Osun state of Nigeria. Osun state is located in the south-western part of the country. It covers an area of approximately 14,875 square kilometers, with an estimated croppable land area of 8,822.55 square kilometers. It shares common boundaries with Kwara, Ogun, Ekiti, Ondo and Oyo states. The indigenes of the state belong to the Yoruba tribe but non-indigene from all parts of Nigeria and foreigners also reside in the state. The major crops grown in the state are cassava, maize, vegetables, cocoa, oil palm, tomatoes etc. This implies that the climate in the state favours both arable and non-arable crops. The states experiences two major seasons, the dry and rainy seasons with August break during the rainy season. The annual temperature varies from 21.1 to 31.1°C, while annual rainfall is within the range of 800 mm in the dry savannah agro-ecology to 1500 mm in the rain forest belt. Traditional land tenure arrangement is still predominant in the state. Rent, share cropping and short term lease arrangements which are often non-legal are popular in Nigeria and some West African Countries.

Data and sampling technique

A multi stage sampling technique was employed in selecting respondents for this study. In the first stage, Osun state was stratified into three based on the state's Agricultural Development Programme (ADP) classification, namely Ife/Ijesha, Iwo Ikire and Osogbo zones. In the second stage, a local government area (LGA) was selected from each of the zones based on the predominance of agricultural practices. In the third stage, three villages were randomly selected from each of the LGA. In the final stage, a minimum of ten respondents were selected per villages. In all, 105 respondents were selected. Data were collected with the aid of structured questionnaire. Data were collected on the socio-economic characteristics of the respondents such as sex, age, level of education and land factors and tenure arrangement such as land size, land ownership type, duration of tenure among others. Data collected were analyzed using the descriptive statistics, budgetary techniques and multiple linear regression technique. Descriptive statistics uses frequency count and percentage to describe the socio-economic variables of respondents in the study area. A total farm budget approach was undertaken to estimate costs and returns accruing to maize, yam and cocoa enterprises in the study area.

Since a budget is the quantitative expression of total farm plan summarizing the income, cost and profit -a residue of total cost from total revenue (Alimi and Manyong, 2002). Gross margin which is the difference between total revenue and total variable cost were analyzed. The total cost component is expressed as:

TC= TFC + TVC

Where: TC = Total cost; TVC = total variable cost; TFC = total fixed cost.

Gross margin = (TR) - (VC)
TVC = TC - TFC
TFC = TC - TVC
TR = Total revenue = price × quantity that is, PQ
VC = Variable cost
Profit = TR - TC
Labour efficiency = Total output/amount of labour used

The efficiency ratios that were analyzed were fixed cost ratios, variable to total cost ratio, labour efficiency amongst others. These were computed to indicate the performance of each of the enterprises. The data collected were analyzed using multiple linear regression models. This model was employed to examine the effects of socio-economic and land tenure factors on the outputs of maize, yam and cocoa. For each of the crops, the dependent variable of the regression model is the output (kg). The postulated model assumed a relationship between the output of the crops and factors affecting crop output(s). The general empirical model is:

$$\gamma = \alpha + \beta_i x_i$$

Where α = constant /intercept; β_i = coefficient of independent variables x_i.

The regression model for each of the production is modeled as:

$$\gamma = \alpha + \beta_1 x_1 + \beta_2 x_2 + \beta_3 x_3 + \beta_4 x_4 + \beta_5 x_5 + \beta_6 x_6 + \beta_7 x_7 + \beta_8 x_8 + \beta_9 x_9 + \beta_{10} x_{10} + \mu$$

Where γ_i is the quantity of individual crop produced in kilogram.

x_1 - x_{10} = explanatory variables
α = constant (intercept)

The fitness of the model was based on the coefficient of multiple determinations (R^2), adjusted R^2 and significance of regression coefficient at a specific level (1, 5 or 10%).

x_1 = Age of the respondent (AGERES) in years
x_2 = Marital status (MARISTAT); (1 if married and 0 otherwise)
x_3 = Farming experience (FARMEXP) in year
x_4 = Household size (HHSIZE)
x_5 = Duration of land use (DURLNDU) in years
x_6 = Land right type (RIGHTYP); (use right only = 0; use and transfer right = 0)
x_7 = Land ownership type (LNDOWSIP); 1 if owned and 2 otherwise
x_8 = Total farm size (FARMSIZ) in hectare
x_9 = Level of education (LEVEDU) in years
x_{10} = Extension visit (EXTVIT); 1 if visited and 0 otherwise
μ_i = error term.

A priori expectation signs of the coefficients

The multiple independent variables included socio-economic and tenural factors that may influence crop output. These variables include age (AGE) of respondents in years, marital status (MARISTAT), farming experience (FARMEXP), household size (HHSIZE), duration of land use (DURLNDU), land rights type (RIGHTYP), land ownership type (LNDOWSIP), farm size (FARMSIZ), level of education (LEVEDU) and extension visit (EXTVIT). The rationale for inclusion of these variables was based on a priori expectation of factors influencing agricultural output. The effect of age (AGERES) on the output may be positive or negative. Previous study shows that the age of individuals affects their output in several ways. Younger farmers have been found to be more agile and would be ready to take on new practices that could improve crop yield. The older the farmer, the less likely may be his output from crops (Amos, 2007). Marital status (MARISTAT) of respondents may have influence on respondents output positively. It is expected that the married individuals have greater number of labour which may increase output. Farming experience is a measure of the number of years a respondent has farmed. We hypothesized that farming experience will positively influence farmers output. It is expected that the more the experience, the greater the resource use and hence the better the output (Amos, 2007; Akinola and Adeyemo, 2008).

Household size (HHSIZE) determines the supplementary man days' of labour that could be produced by the family (Amos, 2007; Yang and Zhang, 1999). We then hypothesized that household with larger size have higher probabilities to acquire more output than smaller household size because the larger the household size, the greater the man power, hence, more labour to work on the farm. Duration of use (DURLNDU) of land is a measure of the length in years that the land would be used. The longer the time period the greater the likelihood those farmers would adopt soil enhancing technology that would increase crop yield (Tenaw et al., 2009). It is expected that the longer the length of use, the greater the tendency that the land occupiers owns to land and the greater the probability of investing in land enhancing technology that would enhance greater (Ogedengbe and Akinbile, 2004). The type of right (RIGHTYP) individual is having over a parcel of land may have positive or negative influence on production. Individual with use right only may not adopt land improving technology and hence, low output while individual with both use and transfer right can adopt new and hence, greater output (Ogedengbe and Akinbile, 2004; Clay, 2008). The Ownership type (LNDOWSIP) defined the land ownership type, which is whether land is owned or otherwise. It is expected that ownership type will influence the rights to hold a parcel of land because the more your income the larger the amount of land you can purchase (Clay, 2008).

Farm size (FARMSIZE), the total farm size owned by respondents is expected to positively influence crop output. The larger the farm size owned, the greater the area that will be put under cultivation and the more the expected output (Clay, 2008). Level of education (LEVEDU) is expected to positively influence crop output. Extension visit (EXTVIT): It is hypothesized that the greater the land allocated for permanent crops, the greater the output from permanent and the less the available land for arable and hence, the lower the yield from arable.

RESULTS AND DISCUSSION

Socio-economic characteristics of respondents

Table 1 showed the socio-economic characteristics of the respondents. The analysis revealed that 75.2% of the respondents were male while 24.8 were female. This implies that farming in the study area were male dominant. The analysis further showed that 57% of the respondents fell between the ages of 41 to 50 years, while about 28.6% of the respondents have age of 51 years and above. This implies that most of the farmers in the study area are still in their active age. The farmers in the area are experience. Analysis revealed that 57 and 22% of the respondents were having 5 to 10 years and 11 to 15 years of experience, respectively. On the duration of land use, 66.7% of the respondents has maximum of 5 years of duration, while just about 33.32% has 6 years duration and above. This implies that the farmers in the area will be reluctant to adopt soil enhancing technology that would improve crop yield. Also, 63.8 and 36.2% of the respondents indicated that they have use rights and transfer rights, respectively. Majority of the respondents (62.9%) do not own land while just 37.1% of the respondents owned land. Majority of the respondents are relatively educated. 39.4 and 48.8% of the respondents finished from primary and secondary schools, respectively. The respondents in the area do not have access to extension service. Analysis revealed 93.3% of the respondents indicated that they do not have access to extension services.

Budgetary analysis for maize, yam and cassava

Results of the budgetary analysis revealed that the average gross revenue for maize, yam and cassava were N104, 875, N583, 846.2 and N438, 208.5, respectively (Table 2). The average variable costs incurred in maize, yam and cassava were N43, 814.9, N107, 414.9 and

N96, 543.9, respectively. The higher cost incurred in yam may probably due to extra cultural practices like staking and mulching involved in yam production. Gross margin values were N60,672.6, N476,431 and N341,664.1 (N is Nigerian currency equivalent to about $0.0067) for maize, yam and cassava, respectively. The rate of returns for maize, yam and cassava were 1.2, 3.4 and 3.1, respectively (Table 3). This implies a better viability of yam enterprise in the study area.

The multiple linear regression result

The results of the multiple linear regressions shown in Tables 4, 5 and 6 revealed that R-square values for maize, yam and cassava were 79.6, 70.7 and 86.6%, respectively, while the adjusted R-squared were 68.3, 56.3 and 76.2, respectively. This implies that 68.3, 56.3 and 76.2% changes in the outputs of maize, yam and cassava were accounted for by the independent variable. The result (Table 4) showed that household size, right type, ownership type, total farm size used for farming and level of education were statistically significant affect the output of maize at 10, 10, 10, 1 and 5%, respectively. This implies that household size, right type, ownership type, total farm size used for farming and level of education were significant determinants of maize production in the study area. The significance of rights type and ownership type indicated the land tenure arrangements (rights) have significant effects in maize production in the area. The result of Table 5 revealed that farming experience, duration of land use, ownership type and total farm size were significant determinants of output of yam in the area. They were significant at 10, 10, 5 and 10%, respectively. This implies that the greater the duration a plot of land, the greater the tendency that farmers output will increase. Also, those who owned land will have better output as the will be willing to adopt output enhancing technology. The result of Table 6 revealed that duration of land use, total farm size and level of education were significant determinants of output of cassava in the area. They were significant at 5, 1 and 5%, respectively. This implies that the greater the duration a plot of land, the greater the tendency that farmers output will increase because he will be willing to adopt output enhancing technologies.

Also, those with larger farm size will have better output as they will enjoy economics of large scale production. It could be seen that land rights have significant effects on the outputs of crops in the study area.

Conclusion

Farmers in the study area were mostly married, middle aged with majority having formal education. The analysis revealed that farming activities in the study area is male dominant as 75.2% of the respondents were male while

Table 1. Socio-economic, demographic and farm characteristics of respondents.

Variable	Frequency	Percentage
Sex		
Male	79	75.2
Female	26	24.8
Total	105	100
Age		
≤30	7	6.7
31-40	11	10.5
41-50	57	54.3
51-60	13	12.4
>60	17	16.2
Marital status		
Single	11	10.5
Married	78	74.3
Others	16	15.2
Total	105	100
Level of experience		
< 5 17	16.2	
5-10 57	54.3	
11-15 22	21	
16 and above 9	8.6	
Total	105	100
Household size		
1-2	43	41
3-4	41	39.5
5-6	9	8.6
7-9	8	7.6
10 and above	4	3.8
Total	105	100
Duration of land use in years		
<2	23	21.9
3-5	47	44.8
6-8	8	7.62
>8	27	25.7
Total	105	100
Right type		
Use right only	67	63.8
Use and transfer right	38	36.2
Land ownership		
Owned	39	37.1
Otherwise	66	62.9
Total	105	100
Farm size		
<1	67	63.8
1.1-2.0	19	18.1
2.1-3.0	11	10.5
3.1 and above	8	7.6

Table 1. Contd.

Total	105	100
Level of education		
None	11	10.5
Primary	41	39.4
Secondary	46	48.8
Tertiary	7	6.7
Total		105
Extension visit		
None	98	93.3
Regular	1	0.95
Occasional	6	5.7
Total	105	100

Source: Field survey (2011).

Table 2. Budgetary analysis for maize, cassava and cassava enterprises.

Items	Maize	Yam	Cassava
(A) Gross revenue (₦)	204,487.5	583,846.2	438,208.5
(B) Variable cost (₦)			
Land clearing	15041.5	22,041.5	21,220.5
Labor (harrowing, ridging)	15,857.9	49357.1	44,543.8
Weeding	19,551.8	23951.8	21,320.5
Harvesting	5,364.5	11,064.5	6,006.7
Haulage	3,000.1	4,000	3,452.4
Planting material	12,470.4	13,670.9	11,665.8
Total variable cost (₦)	66,284.3	114,085.8	96,543.9
(C) Fixed cost (₦)			
Rent	28,681.7	30,681.3	31,223.5
(D) Total fixed cost (₦)	41,151.7	44,351.7	42,889.3
E) Total cost (B+C) (₦)	84,966.6	151,766	139,433.2
(F) Net farm income (₦)	119,908.4	432,079	298,775.3

Source: Field survey (2011).

Table 3. Profitability and efficiency measures for maize, cassava and cocoa enterprise.

Description	Maize	Yam	Cassava
Profit (₦)	19,908.4	432,079	298,775.3
Gross margin (GM) (₦)	60,672.6	476,431	341,664.6
Rate of return (₦)	1.2	3.4	3.1
Cost ratio	1.1	2.4	2.2

just 24.8% farmers in the area are well experienced. Analysis revealed that 57 and 22% of the respondents were having 5 to 10 and 11 to 15 years of experience, respectively. Majority has short duration of land use. About 66.7% of the respondents has maximum of 5 years of duration, while just about 33.32% has 6 years duration and above. Also, 63.8 and 36.2% of the respondents indicated that they have use rights and transfer rights, respectively. Majority of the respondents (62.9%) do not own land while just 37.1% owned land. Budgetary analysis

Table 4. Result for linear regression for production of maize.

Variable	Coefficients	Standard error	T-ratio
Age	116.261	241.887	0.762
Marital status	-942.137	1684.930	0.481
Farming experience	35.15	148.832	0.236
Household size	-2254.999	1078.482	-2.091*
Duration of land use	-4080.7087	3375.907	-1.209
Right type	16139.234	8491.657	1.901*
Ownership type	3928.653	2025.114	1.940*
Total farm size	1799.067	325.348	5.530***
Level of education	1388.818	443.899	3.129**
Extension visit	-440.207	668.776	-0.658
Constant	9703.072	12737.720	0.762

*** = significant at 1%, ** = significant at 5% and * = significant at 10%, R square 79.6; adjusted R square 68.3. Sources: Survey data (2011).

Table 5. Result for linear regression for production of yam.

Variable	Coefficients	Standard error	T-ratio
Age	-68.841	271.870	-0.253
Marital status	629.863	1659.655	.380
Farming experience	290.228	139.932	2.074*
Household size	-1938.522	1159.615	-1.672
Duration of land use	7006.879	3752.936	1.867*
Right type	-6450.729	8504.342	-0.759
Ownership type	3285.376	2337.540	1.405**
Total farm size	549.199	300.935	1.825*
Level of education	195.062	418.985	0.466
Extension visit	-581.645	600.349	-0.969
Constant	24049.714	13769.053	1.747

*= significant at 10%, R square 70.7; Adjusted R square 56.3. Sources: Survey data (2011).

Table 6. Result for linear regression for production of cassava.

Variable	Coefficients	Standard error	T-ratio
Age	101.163	228.776	0.442
Marital status	-1571.179	1552.849	-1.012
Farming experience	-171.979	145.846	-1.179
Household size	-1254.575	1048.515	-1.197
Duration of land use	8194.197	3005.352	2.727**
Right type	-7248.648	10502.295	-0.690
Ownership type	-352.914	2151.264	-0.164
Total farm size	1634.392	297.457	5.495***
Level of education	1085.750	406.419	2.672**
Extension	-544.257	691.718	-0.787
Constant	31171.665	13209.550	2.360

*** = significant at 1%, ** = significant at 5%, R square 86.6; Adjusted R square 76.2. Sources: Survey data (2011).

revealed highest values of gross margin and net income were recorded for yam compared to other enterprises. The average total revenue for yam, cassava and maize were 583,846.2, 438,208.5 and 104,875, respectively.

The average total cost incurred in yam, cassava and maize enterprises were 151,766, 139,433.2 and 84,966.6, respectively. The rate of returns to investments for yam, cassava and maize were 3.4, 3.1 and 1.2, respectively. The result of the multiple linear regression model and its implications revealed that household size, right type, ownership type, farm size and level of education significantly influence maize output. This implies that farmers with defined rights and owned land would have better output. The analysis further revealed that farming experience, duration of land use and ownership also affects yam output. This implies that farmers that owned land can adopt output enhancing technology than those who rent or engage in share cropping. Regression analysis on the factors influencing maize cassava output revealed that duration of land use, farm size and level of education significantly affect cassava output. This implies that the longer the duration, the larger the size and the more educated a farmer is, the greater the output. Therefore, government at all levels and her agencies should put machineries in place that would formulate policies and programmes that would enhance land distribution and ownership in this part of the country and in other regions where the same practices operate.

REFERENCES

Afeikhena J (2000). Land Rights and Investment Incentives in Western Nigeria A paper Presented at the Beijer Research Seminar on Property Rights Structures and Environmental Management, South Africa.

Akinola AA, Adeyemo R (2008). Adoption and Productivity of Improved Rice Varieties in Osun State, Nigeria". Ife J. Agric. 23(1):104-116.

Alimi T, Manyong VM (2000). Partial Budget Analysis for on-farm Research.

Amos TT (2007). Resource Use in Tilapia production among Small Scale Tilapia Farmers in the Savanna Zone of Northern Nigeria. J. Fish. Int. 2(1):42-47. (missing the dot)

Bamire AS, Fabiyi YL (2002). Adoption pattern of fertilizer technology among farmers in the ecological zones of South-Western Nigeria: A Tobit analysis. Aust. J. Agric. Res. 53:901-910.

Besley T (1995). Property Rights and Investment Incentives: Theory and Evidence from Ghana. J. Polit. Econ. 103(5):903-937.

Clay K (2008). Property Rights and Agricultural Production. H. John Heinz III School of Public Policy and Management Carnegie Mellon University Pittsburgh.

Davison J (1988). "Agriculture, Women, and Land" The African Experience. Boulder, Colorado (USA): (ed.) West view Press. pp. 74-75.

Deininger K, Binswanger H (1999). "The Evolution of the World Bank's and Land policy: Principles, Experience and future challenges. World Bank Res. Obser. 14:247- 276.

Fajemirokun B (2000). Land and Resource Rights: Issues of Public Participation And Access to Land in Nigeria. Potential decentralization models are discussed in C. Toulmin, "Decentralization and Land Tenure" in C. Toulmin & J. Quan (eds.), Evolving Land Rights, Policy and Tenure in Africa, (London,). pp. 229-245.

Feder G, Ohdaan T, Chalaniwong Y, Hongladaron C (1988). "Land Policy and Farm Productivity in Thailand". Baltimore: Johns Hopkins University press.

Lastarria-Cornhiel S (1997). "Impact of Privatization on Gender and Property Rights in Africa". World Dev. 25(8):1317-1333.

Ogedengbe K, Akinbile CO (2004). Determination of Consumptive Water Use of Corchorus oliforius (ewedu) Using Blaney-Morin Simulation Model. J. Sci. Eng. Technol. 11(3):5653-5663.

Tenaw SKM, Zahidul I, Parviainen T (2009). Effects of land tenure and property rights on agricultural productivity in Ethiopia, Namibia and Bangladesh. University of Helsinki Department of Economics and Management Discussion Papers no 33 Helsinki 2009.

8

Transaction costs and smallholder household access to maize markets in Zambia

Richard Bwalya[1], Johnny Mugisha[2] and Theodora Hyuha [2]

[1]Institute of Economic and Social Research, University of Zambia, Lusaka, Zambia,
[2]Department of Agribusiness and Natural Resource Economics, School of Agricultural Sciences, Makerere University,
P. O. Box 7062, Kampala Uganda.

After liberalization of the Zambian economy, farmers were faced with the responsibility of finding the right buyers, negotiating prices and delivering produce leading to them incurring transaction costs. This study aimed at identifying and quantifying transaction costs factors and their impact on maize market participation for small holder farmers in Zambia. The study used primary data collected from a sample of 240 randomly selected households from Zambia's central Province. The Heckman's procedure was used to analyze factors affecting the likelihood and extent of participation in maize markets. The logit results (from the Heckman's two-stage process) show that ownership of assets such as radios and having access to alternative marketing channels increased the likelihood of market participation while the heckit results (OLS corrected for selectivity bias) shows that ownership of ox-carts, increased family size and experience in maize marketing were the factors that increased quantities of maize marketed. The study recommends provision of market information, improving accessibility to markets as well as increasing access to productive assets as means of alleviating impact of transaction costs.

Key words: Transaction costs, maize, market access, Zambia.

INTRODUCTION

Maize is one of the most important crops in Zambia. According to the Regional Agricultural Trade Expansion Support (2003), as a staple food, it comprises of up to 55% of the total dietary energy supply and affects food security and incomes of about 80% of the population. It also accounts for between 50 and 67% of the total area under cultivation (Central Statistical Office [CSO], 2002) and it is the single most important crop in the small scale sector in terms of gross value of production and crop sales. Although about 900,000 small-scale farmers account for over 65% of the total national production, they only contribute about 30% to the marketed surplus (Zulu et al., 2007). The smallholder maize market is also highly concentrated with more than 80% of the sales attributed to less than 30% of the sellers (Nijhoff et al., 2003). These low levels of market participation have been attributed to high transaction costs that make access to markets difficult (Kahkonen and Leathers, 1999). Due to differential access to assets, markets and information, transaction costs tend to be household specific and affects households differently leading to some being completely excluded from the markets.

The problems faced by smallholder farmers in marketing their produce have been linked to the liberalization of agricultural markets. For instance, Simatele (2006) argues that despite liberalization of the agricultural markets, the small-scale agricultural sector has been facing problems which are attributed to inadequacies in the marketing system for staples and agricultural inputs. Major among them, are the low prices of staples leading to problems of low real incomes for smallholder households and also food shortages (Nijhoff et al., 2003). Similarly, using historical trends in agricultural productivity, Yambayamba (2009) shows that since the market reforms of 1991, there has been a decline in absolute maize production, which they attribute to removal of fertilizer subsidies, the abolishment of pan-territorial pricing and the closure of maize collection depots in remote areas. These authors show that over the 12-year period between 1990/1991 and 2002/2003 seasons, the share of maize in total smallholder crop output declined from 76 to 55%. Similarly, Seshamani (1999) shows that the main adverse impact witnessed as a result of agricultural market liberalization were the negative supply response of the smallholder farmers due to the adverse impact on their incomes. This author shows that the index of maize production dropped from 145 in the 1989/90 growing season to 54 in the 1994/95 growing season. However, even though the area under maize cultivation fell by 4% in the 1996/1997 season compared to the previous (1995/1996) season, maize production fell by 32%, while maize sales fell by 53%. These declines in maize production and marketing are partly attributed to the fact that smallholder farmers experienced difficulties in accessing adequate and timely inputs, marketing of produce as well as in getting a fair price for their produce (Seshamani, 1999).

The above statistics show that low sales and non-participation in maize markets can be explained by both low production and reduced access to markets due to government withdrawal from providing support to smallholder farmers. For instance, whereas there have been several highly committed and well-funded efforts aimed at kick-starting a "green revolution" based on the understanding that agricultural productivity is a pre-condition for sustainable poverty reduction and improved living standards, they have been thwarted by their inability to anticipate and address downstream issues of marketing and governance (Jayne et al., 2007). Zambia's agricultural sector is also characterized by an inherent dichotomy in agricultural marketing, with smallholder traders facing an underdeveloped informal marketing system, and the more advanced large-scale traders and processers being part of a formal marketing system (Yambayamba, 2009).

Whereas the problem of low productivity has been extensively explored (Yambayamba, 2009; Zulu et al., 2007), the role that market access plays in leading to low maize productivity and sales has not received much attention, leading to misguided policies by government. For instance, government policies aimed at increasing the production of the national staple food (maize) have mostly revolved around increasing productivity through provision of subsidized inputs. To this effect, about 50%

of the national agricultural budget has always been dedicated to provision of subsidized maize seed and fertilizer over the last eight years. This has been coupled with provision of extension services that are biased towards maize production. However, despite all these efforts aimed at increasing production, not much effort has been spent on assessing the role that access to markets play in stimulating production as well as market participation. This is despite some earlier studies (Kahkonen and Leathers, 1999) indicating that Zambian maize markets are riddled with high transaction costs leading certain potential participants being excluded from participating. As Seshamani (1999) points out, faced with a situation where government agents do not come to purchase his produce, the smallholder farmer has to go to the market centres to sell to them, which is not easy in view of the lack of transport to reach the markets. The author also shows that in the event that the farmer reaches the markets, he finds them to be buyers' markets where the prices are not in his favor.

The fact that farmers do not only have to produce but also have to find the right buyers, they negotiate on prices and deliver their produce which leads them to incur transaction costs. According to Eggertson (1990), these are costs that arise when individuals exchange ownership rights for economic assets and enforce their exclusive rights[1]. They originate from activities such as searching for trading partners, screening partners, bargaining, monitoring, enforcement and transferring product (Key et al., 2000). These transaction costs may also include the costs associated with reorganizing of household labor and other resources in order to produce enough for the market (Makhura et al., 2001; Zaibet and Dunn, 1998). This paper attempts to explain the impact of transaction costs on maize market participation among the smallholder farmers in the Central Province of Zambia.

Transaction costs theory has been used to explain farmers' behavior in both input and output markets. A study by de Janvry et al. (1991) showed that high transaction costs lead to missing markets for certain commodities. They concluded that in the absence of food markets households must be self-sufficient in terms of food, which confines their ability to reallocate land and labor to cash crops. These households tend to face wide margins between low selling price and high buying price. They also showed that the poorer the infrastructure, the less competitive the marketing systems, the less information is available, and the more risky the transactions which reduce the incentives.

In a study of household food marketing behavior in Senegal, Goetz (1992) used a range of factors to reflect the effect of transaction cost factors on the market participation in grain, both for buying and selling. For exogenous regressors, variables theoretically expected to

---

[1]Exclusive rights being defined as the power or in a wider sense, the right to perform an action or acquire a benefit and to permit or deny others the right to perform the same action or to acquire the same benefit.

affect quantities purchased and sold, as well as specific proxy variables for fixed transaction costs were used. These included ownership of carts for transportation to market, physical distance from market, number of persons in the household and a regional dummy variable separating study area into two regions with region being well integrated into the transport and communication infrastructure hence facing low information gathering costs while the other one was not. Other variables used included age of household head with older and more experienced heads expected to have greater contacts, which allow them to discover trading opportunities at low cost. An interaction term for information was also included. The study found that in the case of effects of fixed cost-type variables on market participation, better information plays an important role. For buyers, adding a person to the household raises the likelihood of market participation while ownership of assets was important in reflecting market access.

Key et al. (2000) extended Goetz's analysis by focusing on participation in maize markets in Mexico. Their study found that both fixed and variable transaction costs play a significant role in explaining household behavior. They also showed that ownership of assets such as transport equipment (pick-up) tends to reduce entry barriers into the market. Omamo (1998) used the transaction costs approach to determine households' decisions to rather devote resources to low-yielding food crops than to cash crops with higher market returns in Kenya. The analytical results show that transport costs are sufficient to explain the cropping choices. This implies that relatively more land is devoted to cash crops and less to food crops the closer the households are to markets. Matungul et al. (2001) used transaction costs theory to determine the determinants of crop marketing in South Africa. Using regression analysis, they found that the level of income generated from food crop sales by small-scale farmers is influenced by transaction costs and certain household and farm characteristics. Still in South Africa, a study to determine the role of transaction costs in participation of smallholder farmers in maize markets (Makhura et al., 2001) found out that transaction costs differ among households due to asymmetries in access to assets, market information, infrastructure and extension.

In Zambia, Kahkonen and Leathers (1999) analyzed changes in transactions costs for evidence of the private sector's ability to fill the vacancy left by government's withdrawal from agricultural marketing. Their assessment of the maize and cotton markets show that although there has been significant success in the private sector's response to liberalization, there are still many conditions that lead to inflated transactions costs especially at the farm level. They concluded that the limited competition among traders at the farm level in remote areas was the source of high transaction costs. Farmers are not well informed about prices in nearby markets, and find it difficult or impossible to search out alternative markets. The factors contributing to these costs are the poor

quality of roads, unavailability of transport, poor quality of communications infrastructure, and unavailability of credit. However, this study focused more on the impact of institutional arrangements (government interventions) on transaction costs, hence the need to study the farmer characteristics that influence the transaction costs they incur as they participate in the markets. This paper complements other studies by examining transaction costs at household level in Zambia. The objectives include identifying key transaction cost factors in the smallholder maize markets, examining their influence on the likelihood of market participation as well as their influence on quantities of maize marketed. In line with the Government's policy of increasing market access for smallholder farmers, this information would be useful to policy makers as an input in the design for interventions to enhance smallholder participation in maize markets.

METHODOLOGY

The study area, data sources and type

The study was carried out in Central province of Zambia. The dominant crops grown are maize, cassava, millet, groundnuts and beans. According to the 2010 population census (CSO, 2012) the population in the province was estimated at 1,307,111 which is about 10% of the national population. The population density is 10.7 persons per square kilometer. By stratifying the households into market participants and non-participants based on the 2005/06 agricultural season, 240 households were sampled using purposive quota sampling. Using a pre-tested structured questionnaire, data on socio-economic characteristics such as household, assets structure and factors like physical location and information access were collected. Household data included variables such as family size, age and education level of household head. Asset structure data comprised of ownership of assets such as bicycles, ox-carts, radios and televisions. These factors were used as proxies for transaction costs to test the main hypothesis that houses facing lower transaction costs had a high probability of market participation.

Theoretical framework

To incorporate transactions costs into an agricultural household model framework, it is convenient to specify market participation as a choice variable (Key et al., 2000). That is, in addition to deciding how much of each good i to consume c_i, produce q_i, and use as an input x_i, the household also decides how much of each good to "market" m_i (where m_i is positive when it is a sale and negative when it is a purchase). If there were no transactions costs, the household's objective would be to maximize the utility function:

$$u(c_a, c_m, c_l; z_u) \tag{1}$$

where: c_a = household staple food (maize in this case); c_m = purchased good; c_l = home time
subject to:

$$\sum_{i=1}^{N} p_i^m m_i + T = 0 \tag{2}$$

(Cash constraint)

$$q_i - x_i + A_i - m_i - c_i = 0, \qquad i = 1,......N$$
(Resource
balance)
(3)

$$G(q, x; z_q) = 0 \quad \text{(Production technology)}$$
(4)

$$c_i q_i, x_i \geq 0 \quad \text{(non-negativity constraint)}$$
(5)

where p_i^m is the market price of good i, A_i is an endowment in good i, T is exogenous transfers and other incomes, z_u and z_q are exogenous shifters in utility and production, respectively, and G represents the production technology.

Considering that in economic terms, transaction costs are costs paid by buyers but not received by sellers, and/or the costs paid by sellers but not received by buyers (Kissel, 2006), they effectively raise the price paid by a buyer and lower the price received by a seller (Minot, 1999). Although these costs are mostly unobservable and cannot be easily recorded (Key et al., 2000), factors that explain them can be observed (Heltberg and Tarp, 2001). Therefore, by introducing and expressing the transaction costs in monetary terms, the cash constraint becomes:

$$\sum_{i=1}^{N} \left[(p_i^m - t_{pi}^s(z_t^s))\delta_i^s + (p_i^m + t_{pi}^h(z^h))\delta_i^b \right] m_i + T = 0$$
(6)

where δ_i^s is equal to one if $m_i > 0$ and zero otherwise, and δ_i^b is equal to one if $m_i < 0$ and zero otherwise. Introduction of transaction costs imply that the price effectively received by the seller is lower than the market price p_i^m by the unobservable amount t_{pi}^s, and the price effectively paid by the buyer is greater than p_i^m by the unobservable amount t_{pi}^b. Transaction costs are expressed as a function of observable exogenous characteristics, z_t^s and z_t^b, that affect these costs when selling and buying. As such, under transaction costs, the household's objective can be expressed by Equations (1) and (3) to (6), while to derive the supply and demand equations, we define the Lagrangian:

$$L = u(c; z_u) + \sum_{i=1}^{N} \mu_i(q_i - x_i + A_i - m_i - c_i)$$
$$+ \phi G(q, x; z_q)$$
$$+ \lambda \left[\sum_{i=1}^{N} [(p_i^m - t_{pi}^s)\delta_i^s + (p_i^m + t_{pi}^b)\delta_i^b] m_i + T \right]$$
(7)

where μ_i, φ, and λ are the Lagrange multipliers associated with the resource balance, the technology constraint, and the cash constraint, respectively. Because the transaction costs create discontinuities in the Lagrangian, the optimal solution cannot be found by simply solving the first order conditions (Key et al., 2000; Minot, 1999). The solution is decomposed in two steps, solving first for the optimal solution conditional on the market participation regime, and then choosing the market participation regime that leads to the highest level of utility. Under the usual assumptions for utility and technology, the conditional optimal supply and demand

are obtained by solving for the first order conditions are as follows:

$$\frac{\partial u}{\partial c_i} - \mu_i = 0, \quad i = \{i | c_i > 0\} \quad \text{(for consumption goods)}$$
(8)

$$-\mu_i + \phi \frac{\partial G}{\partial q_i} = 0, \quad i = \{i | q_i > 0\} \quad \text{(for outputs)}$$
(9)

$$-\mu_i + \phi \frac{\partial G}{\partial x_i} = 0, \quad i = \{i | x_i > 0\} \quad \text{(for inputs)}$$
(10)

$$-\mu_i + \lambda \left[(p_i^m - t_{pi}^s)\delta_t^s + (p_i^m + t_{pi}^b)\delta_i^b \right] = 0 \quad \text{(for traded goods)}$$
(11)

The decision price p_i is given as: $p_i = p_i^m - t_{pi}^s$, if $m_i > 0$, for sellers; $p_i = p_i^m + t_{pi}^b$, if $m_i < 0$, for buyers; $\tilde{p}_i = \mu_i / \lambda$, if $m_i = 0$, For self-sufficient where $\tilde{p}_i$ is the autarky shadow price (ASP). Using the decision prices p_i and the first order conditions, utility maximization subject to the technological constraint leads to a system of output supply equations $q(p, z_q)$ and input demand equations $x(p, z_q)$. Utility maximization subject to the income constraint leads to a system of demand equations for consumer goods $c(p, y, z_u)$.

$$\sum_{i=1}^{N} p_i c_i = y = \sum_{i=1}^{N} [p_i(q_i - x_i + A_i) - t_{fi}^s \delta_i^b] + T$$
(12)

To derive the household supply curves for home produced goods as a function of the market price under fixed transaction costs (FTCs) and proportional transaction costs (PTCs)[2] (Figure 1), let $q(p^m, z_q)$ be the supply curve without transaction costs. Then with transaction costs, the supply curve is:

$$q^s = q(p^m - t_p^s, z_q) \quad \text{for sellers}$$
(13)

$$q^b = q(p^m + t_p^b, z_q) \quad \text{for buyers}$$
(14)

$$q^a = q(\tilde{p}, z_q) \quad \text{for autarky}$$
(15)

Showing transaction costs shift the supply curve upward for sellers and downward for buyers. The supply curve is discontinuous with three distinct regions:

q^b = buyers supply curve for market prices below $\tilde{p} - t_p^b$
(16)

q^s = sellers supply curve for market prices below $\tilde{p} + t_p^s$
(17)

q^a = autarky price between the two thresholds
(18)

---

[2]FTCs do not vary with the level of sales, while PTCs are those that vary with the level of sales.

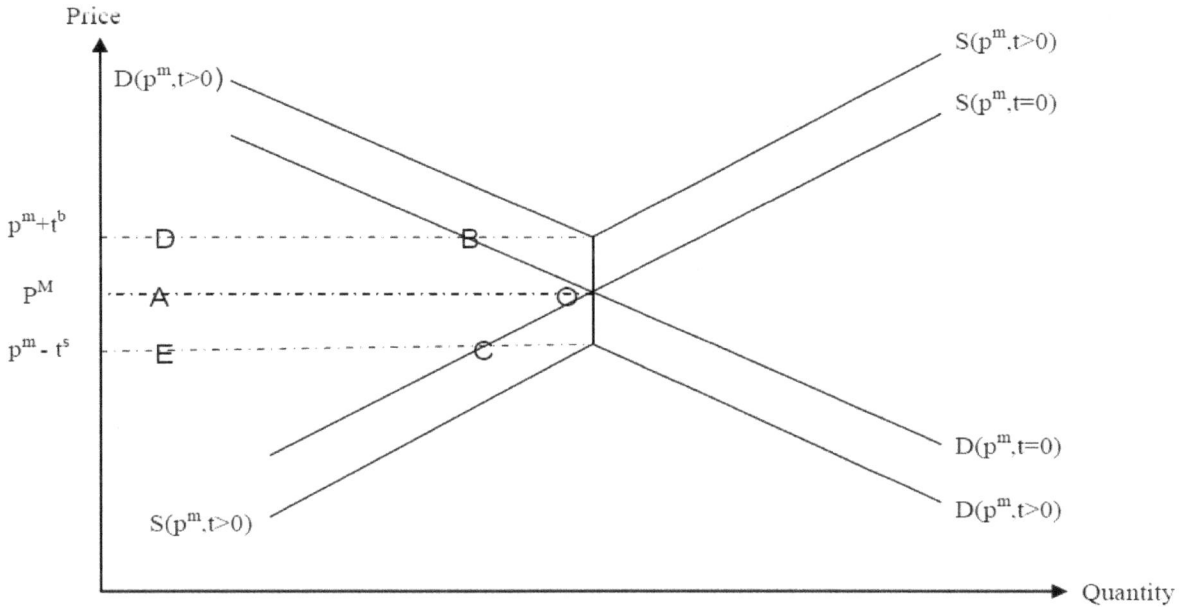

Figure 1. Household demand and supply under transaction costs. Source: Minot (1999).

This implies that fixed transaction costs delay entry into a market as a seller until market price reaches the higher level of $\tilde{p}+t^s_p$. Similarly, they delay entry into a market as a buyer until market price is as low as $\tilde{p}-t^b_p$. The household remains self-sufficient between these two thresholds. A household will switch from autarky to selling when the price that it receives is high enough to compensate for transaction costs.

Empirical model and estimation procedure

Assuming linear expressions:

$$q(p,z_q) = p\beta_m + z_q\beta_q \quad \text{(for supply functions)} \quad (20)$$

$$t^s_p = -z^s_t\beta^t_p \quad \text{(for PTCs for sellers)} \quad (21)$$

$$t^b_p = -z^b_t\beta^b_p \quad \text{(for PTCs for buyers)} \quad (22)$$

This leads to linear expressions for the supply by sellers (q^s):

$$q^s = p^m\beta_m + z^s_t\beta^s_t + z_q\beta_q \quad (23)$$

and by buyers (q^b):

$$q^b = p^m\beta_m + z^b_t\beta^b_t + z_q\beta_q \quad (24)$$

and for autarky households supply (q^a):

$$q^a = z_q\beta^a_q + z_c\beta^a_c \quad (25)$$

For production thresholds, linear expressions for (q^s) are used such that:

$$\underset{-}{q^s} = z^s_t\alpha^s_t + z_q\alpha^s_q + z_c\alpha^s_c \quad (26)$$

and for (q^b) such that:

$$\underset{-}{q^b} = z^b_t\alpha^s_t + z_q\alpha^b_q + z_c\alpha^b_c \quad (27)$$

The econometric specification is obtained by adding error terms to the supply equations:

$$q^s = p^m\beta_m + z^s_t\beta^s_t + z_q\beta_q + u_i \quad \text{(seller supply equation)} \quad (28)$$

$$\equiv x_i\beta_i + u_i$$

$$\underset{-}{q^s} = z^s_i\alpha^s_i + z_q\alpha_q z_c\alpha_c + u_2$$

(seller threshold equation) $\quad (29)$

$$\equiv x_2\alpha_2 + u_2 \quad (30)$$

Where x_i is a vector of exogenous explanatory variables such as household characteristics and location characteristics that influence market participation. The market participation indicator variable (q^s) for the commodity is defined as:

$$q^s = 1 \text{, if } p^m \geq \tilde{p}+t^s_f \text{ or } p^m \leq \tilde{p}-t^s_f \quad \text{(when a household sells)} \quad (31)$$

Table 1. Variables used and hypothesized relationships.

Variable description	Variable	Hypothesized relationship	
		Participation decision	Participation level
Ownership of bicycle	D2	+	+
Ownership of ox-cart	D3	+	+
Ownership of radio	D4	+	
Availability of alternative channels	D5	+	+
Listening to agricultural programs	D6	+	+
Ownership of television	D7	+	
Membership to farmer associations	D8	+	+
Size of harvest	QHST	+	
Age of household head	AGE	+	+
Distance to nearest maize markets	DIST	-	-
Education level of household head	EDU	+	
Household size (number of adults)	HHS	+	+
Frequency of listening to radio	FRR	+	+
Experience in maize marketing	EXP	+	+

$$q^s = 0 \text{ , if } \tilde{p} - t^s_f \leq p^m < \tilde{p} + t^s_f$$

(when a household does not sell) (32)

Data analysis

Under transaction costs, households face a two-stage decision problem (Winter-Nelson and Temu, 2003; Key et al., 2000; Makhura et al., 2001; Goetz, 1992). The first decision, is whether to trade or not and the second is how much to trade and is conditional on participation as a buyer or seller. Because some households participate in the market while others do not, if ordinary least squares regression (OLS) is estimated, the non-participants will be excluded introducing a sample selection bias in the model (Gujarati, 2004). Therefore, in order to analyze the factors affecting the probability and extent of participation in maize markets, a two-step Heckman's procedure (Heltberg and Tarp, 2001; Makhura et al., 2001; Nkonya et al., 1998; Goetz, 1992) was used. This involved two estimation steps. In step one, a logistic regression model was estimated to give the estimated probability that a house i purchased or sold maize. In step two, the intensity of participation was estimated by running a heckits that is OLS corrected for selectivity bias. This was run on observations for which sales were greater than zero.

$$P(SAI) = \alpha_1 + \alpha_2 D_{2i} + \alpha_3 D_{3i} + \alpha_4 D_{4i} + \alpha_5 D_{5i} + \alpha_6 D_{6i} + \alpha_7 D_{7i} + \alpha_8 D_{8i} + \beta_2 QHS + \beta_3 AGE + \beta_4 HHS + \beta_5 EDU + \beta_6 DIST + \beta_7 FRR + \beta_8 EXP + U_1$$

(33)

The results from Equation (33) showed the influence of independent variables on the probability of maize marketing ($\delta Pr/\delta x$). The second model (step two) was used to identify the factors affecting the quantities of maize sold and was expressed as:

$$QTY = \alpha_1 + \alpha_2 D_{2i} + \alpha_3 D_{3i} + \alpha_5 D_{5i} + \alpha_9 D_{9i} + \beta_3 AGE + \beta_4 HHS + \beta_6 DIST + \beta_7 FLR + \beta_8 EXP + \sigma IMR + U_2$$

(34)

Where QTY = Quantity of maize sold while all the other independent variables are the same as those used in step one except for dummies for radio and television as well as quantity

harvested and education variables. This model was run using data from market participants only and included an inverse mills ratio (IMR) to correct for selectivity bias. It was used to estimate the impact of exogenous variables on quantities of maize sold. Table 1 shows the hypothesized relationships between the explanatory variables and probability of maize market participation as well as quantities of maize marketed.

RESULTS AND DISCUSSION

Quantitative factors affecting transaction costs

Comparisons show that the mean harvest size, mean asset value and mean land holding for market participants, were significantly higher ($P \leq 0.05$) than for non-participants (Table 2). The mean distance from commercial centres and main roads for participating households was also significantly lower ($P \leq 0.05$) than for non-participants. However, the average household age, mean household size and years of formal education attained by the household head were not significantly different between the two groups.

Effects of transaction cost on decision to participate in maize markets

Table 3 presents the results of the logit estimations of factors influencing the decision to sell maize. The model x^2 (14) was 132.544 (and significant at the five percent level) implying that the model was predicting decision to sell better than if only the constant had been used. The R-Square of 0.708 indicates that 70.8% of the variation in the decision to sell maize can be explained by the independent variables in the model. The significant transaction costs factors influencing decisions to

Table 2. Comparison of quantitative transaction cost factors between market participants and non-market participants.

Variable	Non-participant (n = 105)	Participant (n = 135)	F-Statistic
Mean size of harvest (50 kg bags)	14.65	87.69	27.57**
Mean value of assets (million Kwacha)	3.15	8.89	15.34**
Mean age of household head (years)	46.60	45.72	0.13
Mean household size (number of adults)	6.45	6.78	0.04
Mean distances from commercial centres (km)	5.84	3.63	16.97**
Years of formal education completed by Household Head	8.05	9.00	0.09
Mean size of land holding (hectares)	5.17	19.07	6.99**

**Significant at 5%; *Significant at 10%.

Table 3. Factors determining households' decisions to participate in maize markets.

Variable	Coefficients	Standard error	Exp(B)
Constant	-1.859	1.664	0.264
Ownership of radio	2.191**	0.799	8.946
Ownership of television	2.479**	0.824	11.933
Own mobile phone	-2.436**	0.834	0.088
Listening frequency programs	-0.114	0.068	0.999
Distance to main markets	-0.372**	0.108	1.449
Ownership of bicycle	-0.185	0.831	0.539
Ownership of ox-cart	1.513*	0.853	4.540
Availability of multiple channels	1.818**	0.543	6.162
Education of household head	0.028	0.047	1.029
Age of household head	0.006	0.021	1.006
Household size (Number of adults)	-0.066	0.089	0.936
Size of maize harvest	0.093**	0.022	1.097
Membership to farmer groups	-0.114	0.575	0.892
Experience in maize marketing	-0.043	0.033	0.958
R^2=0.708 (Cox and Snell) $\chi^2(11) = 132.544^{**}$			

**p < 0.05, *p<0.10; Dependent variable: Sold maize in 2005/6 season; sample size: n= 220.

participate in maize markets were ownership of radio, ownership of television, availability of multiple maize marketing channels, distance to maize markets, ownership of ox-carts and the harvest size. Ownership of assets such as radio and television enables households to acquire market information at a lower cost thus reducing expenditure on search, negotiation and screening costs (Key et al., 2000; Goetz, 1992). This reduces the magnitude of the transaction costs thus increasing the probability of market participation for the household.

Presence of alternative marketing channels increases the efficiency of the marketing system through prevention of monopolistic tendencies (Minten, 1999; Kirsten and Vink, 2005) where short distance to markets reduces the magnitude of the transaction costs by reducing the amount of time and money spent in search for information. By reducing information asymmetry between buyers and sellers, these factors reduce the magnitude of

transaction cost thus increasing the probability of maize market participation. Size of the harvest was found to significantly increase household's probability of maize marketing. This has been explained by the fact that those smallholder farmers who were faced with challenges in maize marketing responded by switching to other crops (Zulu et al., 2000; Seshamani, 1999). Similar results have been reported in South Africa (Matungul et al., 2001; Makhura et al., 2001) where households with larger maize harvests were likely to have surpluses for sale. Age and education level of the household head, maize marketing experience and membership to farmer organizations were not significant.

Effect of transaction cost factors on level of maize sales

Table 4 presents the results of the factors determining

Table 4. Factors influencing the quantities of maize sold by households.

Variable	Coefficient	Std. error	t-statistic
Constant	35.180	25.274	1.392
Experience in maize marketing	0.577*	0.323	1.777
Age of household head	-0.327	0.328	-0.995
Household size (Number of adults)	3.480**	1.515	2.298
Membership to farmer associations	-9.903	8.702	-1.046
Availability of alternative channels	-4.539	1.326	-0.440
Distance to commercial centers	-2.339	1.848	-1.265
Frequency of listening to radio	3.331**	1.139	2.925
Ownership of ox-carts	44.243**	10.272	4.304
Ownership of bicycles	1.398	9.264	-0.151
LAMBDA (IMR)	-26.145**	12.189	2.145
R^2	0.445		
Adjusted R^2	0.395		
S.E. of estimate	44.25		
F-Statistic	8.823		
Prob. (F-Statistic)	0.000		

** $P < 0.05$, *$P<0.10$. Dependent variable: Number of bags of maize sold; Sample size: n = 90.

the quantities of maize sold by the households. The R^2 and adjusted R^2 were quite low (0.445 and 0.395 respectively) which is not unusual for cross sectional data, while the overall significant fit (F) was 8.823 indicating that the data correctly fits the model. The coefficient on the inverse mills ratio (lambda) was significant at five percent level indicating that correlation between the error terms of the decision to sell (u_1) and level of market participation (u_2) was different from zero, $\sigma_{\varepsilon u} \neq 0$. This implies that sample selection bias would have resulted if the level of maize sales had been estimated without taking into account the participation decision.

The significant transaction costs factors influencing the quantities of maize marketed were household size, experience in maize marketing, frequency of listening to agricultural programs on the radio and ownership of ox-carts. As the household size increased by one adult, the quantity of maize sold by the household would increase. Although family size has two opposing effects with large family size implying large food demand thus reducing marketable surplus, large family size also implies increased labor supply (Makhura et al., 2001). Considering that the sampled households depended on family members for labor supply, the larger the number of adults in the household, the more labor they had and the more maize they were likely to produce. An increase in maize marketing experience also increased the quantities of maize sold. Experience in maize marketing makes certain information and search costs low (Goetz, 1992; Makhura et al., 2001) due to prevalence of social networks. Experienced households may also have greater contacts and increased trust gained through repeated exchange with the same parties (Kirsten and

Vink, 2005) allowing them to discover trading opportunities at lower costs.

By reducing the unit cost of production and delivering produce to the market, assets such as oxen reduces variable transaction costs faced by households leading to higher levels of market participation (Key et al., 2000). The regression results show that households that owned ox-carts marketed 2,200 kg more than those that did not own ox-carts. This observation may be explained by the fact that most transactions were being conducted either at the market centers or trader's premises with farmers bearing the cost of delivering the produce. Similar results have been reported in Mozambique (Heltberg and Tarp, 2001), Mexico (Key et al., 2000) and South Africa (Makhura et al., 2001).

CONCLUSION AND RECOMMENDATIONS

The results show that high transaction costs negatively influence the decision to participate in maize markets as well as the quantities marketed in Zambia. Based on these findings, it is recommended that information be provided for farmers, through existing government agencies such as the National Agricultural Information Services (NAIS) on who is buying maize, at what prices they are buying and the location of these buyers using mass media such as radio and television. To increase the likelihood of market participation, action should be taken to increase farmers' access to marketing channels through increased access to transport which also minimises the impact of distance on those farmers located far away from major maize trading centres. This can be achieved by improving on the quality of rural roads by rehabilitating feeder roads connecting villages to major trading centres and highways so as to encourage private transporters to venture into these rural areas. Furthermore, public investments that raise smallholders' productivity, such as improved seeds availability and innovative extension programs should be intensified while actions aimed at increasing household's productive asset base such as ox-carts should also be intensified through provision of affordable loans as well as work-for-asset programmes which are already being implemented in some areas.

REFERENCES

Central Statistical Office (2002), Agriculture Analytical Report for the 2000 Census of Population and Housing, Lusaka Zambia.

Central Statistical Office (2012), Zambia 2010 Census of Population and Housing: Population Summary Report, Lusaka Zambia.

de Janvry A, Fafchamps M, Sadoulet E (1991). Peasant Household Behaviour with Missing Markets: Some Paradoxes Explained. Econ. J. 101(409):1400-1417.

Eggertson T (1990). Economic Behavior and Institutions, Cambridge: Cambridge University Press.(Book, pp.14-15)

Goetz SJ (1992). A selectivity model of household food marketing behavior in Sub-Saharan Africa". Am. J. Agric. Econ. 74(2):444-452.

Heltberg R, Tarp F (2001). Agricultural Supply Response and Poverty in Mozambique". Institute of Economics, University of Copenhagen, Paper presented at the conference on "Growth and Poverty" at the World Institute for Development Economics Research in Helsinki on 25th -26th May, 2001. Discussion Paper No. 2001/114.

Jayne TS, Chapoto A, Govereh J (2007). Grain Marketing Policy at the Crossroads: Challenges for Eastern and Southern Africa: Paper Prepared for the FAO Workshop on "Staple Food Trade and Market policy Options for Promoting Development in Eastern and Southern Africa", Rome, Italy.

Kahkonen S, Leathers H (1999), Transaction costs analysis of maize and cotton marketing in Zambia and Tanzania. SD Publications Series, P. 105.

Key N, Sadoulet E, de Janvry A (2000), "Transactions costs and agricultural household supply response Am. J. Agric. Econ. 82(2):245-259.

Kirsten J, Vink N (2005), The Economics of Institutions: Theory and Applications to African Agriculture. University of Pretoria, South Africa.

Kissel R (2006). The expanded implementation shortfall: "Understanding transaction costs components. Forthcoming, J. Trading summer.

Makhura M, Kirsten J, Delgado C (2001). Transaction costs and smallholder participation in the maize market in Northern Province of South Africa". Seventh Eastern and Southern Africa Regional Maize Conference, 11th -15th February, 2001, pp. 463-467 (Place missing on paper)

Matungul PM, Lyne MC, Ortmann GF (2001). Transaction Costs and Crop Marketing in the Communal Areas of Impendle and Swayimana, KwaZulu-Natal. Development Southern Africa, 18(3):347-363.

Minten B (1999). Infrastructure, market access, and agricultural prices: evidence from Madagascar". MSSD Discussion Paper No. 26 Market and Structural Studies Division International Food Policy Research Institute. Washington D.C.

Minot N (1999). Effects of Transaction Costs on Supply Response and Marketed Surplus: Simulations Using Non-Separable Household Models. MSSD Discussion Paper No.36. Int. Food Pol. Res. Inst. Washington D.C.

Nijhoff JJ, Tembo G, Shaffer J, Jayne TS, Shawa J (2003), How will the Proposed Crop Marketing Authority Affect Food Market Performance in Zambia: An Ex Ante Assessment to guide Government Deliberation. Working Paper No. 7, Food Security Research Project, Lusaka, Zambia.

Nkonya E, Xavery P, Akonaay H, Mwangi W, Anandajayasekeram P, Verkuijl H, Martella D, Moshi A (1998). "Adoption of maize production technologies in Northern Tanzania". Mexico D.F: International Maize and Wheat Improvement Center (CIMMYT), The Republic of Tanzania and the Southern Africa Center for Cooperation in Agricultural Research (SACCAR).

Omamo SW (1998). Transport Costs and Smallholder Cropping Choices: An Application to Siaya District, Kenya. Am. J. Agric. Econ. 80(1):116-123.

Regional Agricultural Trade Expansion Support (2003). Maize market assessment and baseline study for Zambia". Prepared by the IMCS Center, Lusaka, Zambia.

Seshamani V (1999). The impact of market liberalization on food security in Zambia. Food Policy, 23(6):539-551

Simatele CM (2006), Impact of Structural Adjustment on Food Production in Zambia. University of Hertfordshire, Hartfield, United Kingdom

Winter-Nelson A , Temu A (2003), Impact of Prices and Transaction Costs on Input Usage in a Liberalizing Economy: Evidence From Tanzanian Coffee Growers . Agricultural and Consumer Economics, University of Illinois, USA.

Yambayamba ES (Eds.) (2009). Inaugural National Symposium on Agriculture under the theme: Harnessing the Potential of Agriculture to meet the Increasing Demands of a Growing Population. University of Zambia

Zaibet LT, Dunn EG (1998), Land tenure, farm size, and rural market participation in developing countries: the case of the Tunisia olive sector. Economic Development and Cultural Change, 46(4):831-848.

Zulu B, Nijhoff JJ, Jayne TS, Negassa A (2000). Is the Glass Fullor Half Empty? An Analysis of Agricultural Production Trends in Zambia. FSRP Working Paper No. 3, Food Security Research Project, Lusaka, Zambia.

Zulu B, Jayne TS, Beaver M (2007). "Smallholder household maize production and marketing behavior in Zambia". Working P. 22, Food Security Research Project. Lusaka, Zambia . http: www.aec.msu.edu/agecon/fs2/zambia/index.htm

9

Measuring the effect of climate change on agriculture: A literature review of analytical models

Maria De Salvo[1], Diego Begalli[1] and Giovanni Signorello[2]

[1]Department of Business Administration, University of Verona, via Della Pieve, 70, San Pietro in Cariano, Verona, Italy.
[2]Department of Agri-food and Environmental System Management, University of Catania, via Santa Sofia, 100, Catania, Italy.

This article provides a short overview of the principal models that can be used to estimate the effects of climate change on agriculture. The models are classified in relation to the following criteria: the specific impacts they aim to assess, their ability to measure production and/or economic losses, and the adoption of social indicators of the effects and responses. The weaknesses and strengths of the models are also identified and discussed. The most relevant factors for the choice of the most appropriate model are analysed. Through a comparative analysis of the literature, an easily adoptable scheme for selecting the most appropriate method to estimate the effects of climate change according to the characteristics of the case study is identified. The adopted classification scheme demonstrates that one model is capable of simultaneously considering many aspects related to climate change and classifying these in different class.

Key words: Climate change, impacts, agriculture, models.

JEL Classification: C50, Q15, Q51.

INTRODUCTION

Agriculture is one of the sectors most affected by ongoing climate change. The wide range of literature on this subject demonstrates that damages caused by climate change can be relevant to both cropping and livestock activities (IPCC, 1990; Adams et al., 1998). Climate change will have a significant effect on the rural landscape and the equilibrium of agrarian and forest ecosystems (Walker and Steffen, 1997; Bruijnzeel, 2004). In fact, climate change can affect different agricultural dimensions, causing losses in productivity, profitability and employment. Food security is clearly threatened by climate change (Sanchez, 2000; Siwar et al., 2013), due to the instability of crop production, and induced changes in markets, food prices and supply chain infrastructure.

Moreover, because of the multiple socio-economic and bio-physical factors affecting food systems and, consequently food security, the capacity to adapt food systems to reduce their vulnerability to climate change is not uniform from a spatial point of view (Gregory et al., 2005).

However, besides its primary role in producing food and fibres, agriculture performs also other functions, such as the management of renewable natural resources, the construction and protection of landscape, the conservation of biodiversity, and the contribution to maintain socio-economic activities in marginal and rural areas. Climate change could affects also this multifunctional role of agriculture (Klein et al., 2013).

The ongoing effects of climate change require the individuation of mitigation policies to reduce greenhouse gas emissions and identify appropriated adaptation strategies that aim to contain agricultural losses both in market goods and environmental services (such as protection of biodiversity, water management, landscape preservation and so on). These strategies can easily be identified and applied if the economic effects of climate change on agriculture are assessed. However, creating models that are able to assess these effects accurately can present difficulties for several reasons. The first is data availability: while data are frequently available, they are often not disaggregated on the necessary temporal and/or spatial scales. Another reason is that research about the effects of climate change involves multidisciplinary skills and competencies because analyses of the effects of climate change involve many factors such as the consideration of (Bosello and Zang, 2005):

1. Climate and other induced climate-change environmental aspects,
2. Biological and plant physiology aspects,
3. Technical and socioeconomic factors,
4. Strategies to coping with the effects of climate change,
5. Impacts on/of the main economic adjustment mechanisms at the national and international level,
6. Feedback of the changed conditions on climate.

Economic and agricultural policies play an important role in such analyses, as does the geographical scale (e.g. local, regional or international) considered for the analysis. In addition to these aspects, it is also important to consider the temporal and spatial variability of the events which in turn causes a difficult predictability of future scenarios.

Considering all these aspects simultaneously is problematic. For this reason the literature proposes several models that are suitable for estimating the effects of climate change on agriculture addressing specific research issues. In light of this the present article offers an overview of the models most used to estimate the effects of climate change on agriculture (section 2) aimed to classify these models and to propose a logical scheme to help researchers in the selection of the model that best suits their research goals (section 3). The fourth section presents the conclusions.

LITERATURE REVIEW

The literature suggests that various models can be employed to assess the effects of climate change on agriculture. Each model has advantages and shortcomings, and presents different levels of complexity and completeness in relation to the specific aspects considered in its analysis. These peculiarities are

discussed below for each models category.

The effects of climate change were evaluated by several scholars with consideration given only to the changes in the production of specific crops (principally maize, rice, cotton and soybean), using the so-called 'crop simulation models'. These models restrict the analysis to crop physiology, and simulate and compare crop productivity for different climatic conditions (Eitzinger et al., 2003; Torriani et al., 2007a). Crop models are considered 'agriculture oriented' because the analysis of these models is focused on the biological and ecological consequences of climate change on crops and soil. In these models, farmers' behaviour is not captured and the management practice is considered fixed. Moreover, they are crop and site specific, and they were calibrated only for the major grains and for a limited number of places (Mendelsohn and Dinar, 2009).

Others scholars estimated the sensitivity of yields to climate using empirical yield models that apply the production–function approach (Terjung et al., 1984; Eitzinger et al., 2001; Isik and Devadoss, 2006; Lhomme et al., 2009; Poudel and Kotani, 2013). The basic idea of this approach is that the growth of agricultural production depends on soil-related and climatic variables that are implemented as explanatory variables in the model for estimating the production function. Changes in climate scenarios are usually simulated using the general circulation model (GCM) (Chang, 1977; Randall, 2000).

In the production function approach, the economic dimension is of secondary importance and is considered in a partial and simplified manner (Bosello and Zang, 2005), even if these models produce important information for larger model frameworks that consider economy, later discussed. Some studies explicitly assess the economic impact of climate change through the estimation of the economic production function (Adams, 1989; Rosenzweig and Parry, 1994). However, other research evaluates the economic effects of climate change by implementing the results of agronomic analyses or of empirical yields models in mathematical-programming models (Kaiser et al., 1993; Finger and Schmid, 2007).

The main weakness of the production–function model is that it is crop and site specific. It endorses the so-called 'dumb-farmer' hypothesis, which excludes from analysis the plausible adoption by farmers of strategies for coping with the effects of climate change, for example, strategies that replace crops that are most sensitive with others that are less so (Rosenzweig et al., 1993; Reilly et al., 1994).

To overcome this limitation, Mendelsohn et al. (1994) proposed the Ricardian model. The principal characteristic of the Ricardian model is that it treats adaptation to climate change as a 'black box'. In fact it estimates the relationship between the outcomes of farms and climate normals using cross-sectional data and including, among regressors, appropriate control variables. As such, it implicitly considers farmer adaptation strategies without the need to implement such strategies as explicit exploratory variables (Mendelsohn and Dinar, 2009).

However, this aspect could also represent a weakness in the model if the aim of the analysis were to estimate the effect of farmer adaptation strategies on climate change. Due to this weakness in analysis, models have been proposed that use mathematical programming to consider specifically farmer adaptation strategies (Adams et al., 1990; Kaiser et al., 1993; Mount and Li, 1994), especially concerning irrigation (Medellín-Azuara et al., 2010). However, these applications often suffer the limitation of considering hypothesised and simulated strategies that can be derived by incorrect simulation of the farmers' goal function.

The latest applications of the mathematical-programming model use positive mathematical programming (PMP) (Qureshi et al., 2010, 2013; Howitt et al., 2012). These surpass the traditional limitations of linear-programming methods, for example, the unavailability of detailed information about the relationship between inputs and yields through the function cost. In the field of the assessment of climate change impacts on agriculture this model is particularly suitable for analysis of the effects of drought on agriculture because it allows different aspects related to the use and availability of water to be explicitly treated. However, given that this model needs to consider data that can be difficult to collect (e.g. water cost by considering the source of water, the water requirements of crops, and the availability of water resources), its applicability is also limited.

More recently, other research has attempted to overcome the limitations of the Ricardian model in considering farmer adaptation strategies[1] by using econometric models estimated on farm survey data. These applications explicitly treat farmer adaptation strategies by using their proxies as explanatory variables (Di Falco and Veronesi, 2013a, b; Oluwasusi, 2013) or by modelling adaptation as the dependent variable (Gebrehiwot and Van Der Veen, 2013). These applications have the advantage of being able to estimate using the available data.

Moreover, they are suitable to be specified through sophisticated models that can consider specific characteristics of the database such as endogeneity, stratified samples, spatial correlation, and panel and time-series data. With such applications, it is also possible to hypothesise different equation functional forms (e.g. linear, log-linear, quadratic, Box Cox) as well as different distributions for the error term (e.g. normal, Weibull, probit, logit) while at the same time, using the most suitable estimator (e.g. ordinary least squares,

maximum likelihood estimator) according to the specific model. However, the predictive ability is strongly connected with the accuracy of the model specification and the data quality. On this last aspect impacts the impossibility to consider strategies that are new. In fact in the past we did not have climate change so in the future new approaches need to be developed.

All the models that have been discussed focus on the agricultural sector, its specific branches, or crops without considering the relationships with other economic sectors. For this reason, further research developed general equilibrium economic models (GEMs) (Darwin et al., 1995; Borsello and Zang, 2005; Calzadilla et al., 2010a, b). GEMs examine the economy as a complex system composed of interdependent components (e.g. industry, factors of production, institutions and international economic conditions). GEMs have the advantages: to capture economy-wide and global changes, and to measure the effects of climate change on other economic sectors. Conversely, they are limited in that they aggregate in a single entity different sector characterised by specific economic and spatial dimensions. For example, agriculture is generally considered as an aggregate sector at the national level without considering its local specificities. Similarly, production factors (including irrigation water) are implemented in the model as undifferentiated commodities. Further, GEMs do not consider farmer adaptation to climate change or all dimensions, skills, and competencies that should be involved in the analysis of the effects of climate change (Mendelsohn and Dinar, 2009).

Consequently, researchers developed integrated assessment models (IAMs)[2] that combine the use of GCM with data on crop growing, soil usage, and economic models (Prinn et al., 1999; Kainuma et al., 2003). IAMs describe the causes and effects of climate change, integrating knowledge from different academic disciplines into a single framework to generate useful information for policymakers (Dinar and Mendelsohn, 2011).

The integration of such varied skills and disciplines means IAMs are often particularly complex. Moreover, interactions between agriculture and land usage with climate are only partially treatable in such models and the accuracy of this model is subject to the treatment of complex interactions (e.g. the availability and the competitive use of water between economic sectors). Another limitation is that productivity is treated as an exogenous variable, even if it is strongly correlated with the climate (Dinar and Mendelsohn, 2011). Tables 1 and 2 summarises the advantages and limitations for each of the models that have been discussed in the literature review.

---

[1] Seo and Mendelsohn (2008) propone a multiple-stage model called the structural Ricardian model that first estimates an adaptation model on farmer choice, and then estimates the conditional income for each choice using a traditional Ricardian formulation.

[2] For more information on IAMs, see: IMAGE (http://www.mnp.nl/en/themasites/image/index.html) or IGSM-MIT (http://globalchange.mit.edu/igsm/).

Table 1. Principal models used to estimate the effects of climate change on agriculture.

Model	Brief description	Advantages	Limitations
Crop simulation	This model restricts the analysis to crop physiology, and simulate and compare crop productivity for different climatic conditions	It is based on a deep understanding of agronomic science It is suitable to integrate effects of carbon dioxide fertilization It is calibrated to local condition	Analysis is focused on the biological and ecological consequences of climate change on crops and soil Economic dimensions are not considered. This model can be coupled with other models to better treat economic dimension. In the traditional formulation adaptation is not considered and the farmer's management practice is considered fixed. Some researchers consider adaptation exogenously. It do not consider crop's switching. It is crop and site specific It was calibrated for the main grains and for a limited number of places
Production Function	Yields sensitivity to climate is estimated assessing a empirical production function that links water, soil, climate and economic input to yields for specific crops.The effect of climate change is assessed by considering yield variations comparing two alternative scenarios. Future climate scenarios are usually simulated using a GCM.	Easy to estimate It is possible to measure the effect of weather on yields over time	Crop specific Social and economic dimensions of agriculture are considered of secondary importance. This model can be coupled with other models to better treat economic dimension. Assumption of the 'dumb-farmer' hypothesis (farmer adaptation strategies are not considered) Calibrated for a specific context; if the location is not representative, can provide biased predictions.
Ricardian	This model treats the full range of farmer adaptation strategies as a black box by performing a cross-sectional regression of land values or net revenues on climate normals and other control variables. Climate normals are calculated as averages in a long-term scenario (usually 30 years). The effects of climate change are assessed in terms of farm outcome variations, comparing the current situation to simulated scenarios.	Does not assume the 'dumb-farmer' hypothesis Easy to estimate Possible to consider spatial correlations and to analyse panel data Possible to elicit farmer adaptation in estimation if a multinomial logit model (e.g. a structural Ricardian model) is used.	Omitted variables, such as unobservable farm and farmer characteristics could lead to bias of unknown sign and magnitude In the traditional formulation, farmer adaptation strategies are considered but not explicitly treated In the traditional formulation, the role of irrigation is not considered. More recently, this variable was included among the regressors. However, it is not treated endogenously and multicollinearity problems are not adequately considered Analysis is focused on the economic dimension of agriculture and only indirectly on other dimensions (e.g. biological and social) Assumes a partial equilibrium model and does not consider relationships with other sectors Assumesthe output and input prices constancy and does not measure adjustment costs.

Table 1. Contd.

PMP	This is an economic management model estimated by solving a mathematical-optimisation problem using farm data. The pay-off function can be formulated considering the profit (to be maximised) or the cost (to be minimised). The latter, known as the Positive Mathematical Programming, surpasses the traditional limitations of linear-programming methods such as the unavailability of detailed information on the relationships between inputs and yields through the dual function cost.	Useful for assessing the economic effects of climate change, especially in the simulation of irrigation-farmer adaptation options and/or water policies, including water markets and irrigation efficiency improvement.	Difficult to estimate. Often difficult to find data on technical coefficients and limiting production factors. Assumes simulated farmer strategies not obtained from observed choices in specific climatic scenarios.
GEM	These look at the economy as a complex of interdependent components (e.g. industry, production factors, institutions).	Assumes a general economic equilibrium, considering all economic sectors. Captures economy-wide and global changes such as those linked to input and output prices. Provides information on the effect of climate change in different regions. Measures the effect of climate change on other economic sectors.	Difficult to estimate. Aggregates into one single entity sectors that are different in economic and spatial characteristics. Production factors, including irrigation water, are considered in the model as undifferentiated inputs. Difficult to analyse farmer adaptation strategies. Does not allow consideration of details of the studied phenomena.
IAM	These are based on the joint use of General Circulation Model, crop growing, soil usage, and economic models. These models integrate different skills and competencies.	Analysis simultaneously considers all agricultural dimensions. Generates useful information for policymakers.	Difficult to estimate. These models can be very complex. In some cases the required data are not available. Interaction between agriculture and land use with the climate are only partially treatable. Accuracy of model is subject to the treatment of the complex interaction between different factors, especially concerning water usage and availability. Productivity is treated as an exogenous variable.

CLASSIFICATION OF MODELS, RESEARCH QUESTIONS TO BE ANSWERED, AND CRITERIA FOR CHOOSING THE MOST SUITABLE MODEL

To assess the effect of climate change on agriculture, the choice of the most appropriate model depends on the following factors:

1. The level at which the analysis needs to be conducted—this could be the agricultural sector; whole, or one crop, or a particular agricultural branch[3]

2. The (temporal or spatial) scale of analysis; as a

---

[3] The literature discusses numerous applications that estimate the effect of climate change on permanent cultivations (Lobell et al., 2006), viticulture (Tate,

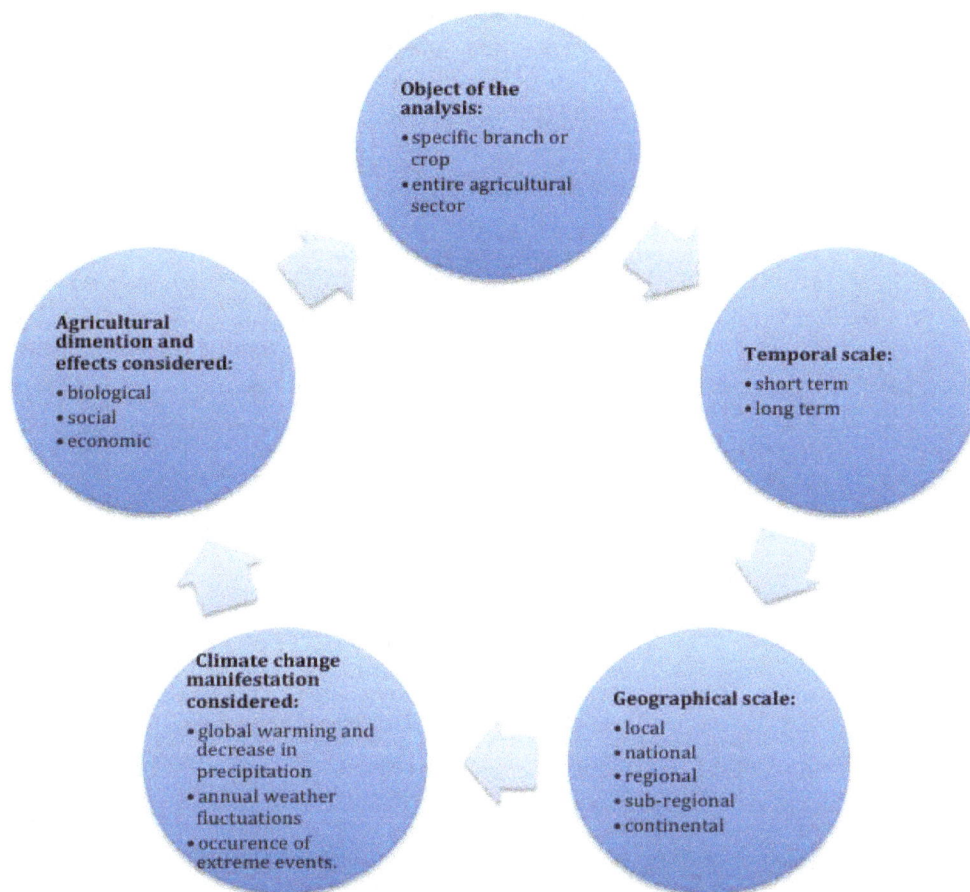

Figure 1. Aspects that influence the choice of model to be used; Source: Authors' elaborations.

whole, or one crop, or a particular agricultural branch[4];
3. The climatic phenomenon used to measure the analysed climate change (Tate, 2001; Bernetti et al., 2012), and livestock (Seo, 2008; Reynolds et al., 2010; Kimaro and Chibinga, 2013);
4. The agricultural dimension (biological, social or economic) with respect to which climate change impacts are assessed.

Figure 1 summarises the hierarchical links between these elements. The first aspect (the level of the analysis) and the fourth aspect (the agricultural dimension to be considered for estimating the effects of climate change) are connected. In fact, the models devoted to the analysis of the biological dimension of agriculture are crop specific; consequently, they concern only a single crop or branch. Conversely, the models devoted to assessing the effect of climate change on the social or economic dimensions of agriculture can consider the agricultural sector as a whole or one of its branches.

In reference to the scale of analysis it can concern cross-sectional, panel, or time-series data. In the latter case the length of the time period to be considered depends on the analysed scenario. The spatial scale can be very significant when the empirical evidence demonstrates that the magnitude of the effect of climate change varies significantly according to the location and the size of the areas studied. Previous research has highlighted that agriculture in warmer areas is more affected by climate change than agriculture in colder areas (Mendelsohn et al., 1994; Schlenker et al., 2005). However, the effects can vary dramatically on international, national and local scale (Bindi and Olesen, 2011). This variation in the effects is due to differences in adaptation strategies, which correlate highly with the local cultural, institutional and environmental conditions.

Another important issue to be considered is the specific manifestation of climate change that the model considers in calculating its effect on agriculture. This issue may concern:

1. A general increase in temperatures, accompanied by a decrease in precipitations characterising a long-term scenario (climate warming and precipitations change);

---

[4] The literature discusses numerous applications that estimate the effect of climate change on permanent cultivations (Lobell et al., 2006), viticulture (Tate,

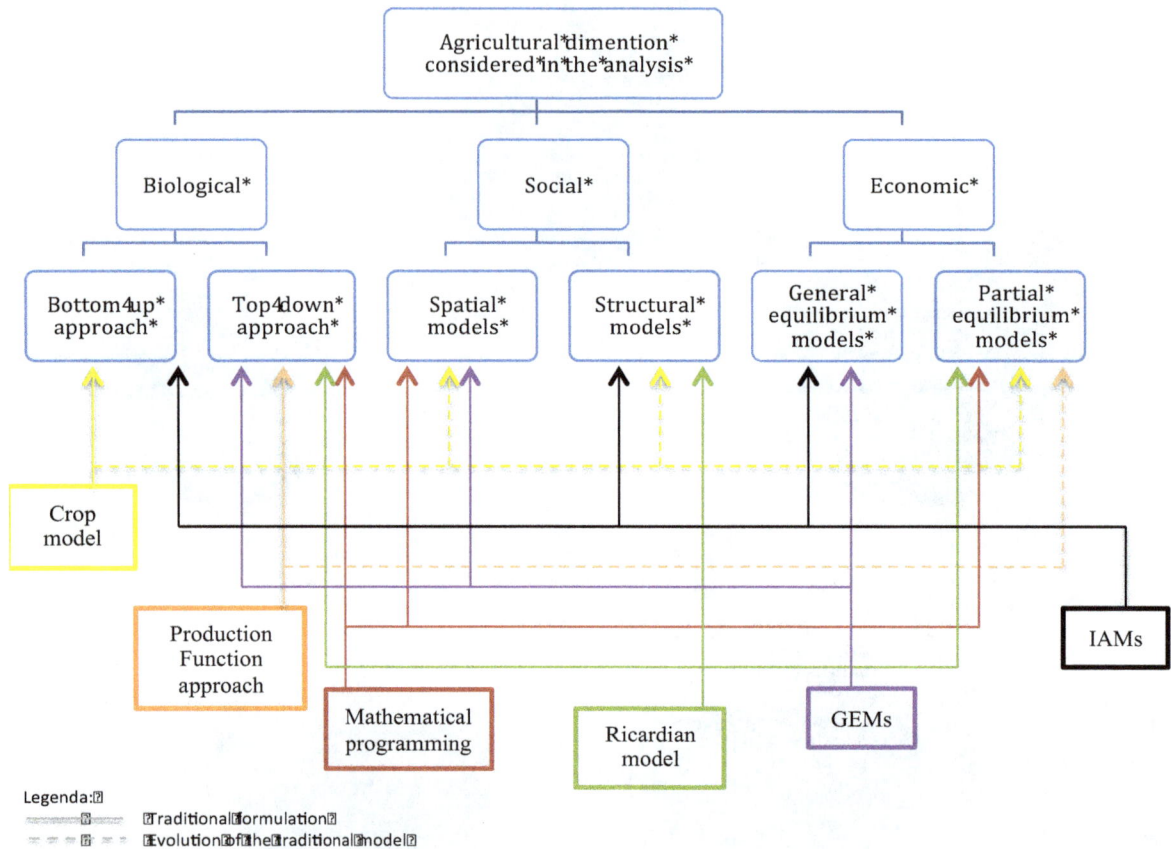

Figure 2. Classification of models by agricultural dimension, Legend: Traditional formulation; Evolution of the traditional model; Source: Authors' elaborations.

2. Annual fluctuations in the weather in terms of temperature and precipitations;
3. The frequency of extreme weather events such as droughts or floods.

Each of these aspects plays a different role and causes different effects on agriculture. The issue that has been the subject of most research is the effect of climate change in a long-term scenario. This has been widely analysed using the Ricardian model. The other two forms of the effects of climate change have been less investigated. Annual fluctuations in the weather were examined by Kelly et al. (2005) and Deschenes and Kolstad (2011). The effects of drought were analysed by Trnka et al. (2010, 2011) and of cyclones by Dasgupta et al. (2011). Figure 2 presents a classification of models that consider the biological, social, and economic dimensions of agriculture.

As demonstrated in Figure 2, if the focus is on the effects in terms of production change, by considering the biological aspects and their dynamics, it is possible to implement plant-physiology models that correlate the production output to climate variables or vegetation distribution behaviours. As such, it is possible to explain the spatial distribution of crops in relation to the climate

scenario. In this case the model adopted is a bottom-up model (Bosello and Zang, 2005). Alternatively, it is possible to use a top-down model (or spatial analogue), which analyses crop reaction to climate change based on the productivity values in different temporal and spatial scenarios.

Further, in the assessment of the social effects, it is possible to distinguish spatial versus structural models (Bosello and Zang, 2005). Through the analysis of choices, strategies, and technologies used in different climatic and geographic scenarios, both of these models provide the possibility of forecasting behaviours will be adopted by farmers to face climate change.

Spatial models analyse variations in a farm's performance when dealing with climate change without considering farmer adaptation. This type of model hypothesises that such variations do not affect the prices of agricultural commodities and inputs. Consequently, this model does not consider the effects of climate change on agricultural demand and supply. Moreover, spatial models implicitly assume the absence of progressive farmer adaptation processes through changes in production cost in the short-term and medium-term scenarios. It follows that it is not possible to differentiate climate-change adaptations endorsed by the

Table 2. Characteristics demonstrated by the most commonly used models to assess the effects of climate change on agriculture.

Model	Object of the analysis	Temporal scale	Geographical scale	Climate change manifestation	Agricultural dimension			References
					Biological	Social	Economic	
Crop simulation	A specific crop	Short time	Local	Weather annual fluctuation	Treated	Not treated in the traditional formulation. It is possible to treat it exogenously.	Not treated in the traditional formulation. However it is possible to couple this model with larger model frameworks that consider economy.	Eitzinger et al. (2003), Torriani et al. (2007)
Production function	A specific crop, a group of crops or a particular ecosystem	Both short term and long term	All possibilities	All possibilities	Not explicitly treated	Treated in a secondary manner.	In the traditional formulation treated in a secondary manner. Some studies estimate and economic production function. Others couple this model with larger model frameworks that consider economy.	Terjung et al. (1984), Isik and Devadoss (2006), Poudel and Kotani (2013)
Ricardian	The whole agricultural sector or a particular branch or crop	Long term	All levels, providing enough climatic variability is assured	Global warming and precipitations decreasing	Not explicitly treated	Not explicitly treated in the traditional formulation but explicitly treated in the structural Ricardian model	Treated	Mendelsohn et al. (1994), Schlenker et al. (2005), Seo and Mendelsohn (2008), De Salvo et al. (2013), Massetti and Mendelsohn (2011)
Econometric model	The whole agricultural sector or a particular branch or crop	Both short term and long term	All levels, especially local, national or regional	All possibilities	This depends on the model formulation	This depends on the model formulation	This depends on the model formulation	Schlenker and Roberts (2006), Deschênes and Greenstone (2007), Di Falco and Veronesi (2013a, b).
PMP	The whole agricultural sector or a particular branch	Both short term and long term	All levels, especially local, national or regional	All possibilities	Not explicitly treated in the traditional formulation. Some researchers treat it explicitly coupling this model with a crop simulation model	Treated	Treated	Quresh et al. (2010), Howitt et al. (2012), Qureshi et al. (2013)

Table 2. Contd.

					Not treated	Not explicitly treated	Treated	
GEMs	The whole agricultural sector or a particular branch if appropriately formulated	Long term	All levels, especially national or higher	All possibilities			Treated	Darwin et al. (1995), Calzadilla et al. (2010a, b), Trnka et al. (2010, 2011)
IAMs	The whole agricultural sector or a particular branch if appropriately formulated	Long term	All levels, especially national or higher	Global warming and precipitations decreasing	Treated	Treated	Treated	Prinn et al. (1999), Kainuma et al. (2003)

agricultural sector from those deployed by the economy as a whole, and neither is it possible to separate these adaptations from those put in place to deal with factors other than climate change (Molua and Lambi, 2007).

The structural models through which the physical, social, and economic responses of agriculture to climate change are analysed overcome these limits. However, the application of these models is sometimes hampered by a need for detailed information on business-management practices.

By focusing only on the economic dimension, applicable models can consider a partial equilibrium or a general equilibrium in sectorial and/or geographical terms. GEMs, or economy-wide models, were used to estimate the economic effect of climate change on agriculture (e.g. Darwin et al., 1995; Borsello and Zang, 2005; Calzadilla et al., 2010a, b). These applications look at the whole economy and consider the relationships between sectors. However, they present some limitations (Table 1) that are overcome by the partial equilibrium models, which focus on a part of the economic system, consisting of a single market or a set of markets or sectors (Deressa, 2007).

The microeconomic partial equilibrium models can omit important aspects of the issue being considered, for example:

1. The re-allocation of production factors,
2. Changes in demand for agricultural products,
3. The interrelation of the economic sectors,
4. The dynamics of international markets,
5. The endogenous nature of market prices for agricultural products and inputs.

Moreover, the partial microeconomic equilibrium models can be divided into two broad categories: models based on the simulation of the crop-growth processes (crop-growth simulation models) and econometric methods (Kurukulasuriya and Rosenthal, 2003; Deressa, 2007) that also include the widely used Ricardian models. The choice of the best model to assess economic effects depends heavily on the specific aspects that the analysis has to consider and on the level of detail (Table 2).

Conclusion

The assessment of the effects of climate change on agriculture and the choice of the model that better suite the research aims remains a complex area for several reasons. First, data are not always available and/or disaggregated on the necessary temporal or spatial scales. Second, such research involves different skills and professional competencies, which means that analyses have to consider biological and physiological aspects; technical and socioeconomic features; and adaptation strategies adopted by farmers and breeders to face climate change. Third, a relevant role is played by aspects related to economic and agricultural policies and to the geographical (local, regional or international) scale of the analysis. Finally, a valid model should consider the temporal and spatial variability of climate; the uncertainty of future climate scenarios; and the feedback of agricultural changes due to climate change.

Consequently, the selection of the most appropriate model should consider different aspects of the research problem, for example:

1. The specific object of the analysis,
2. The temporal and geographical scales,
3. The specific forms of climate change that are being considered (e.g. climate warming, weather fluctuations or extreme climatic events),
4. The magnitude of the effects expressed according to the agricultural dimensions (biological, social and/or economic) that the analysis aims to consider.

The choice of the model to be implemented is one of the most important steps in a assessment project. In the analysis of the effects of climate change on agriculture, the literature offers a multitude of applicable methods and tools, each of them with specific advantages and disadvantages. Consequently, the choice of the best model can be difficult due to a lack of perfect knowledge of all the possible alternatives. The choice of the model to apply for analysis often follows the trend of the moment, and is applied without detailed analysis of all the assumptions and hypotheses underlying the model. Choosing incorrect models causes a bias of results and an increase in unexplained variability that worsens the analytical framework of an already very complex area issue.

This article attempts to address this lack of information by offering to researchers a useful tool with which to identify all the possible alternatives of models analysing the effects of climate change on agriculture. This article has reviewed the literature and discussed the most popular analytical methods that are presented in the literature, and that are: the Crop Simulation Models, the Production-Function Model, the Ricardian Model, the Mathematical Programming, the General Equilibrium Model (GEMs) and the Integrated Assessment Models (IAMs). It has classified methods of analysis according to the principal aspects that have to be considered in when selecting a model, with particular emphasis on the dimensions under which the effects of climate change should be expressed. The adopted classification scheme demonstrates that one model is capable of simultaneously considering many aspects related to climate change and classifying these in different classes.

REFERENCES

Adams RM (1989). Global climate change and agriculture: An economic perspective., Am. J. Agric. Econ. 71(5):1272–1279.

Adams RM, Rosenzweig C, Peart R, Ritchie J, McCarl B, Glyer J, Curry B, Jones J, Boote K, Allen L (1990). Global climate change and US agriculture. Nat. 345:219–224.

Adams RM, Hurd BH, Lenhart S, Leary N (1998). Effects of global climate change on agriculture: An interpretative review. Clim. Res. 11(1):19–30.

Bernetti J, Menghini S, Marinelli N, Sacchelli S, Alampi Sottini V (2012). Assessment of climate change impact on viticulture: Economic evaluations and adaptation strategies for the Tuscan wine sector. Wine Econ. Pol. 1:73–86.

Bindi M, Olesen JE (2011). The responses of agriculture in Europe to climate change, Reg. Envion. Change 11(1):151–158.

Bosello F, Zang J (2005). Assessing Climate Change Impacts: Agriculture, FEEM Nota di Lavoro 94.2005, Fondazione Eni Enrico Mattei.

Bruijnzeel LA (2004). Hydrological functions of tropical forests: Not seeing the soil for the trees. Agric. Ecosys. Environ. 104(1):185–228.

Calzadilla A, Rehdanz K, Tol RSJ (2010a). The economic impact of more sustainable water use in agriculture: A computable general equilibrium analysis. J. Hydr. 384(3–4):292–305.

Calzadilla A, Rehdanz K, Betts R, Falloon P, Wiltshire A, Tol RSJ (2010b). Climate change impacts on global agriculture, Kiel Working. P. 1617.

Chang J (eds) (1977). General Circulation Models of the Atmosphere, Method in Computational Physics, P. 17, Academic Press, New York.

Darwin RF, Tsigas M, Lewandrowski J, Raneses A (1995), World Agriculture and Climate Change—Economic Adaptations, US Department of Agriculture, Washington, DC.

Dasgupta S, Huq M, Khan ZH, Zahid Ahmed MM, Mukherjee N, Malik Fida Khan MF, Pandey K (2011). Cyclones in a Changing Climate: The Case of Bangladesh (http://www.gwu.edu/~iiep/adaptation/docs/Dasgupta,%20Cyclones%20in%20a%20Changing%20ClimateThe%20Case%20of%20Banglad esh%20(updated).pdf).

De Salvo M, Raffaelli R, Moser R (2013). The impacts of climate change on permanent crops in an Alpine region: A Ricardian analysis. Agric. Syst. 118:23–32.

Deressa TT (2007). Measuring the economic impact of climate change on Ethiopian agriculture: Ricardian approach, World Bank Policy Research Working. P. 4342.

Deschênes O Greenstone M (2007). The Economic Impacts of Climate Change: Evidence from Agricultural Output and Random Fluctuations in Weather. Am. Econ. Rev. 97(1):354-385.

Deschenes O, Kolstad C (2011). Economic impacts of climate change on California agriculture, Clim. Ch. 109(1):365–386.

Dinar A, Mendelsohn R (eds) (2011). Handbook on Climate Change and Agriculture, Edward Elgar, Cheltenham.

Di Falco S, Veronesi M (2013a). How African agriculture can adapt to climate change? A counterfactual analysis from Ethiopia, Land Econ forthcoming in November.

Di Falco S, Veronesi M (2013b). Managing environmental risk in presence of climate change: The role of adaptation in the Nile Basin of Ethiopia. Environ. Res. Ec., article in press.

Eitzinger J, Žalud Z, Alexandrov V, Van Diepen CA, Trnka M, Dubrovský M, Semerádová D, Oberforster M (2001). A local simulation study on the impact of climate change on winter wheat production in north-eastern Austria. Bodenkultur 52(4):199–212.

Eitzinger J, Stastna M, Zalud Z, Dubrovski M (2003). A simulation study of the effect of soil water balance and water stress on winter wheat production under different climate change scenarios. Agric. Water Man. 61:195–217.

Finger R, Schmid S (2007). The impact of climate change on mean and variability of Swiss corn production, Info Agrar Wirtchaft, Schriftenreihe 2007/1 (http://www.cer.ethz.ch/resec/research/workshops/Nachwuchsworksh op/Finger_Paper.pdf)

Gebrehiwot T, Van Der Veen A (2013). Farm level adaptation to climate change: The case of farmers in the Ethiopian highlands, Env. Man. 52(1):29–44.

Gregory PJ, Ingram JSI, Brklacich M (2005). Climate change and food security. Philos. Trans. R Soc. London B Biol. Sci. 29:360(1463):2139–2148. (http://www.ncbi.nlm.nih.gov/pmc/articles/PMC1569578/)

Howitt RE, Medellín-Azuara J, MacEwan D, Lund JR (2012). Calibrating disaggregate economic models of agricultural production and water management., Env. Mod. Soft. 38:244–258.

Intergovernmental Panel on Climate Change (1990): IPCC First Assessment Report 1990 (FAR), (http://www.ipcc.ch/publications_and_data/publications_and_data_re ports.shtml).

Isik M, Devadoss S (2006). An analysis of the impact of climate change on crop yields and yield variability. Ap. Econ. 38(7):835–844.

Kainuma M, Matsuoka Y, Morita T (eds) (2003). Climate Policy Assessment Asia-Pacific Integrated Modeling, Springer-Verlag, Tokyo.

Kaiser HM, Riha SJ, Wilks DS, Rossiter DG, Samphat R (1993). A farm level analysis of economic and agronomic impacts of gradual warming. Am. J. Agric. Econ. 77(2):387–398.

Kelly DL, Kolstad CD, Mitchell GT (2005). Adjustment costs from environmental change, J. Env. Econ Man. 50(3):468–495.

Klein T, Holzkämper A, Calanca P, Fuhrer J (2013). Adaptation options under climate change for multifunctional agriculture: A simulation study for western Switzerland, Reg. Env. Change, article in press.

Kimaro EG, Chibinga OC (2013): Potential impact of climate change on livestock production and health in East Africa: A review, Livestock Res.   Rural   Develop.   25(7):116. (http://www.lrrd.org/lrrd25/7/kima25116.htm).

Kurukulasuriya P, Rosenthal S (2003). Climate change and agriculture: A review of impacts and adaptations, Climate Change Series Paper No. 91, Environment Department and Agriculture and Rural Development Department, The World Bank, Washington DC.

Lhomme JP, Mougou R, Mansour M (2009): Potential impact of climate change on durum wheat cropping in Tunisia. Clim. Change. 96(4):549–564.

Massetti E, Mendelsohn R (2011). Estimating Ricardian Functions with Panel Data. Clim. Change Econ. 2(4):301-319.

Medellín-Azuara J, Harou JJ, Howitt RE (2010). Estimating economic value of agricultural water under changing conditions and the effects of spatial aggregation. Sci. Tot. Environ. 408(23):5639–5648.

Mendelsohn R, Dinar A (eds) (2009). Climate Change and Agriculture— An Economic Analysis of Global Impacts, Adaptation and Distributional Effect, New Horizons in Environmental Economics, Edward Elgar, Cheltenham.

Mendelsohn RO, Nordhaus WD, Shaw D (1994). The impact of global warming on agriculture: A Ricardian analysis. Am. Econ. Rev. 84(4):753–771.

Molua EL, Lambi CM (2007). The economic impact of climate change on agriculture in Cameroon, The World Bank, Development Research Group, Sustainable Rural and Urban Development Team, Policy Research Working P. 4364.

Mount T, Li Z (1994). Estimating the Effects of Climate Change on Grain Yield and Production in the US, USDA Economic Research Services, Washington, DC.

Oluwasusi JO (2013). Farmers adaptation strategies to the effect of climate variation on yam production in Ekiti state, Nigeria. J. Food Agric. Environ. 11(2):724–728.

Poudel S, Kotani K (2013). Climatic impacts on crop yield and its variability in Nepal: Do they vary across seasons and altitudes. Clim. Change 116(2):327–355.

Prinn R, Jacoby H, Sokow AC, Wang XX, Yang Z, Eckaus R, Stone P, Ellerman D, Melillo J, Fitzmaurice J, Kicklighter D, Holian G, Liu Y (1999). Integrated global system model for climate policy assessment: Feedback and sensitivity studies. Clim. Change. 41(3/4):469–546.

Qureshi ME, Schwabe K, Connor J, Kirby M (2010). Environmental water incentive policy and return flows. Water Res. Res. 46:W04517.

Qureshi ME, Whitten SM, Mainuddin M, Marvanek S, Elmahdi A (2013). A biophysical and economic model of agriculture and water in the Murray-Darling Basin, Australia. Env. Mod. Soft. 41:98–106.

Randall DA (ed) (2000). General Circulation Model development: Past, present, and future. International Geophysical Services. P 70.

Reynolds C, Crompton L, Mills J (2010). Livestock and climate change impacts in the developing world. Outlook Agric. 39(4):245–248.

Rosenzweig C, Parry ML, Fischer G, Frohberg K (1993). Climate change and world food supply, Research Report. Environmental Change Unit, University of Oxford, Oxford. P. 3.

Rosenzweig C, Parry ML (1994). Potential impacts of climate change on world food supply. Nature 367:133–138.

Sanchez PA (2000). Linking climate change research with food security and poverty reduction in the tropics Agriculture. Econ. Environ. 82(1–3):371–383.

Schlenker W, Hanemann WM, Fisher AC (2005). Will US agriculture really benefit from global warming. Accounting for irrigation in the hedonic approach. Am. Econ. Rev. 95(1):395–406.

Schlenker W, Roberts MJ (2006). Estimating the Impact of Climate Change on Crop Yields: The Importance of Non-Linear Temperature Effects (Available at SSRN: http://ssrn.com/abstract=934549 or http://dx.doi.org/10.2139/ssrn.934549).

Seo SN (2008). A microeconomics analysis of climate change impacts on livestock management in African agriculture, Yale University, MPRA. P. 6903.

Seo SN, Mendelsohn R (2008). Measuring impacts and adaptations to climate change: A structural Ricardian model of African livestock management. Agric. Econ 38(2):151–165.

Siwar C, Ahmed F, Begum RA (2013). Climate change, agriculture and food security issues: Malaysian perspective. J. Food Agric. Environ. 11(2):1118–1123.

Tate AB (2001). Global warming's impact on wine. J. Wine Res. 12:95–109.

Terjung WH, Hayes JT, O'Rourke PA, Todhunter PE (1984). Yield responses of crops to changes in environment and management practices: Model sensitivity analysis. I. Maize, Int. J. Biomet. 28(4):261–278.

Torriani D, Calanca P, Schmid S, Beniston, M. Fuhrer J (2007). Potential effects of changes in mean climate and climate variability on the yield of winter and spring crops in Switzerland. Clim. Res. 34:59-69.

Trnka M, Eitzinger J, Dubrovský M, Semerádová D, Štěpánek P, Hlavinka P, Balek J, Skalák P, Farda A, Formayer H, Žalud Z (2010). Is rainfed crop production in central Europe at risk? Using a regional climate model to produce high resolution agroclimatic information for decision makers. J. Agric. Sci. 148(6):639–656.

Trnka M, Olesen JE, Kersebaum KC, Skjelvåg AO, Eitzinger J, Seguin B, Peltonen-Sainio P, Rötter R, Iglesias A, Orlandini S, Dubrovský M, Hlavinka P, Balek J, Eckersten H, Cloppet E, Calanca P, Gobin A, Vučetić V, Nejedlik P, Kumar S, Lalic B, Mestre A, Rossi F, Kozyra J, Alexandrov V, Semerádová D, Žalud Z (2011). Agroclimatic conditions in Europe under climate change. Global Change. Biol. 17:2298–2318.

Walker B, Steffen W (1997). An overview of the implications of global change for natural and managed terrestrial ecosystems, Cons. Ecol. 1(2):2 (http://www.consecol.org/vol1/iss2/art2/).

10

Gender discrimination in Agricultural land access: Implications for food security in Ondo State, Nigeria

A. G. Adekola[1], F. O. Adereti[2]*, G. F. Koledoye[2] and P. T. Owombo[3]

[1]Department of Agricultural Economics and Extension, Igbinedion University, Okada, Edo State, Nigeria.
[2]Department of Agricultural Extension and Rural Development, Obafemi Awolowo University, Ile Ife, Osun State, Nigeria.
[3]Department of Agricultural Economics, Obafemi Awolowo University, Ile Ife, Osun State, Nigeria.

This study assessed gender discrimination in agricultural land access: Implications for food security in Ondo State Nigeria. Specifically, it analysed men and female accessibility to forms of land holding and the factors affecting agricultural land accessibility in the study area. Multistage sampling technique was used in selecting 240 respondents used for this study. Data collected were summarized using descriptive statistics such as frequency counts and percentages and correlation analysis was used to test the hypothesis stated. The results revealed that the mean age of the male respndents was 48.3 while that of female was 43.7 with the standard deviation of 14.9 and 11.3, respectively. Also, at $p \leq 0.05$, there was significant relationship between accessibility to agricultural land and male and female socio-economic characteristics such as age ($r = 0.484$), marital status ($r = 0.568$), farm size ($r = 0.504$), farming experience ($r = 0.479$), household ($r = -0.668$), access to credit facility ($r = 0.476$), and membership of social organization ($r = 0.593$). This study therefore concluded that gender differentials, especially with regards to land favour the males. It is therefore recommended that redesigning and redeveloping the structure of land policies to be more gender sensitive and inclusive.

Key words: Gender, land acess, food security, correlation, discrimination.

INTRODUCTION

In a rapidly changing world, food and agricultural land holding systems in developing countries are facing new and increasingly complex challenges (Derman et al., 2007). Fighting poverty, ensuring food and nutrition security while protecting the environment still remains a major challenge facing global development practitioners today. In discussing agriculture, two key factors that cannot be ignored are land and labour. Land ownership, accessibility, and the sustainability of this access are very crucial for any meaningful agricultural development. Either by design or circumstance, women constitute a large proportion of farm labour and agricultural workers. These women usually control very little or no amount of land, which is very important in determining farm productivity and labour welfare (Afonja et al., 2002).

Land is the most valuable form of property in agrarian societies because of its economic, political symbolic and ritual importance (Bioye et al., 2006). It is the basis of political power and social status in most societies of the world. It is a productive wealth-creating and life sustaining asset which every human being craves for and provides a sense of identity and rootedness within a community (Argwal, 1994). Land is used for production of biomass, ensuring food, fodder, renewable energy and raw materials for existence of human and animal life. It is a base for settlement and industrial use and a store of our cultural heritage and is actually a source of raw materials like minerals, clay, energy and water (Blum, 1998). Land stands for continuity of ownership since it is a burial ground where all clansmen are buried and consequently a central place for the spirits of their ancestors for example in African countries.

*Corresponding author. E-mail: fadereti@yahoo.com.

In most developing countries, land is not only the primary means for generating livelihoods but often the main vehicle for investing, accumulating wealth, and transferring it between generations. Thus, the ways in which access to land is regulated, property rights are defined, and ownership conflicts are resolved has broad implications for food security (Deininger and Binswanger, 1999, FAO, 2006, 2008, 2009). Access to land can be through right of ownership, through informal concessions granted by individuals to kin or friends. Legal ownership is normally accompanied by legal restrictions on disposal, that is, there is no effective control here. In most African societies, women have land use priorities from husbands but have no independent rights which allow them control or produce from the land. The direct advantage of land rights are that a woman can use it to grow food crops, fodder for animals, keeping livestock, practicing sericulture, growing trees and vegetable gardening (Cousins and Claassens, 2006). Land rights facilitate access to credit and strengthen support that the women receive from relatives. Access to land means reliable food supply, better healthcare, better housing and reliable income in most cases.

The percentage of men and women employed in agricultural sector decreased between 1997 and 2008 (due to the increasing industrialization of the considered countries), the percentage of women employed in agriculture is still higher in almost all developing regions as shown in apendix 1. In the last years, migration of men towards the cities led to a gradual feminization of small-scale agriculture, with an increasing percentage of women-headed households in rural areas (FAO, 2008). The relevance of women's agricultural labour can be appreciated if we consider that, for instance, the agricultural sector in Sub-Saharan Africa contributes about 30% of the GNP of the continent, employing from 60 to 90% of the population and producing from 25 to 90% of the income deriving from exports (FAO, 2009) and FAO (2002)

Lawanson (2010) opined that Nigeria is a typical patriarchic society where male superiority and dominance originated from historically rooted culture and religion. In Pre-colonial times, females generally were accorded less value and lower social status. Western culture reinforced this anomaly. However, Bruce and Lloyd (1991) stated that the western culture has failed to address gender inequality in access to land, in spite of the role that land plays in the lives of the women who are increasingly being saddled with the responsibility of heading and maintaining households, especially in developing countries.

Women constitute over 50% of the Nigerian population, make up about 37% of the formal sector (World Bank, 2001), and dominate the informal economic sector which is principally made up of home-based enterprises (Soetan, 2002); Moreso, Lawanson (2010) stated that their relative powerlessness both economically and politically, are unable to exercise control over resources, particularly land. Some cultures in Nigeria, through marriage and inheritance practices, prohibit women from owning land. However, Soetan (2002) posited that marital status increases the women ability to own land.

At rural level, women work mainly on their own, linking their activities to the family needs, and just a small percentage of them, everywhere lower than men's receives a wage. In Latin America, for instance, only 2.3% of women in agriculture get a wage against 20.9% of men; in Southern Asia salaried rural women are 11.9%, while men are 21.8%7 (Alice, 2008). With the increasing roles of rural women in agriculture and contributions to food security and the consequence inequality on land access across gender, there is therefore the need to access the gender discrimination in land accessibility and implications to food security in Nigeria

The specific objectives for the study were to:

1. Examine the socio-economic characteristics of the respondents
2. Analyse male and female accessibility to forms of land holdings in the study area
3. Assess the factors affecting male and female accessibility to land.
4. Profile the security of tenue over land across gender.

Hypothesis for this study was stated in the null form as follows: there is no significant relationship between selected socio-economic characteristic of male and female respondents and their accessibility to land in the study area.

The study area

The study was conducted in Ondo State of Nigeria. The state was carved out of the old Oyo state on the 3rd February, 1976 with the capital in Akure. The state covers an area of approximately 15,500 km^2 and it is bounded in the south by the Bight of Benin and Atlantic Ocean; north by Ekiti and Kogi States; east by Edo and Delta States and west by Osun and Ogun states (Figure 1). The state lies between longitude 5°45' and 7°52' on the North – South Pole, and longitude 4°20' and 6°5' on the East – West Pole. According to analytical report of the National Population Commission (NPC) (2006), Ondo State has 3,441,024 million people with eighteen (18) Local Government Areas.

The tropical climate of the state is broadly of two seasons: rainy season (April to October) and dry season (November to March). A temperature throughout the year ranges between 21 to 29°C and humidity is relatively high. The annual rainfall varies from 2,000 mm in the southern areas to 1,150 mm in the northern areas. The soils derived from the basement complex rocks are mostly well drained, with a medium to fine texture. The state enjoys luxuriant vegetation with high forest zone (rain forest) in the south and sub-savannah forest in the northern fringe. The indigenes of the state belong to the Yoruba ethnic group

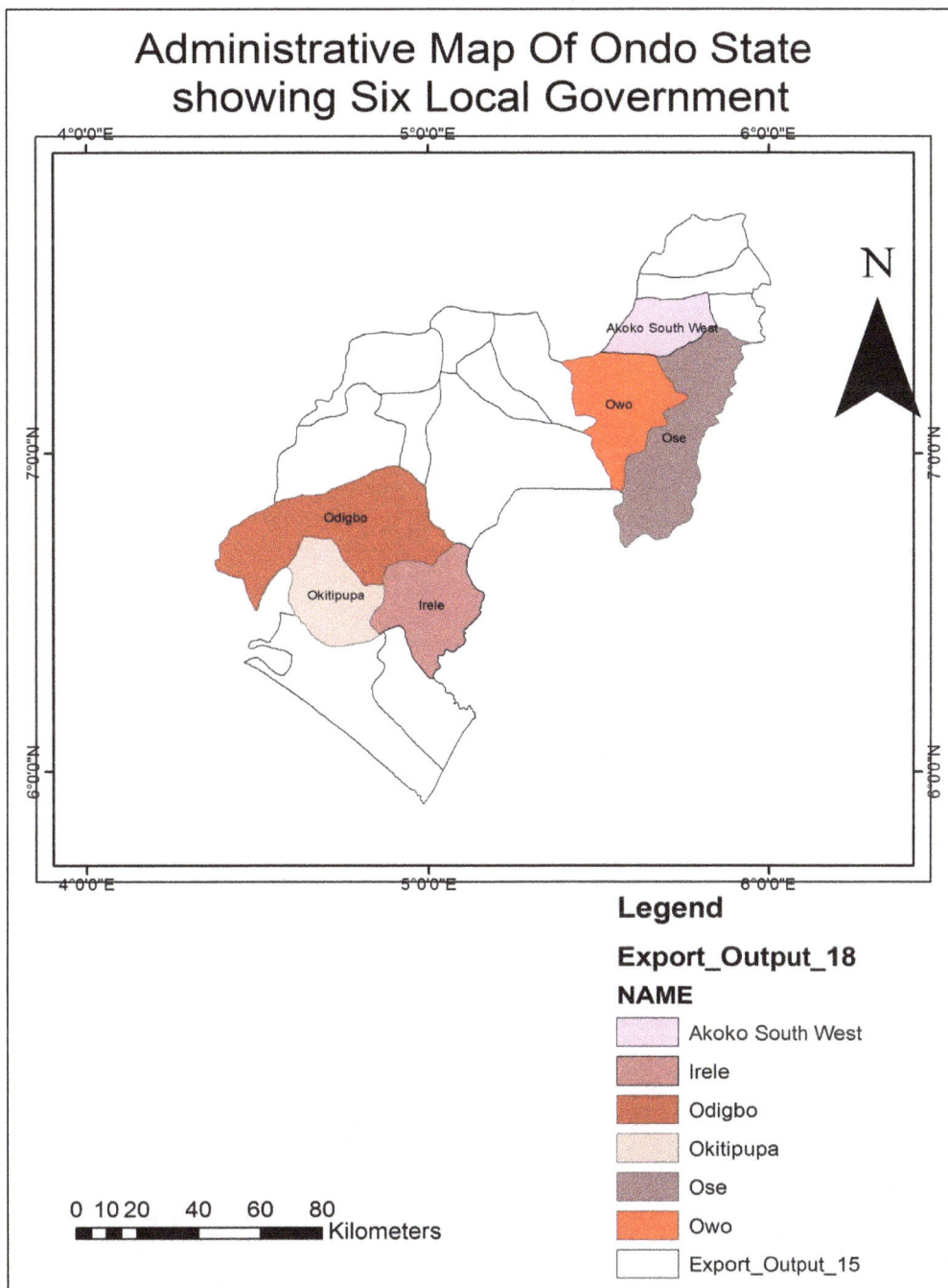

Figure 1. Map of the study area.

and are composed of the Akokos, the Ondos, the Ikales/Ilajes and the Apoi/Ijaw Arogbos. However, non – indigenes from every part of the country and outside reside in the state. Yoruba and English are the languages of the people for official and business transactions. The State is basically agrarian with large scale production of cocoa, palm produce and rubber. Other crops like maize, yam and cassava are produced in large quantities. Sixty-five percent of the state labour force is in the agriculture sub-sector

(Coastalnews, 2012). The state is also blessed with very rich forest resources where some of the most exotic timber in Nigeria abounds. Ondo State is equally blessed with extensive deposits of crude oil, bitumen, glass sand, kaolin, granites and limestone. Therefore, the state has great potentials for rapid industrial growth in view of its raw materials base. Reasonable segment of the populace are also traders and artisans. Other occupations of the people include weaving, mat – making, dying, soap making, wood

Table 1. Distribution of respondent's demographic characteristics.

Variable	Male (N= 160)		Female (N=160)	
	Frequency	Percentage	Frequency	Percentage
Age				
>30	42	27.2	34	21.3
31-60	90	56.3	102	63.8
60 years and above	28	16.5	24	14.9
Level of education				
No formal education	19	11.9	31	19.4
Primary school	56	35.0	61	38.1
Secondary education	63	39.3	50	31.2
Tertiary education	22	13.8	18	11.3
Farming experience				
< 5	10	6.3	5	3.1
6-10	32	20.0	40	25.0
11-20	68	42.5	79	49.4
>20	50	31.2	36	22.5
Household size				
>2	15	9.4	12	7.5
2-5	35	21.9	89	55.6
6-9	84	52.5	40	25.0
>10	26	16.2	19	11.9
Farm size				
<1	12	7.5	62	38.8
1-3	91	56.8	83	51.9
>3	51	35.7	15	9.3

Source: Field survey, 2010.

carving, among many others. The state lies entirely within the tropics of 162 mm per annum.

SAMPLING PROCEDURE

A multi-stage sampling technique was adopted to select respondents for this study. The first stage involved purposive sampling of six local government areas based on their degree of involvement in farming. The local government areas were Irele, Okitipupa, Odigbo, Owo and Akoko, Ose and Akoko South-West. In the second stage, the local government areas were grouped into communities. In the third stage, two communities were randomly selected from each of the local government areas. In the final stage, twenty respondents with equal number of male and female were sampled from each of the communities

Primary data were used for this study. Data were collected using structured questionnaire and interview schedule for both the literates and illiterates respondents respectively. A total of 240 respondents were used for this study.

Analytical technique

Data were analyzed using descriptive statistics and inferential statistics. The descriptive statistics employed were frequency counts

and percentage while the inferential statistics were correlation analysis and Chi-Square. Data were analysed with the aid of Statistical Package for Social Sciences (SPSS) version 14.0 and CoStat analytical software.

RESULTS AND DISCUSSION

The result of the socio-economic characteristics as shown in Table 1 revealed that 27.2% of male respondents were less than 30 years, 56.3% were found within the age bracket of 31 to 60 years while only 16.5% were 61 year sand above. In the female category, 21.3% of the respondents were less than 30 years, 63.8% were within the age group of 31 to 60 years while just 14.9% of the respondents were 60 years and above. The mean ages of male and female respondents were 48.3 and 43.7 years respectively. The findings revealed that majority (56.3 and 63.8%) of respondents were in their middle active ages, an indication that they will still be active to access land. Furthermore, 35.0% of male respondents had primary education, 39.3% had secondary education and only 13.8% had tertiary education while 11.9% did not have

Table 2. Distribution of respondents' accessibility to forms of land holdings.

Gender	Male (N=160)		Female (N=160)	
Variable	Frequency	Percentage	Frequency	Percentage
Access to community land	152	95.0	38	23.8
Access to inherited land	138	86.3	52	32.5
Access to land by purchase	160	100.0	72	45.0
Access to land by lease	160	100.0	61	38.1
Access to land by gift	115	71.9	29	18.1

Source: Field survey, 2010. Multiple responses were given.

Table 3. Chi-Square analysis showing the difference between gender and land accessibility.

Variable	X^2_c	X^2_t	Df	C
Gender	5.83*	3.84	1	0.57

P-Values at 0.05 level of significance; C= Contingency coefficient; $X2c$ = Chi- square calculated; $X2t$ = Chi-square

formal education. In the female group, 38.1% had primary education and only few (11.3%) had tertiary education. This implied that male respondents were more educated than their female counterparts in the study area. For farming experiences in years, majority (42.5%) of male respondents had between 11 to 20 years and 31.2% had greater than 20 years. More so, 49.4% of female respondents had 11 to 20 years of farming experience while only 22.5% were found to have more than 20 years of farming experience as shown in Table 1.

In addition, majority (52.5%) of male respondents had between 6 to 9 household sizes while majority (55.6%) of female respondents had between 2 to 5 household sizes. This implied that male respondents had higher household sizes than the female. This could be as a result of male having more than one wives. Also, 56.8% of male respondents had between 1 to 3 hectares farm size, 35.7% had more than 3 hectares farm size while 51.9% of female respondents had between 1 to 3 hectares and a few (9.3%) cultivated more than 3 hectares of land in the study area. This analysis shows that male respondents had higher farm size than their female counterparts in the study area.

Results in Table 2 revealed that majority (95.0%) of male respondents had access to communal land while few (23.8%) of female respondents had access to community land. Also, 86.3% of male had access to inherited land while only 32.5% of female respondents had access to inherited land. More so, 100.0% of male respondents were found to have access to land by purchase while only 45.0% of female could access land by purchase. In addition, majority (71.9%) of male could access land through gift. The above analyses indicated that male had more access to any form of land than their female counterparts. Also, the result of Chi-Square analysis revealed that there was a significant difference between gender and accessibility to land at 0.05 level of significance as shown in Table 3.

Gender was found to exert a little above avarage (57%) strenght of association on land accessibilty. This significant difference further revealed that male have more access to tabulated; Df = degree of freedom; Source: Field survey, 2010. land than female in the study area. This finding was in conformity with Lawanson (2010), Bruce and Lloyd (1991) and Afonja et al. (2002). The low accessibility of female to agricultural land could be as result of socio-cultural factors that could hinder female from owning land (Lawanson, 2010).

Results in Table 4 revealed that 70.0% of male respondent indicated that income level affects land holding, while majority (100.0%) of female indicated that marital status is the major determinant to hold land in the study area. More so, about 84.5 percent of the female respondents viewed cultural belief as a factor that affects land holding while a few (38.8%) of the male respondents indicated that cultural belief is a factor that affects land holding among men in the study area. This finding revealed that men were less affected by the various factors indicated, an indication that female were not having equal access to agricultural land despite their contributions to the food production in Sub-Saharan Africa. This affirms the position of Alice (2008) in the study "effect of land tenure system on women's knowledge-base and resource management in Manjiya County, Uganda". The study indicated that a very low percent of women in agriculture gets a wage as aginst the percentage of men. The Pearson correlation analysis revealed that marital status was positively correlated with land accessibility. This conforms with Soetan (2002).

Multiple responses were given

On the security of tenure over land, results in Table 5 indicated that about 83.1 percent of male owned land while only few (17.5%) of women owned land in the study area. More so, about 74.4 percent of male had been in possession of their land for more than 5 years while about 1.3% of women had held land for the same duration with men. In addition, about 88.1% of male had the rights to transfer land while none of the female respondents had the same priviledge of transferring land like their male counterparts. This analysis reveals that gender differential

Table 4. Factors affecting land ownership.

Variable	Male (N=160)		Female (N=160)	
	Frequency	Percentage	Frequency	Percentage
Income	112	70.0	129	80.6
Marital status	70	43.8	160	100.0
Access to credit facility	132	82.5	108	67.5
Age	125	78.1	115	71.9
Cultural belief	62	38.8	135	84.5

Source: Field survey, 2010.

Table 5. Security of tenure over land.

Variable	Male		Female	
	Frequency	Percentage	Frequency	Percentage
Land ownership				
Owned	133	83.1	28	17.5
Otherwise	27	16.9	132	82.5
Duration of land use				
<2 years	-	-	39	24.3
3-5 years	41	25.6	118	73.8
>5 years	119	74.4	3	1.9
Rights to land				
Use only	19	11.9	160	100
Transfer	141	88.1	-	-

Source: Field survey, 2010.

Table 6. Results of correlation analysis showing the relationship between male and female socio-economic characteristics and accessibility to agricultural land.

Variable	Correlation coefficient (r)	Coefficient of determination (r^2)
Age	0.484*	0.064
Marital status	0.568*	0.059
Level of education	0.115	0.013
Farm size	0.504*	0.039
Farming experience	0.479*	0.121
Household size	-0.668*	0.072
Access to credit facility	0.476*	0.141
Membership of social organization	0.593*	0.154

Source: Field survey, 2010. Critical value of r = 0.427, Level of significance= 0.05.

exists in security of tenure over land in the study area. This findins conform with Lawanson (2010), who stated that women relative powerlessness both economically and politically in a typical African setting limits their control over resources, particularly land and also that male superiority and dominance over resources in Nigeria originated from historically rooted culture and religion.

The results of the hypothesis stated and tested on Table 6 revealed that there were positive and significant correlation between accessibility to agricultural land and age (r = 0.484); marital status (r = 0.568); farm size (r = 0.504); farming experience (r = 0.479); access to credit (r = 0.476) and membership of social organization (r = 0.593). The r values were compared with the critical value of r = 0.427. Household size had negative but significant relationship with respondents' socio-economic characteristics and accessibility to agricultural land. Only level of education was found to have non significant relationship. The above information established that an increase in the value of the independent variables would result in corresponding relationship between respondents' socio-economic characteristics and accessibility to agricultural land. The values of coefficient of determination (r^2) further revealed the percentage contribution of the independent variables to agricultural land accessibility. The higher the values of r^2, the stronger the influence as reflected in the percentage contribution of the significant variables. This information therefore justified these variables as highly significant between respondents' socio-economic characteristics and accessibility to agricultural land.

CONCLUSION AND RECOMMENDATIONS

Based on the findings of this study, most of the respondents were within their productive age of 31 to 60 years. There was gender differences in land accessibility in the study area as male were found to have more access to agricultural land than their female counterparts. Female were more affected by the factors affecting land holding in Ondo State. Age, marital status, farm size, farming experience; access to credit facility and cultural beliefs were found to determine accessibility to agricultural land in the study area. There is therefore the need for an intensive effort and emphasis on mainstreaming gender in agricultural programmes. This will facilitate the entry of women as active decision makers on issues that relate to food security and income generation. Women are important links to development: they maintain food security and the general well-being of their families/households. Therefore, improving women's status and control over land should be considered as strategically important efforts at all levels in combating the state of food security in the country.

It is therefore recommended that society and government should consider gender in agriculture by restructuring the system of land holding in order to include the vulnerable group as this will have a significant effect on food production in Ondo State and in Nigeria at large. Also, land tenure policies should be restructured to ensure that farmers (male and female) have equal access to agricultural land especially for perennial crops and are able to obtain land on a more permanent basis would be helpful in combating the state of food insecurity in the country.

REFERENCES

Afonja S, Mills-Tettey R, Amole D (2002). Gender differentials in access to land and housing in a Nigerian city, being monograph of the Center for Girder and Social Policy Studies, Obafemi Awolowo University, Ile-Ife.

Alice MK (2008). The effect of land tenure system on women's knowledge-base and resource management in Manjiya County, Uganda.

Argwal B (1994). A field of one's own: Cambridge University Press. FAO corporate document repository: current and emerging issues for economic analysis and research, 2010

Blum CL (1998). Land Tenure and Administration in Africa: Land Tenure and resource Access in Africa, IIED/FAO, London.

Bioye TA, Abdul RA, Joseph BO (2006). Women and Land Rights Reforms in Nigeria: Promoting Land Administration and Good Governance 5th FIG Regional Conference.

Bruce J, Lloyd C (1991). Family Research and Policy Issues for the 1990's', in Understanding how resources are allocated within Households.

Cousins B, Claassens A (2006). 'More than simply 'socially embedded': recognizing the distinctiveness of African land rights'. Keynote address at the international symposium on the frontier of land issues: social embeddedness of rights and public policy' Montpellier, May 17-19.

Coastalnews (2012). Profile of Ondo state, Nigeria.

Deininger K, Binswanger H (1999). The Evolution of the World Bank's Land Policy: principles, experience and future challenges', The World Bank Research Observer 14:247-276.

Derman BR Odgaard O, Sjaastad E (2007). Conflicts over Land and Water in Africa. Oxford: James Currey. The World Bank Research Observer 14:247-270.

Food and Agricultural Organization (2002). Assessment of the World Food Security and Nutrition Situation. Background document for the 38th Session of the FAO, Rome, pp. 34-39.

Food and Agricultural Organization, (2006). The State of Food Insecurity in the World Rome.

Food and Agricultural Organization (2008). Gender, Property Rights and Livelihoods in the Era of AIDS.

Food and Agricultural Organization (2009). Gender Equity in Agriculture and Rural Development. A quick guide to gender mainstreaming in FAO's new Strategic Framework.

Lawanson TO (2010). Gender differentials in Nigeria: Implications for sustainable development. J. Extension. Syst. 21(1):46-57.

Soetan R (2002). 'Women, Small Scale Enterprises and Social Change: Implications for Changes in Industrialization Strategy", in Afonja.S and Aina O (eds) Gender and Social Policy Studies. Obafemi Awolowo University, Ile Ife

The World Bank (2001). Integrating Gender into the World Bank's Work: A Strategy for Action. www.worldbank.org

APPENDIX

Appendix 1. Percentage of women and men employed in the primary sector.

Region	1998		2007	
	Male	**Women**	**Male**	**Women**
World	39.4	42.9	33.1	36.4
Eastern Asia, South-Eastern Asia and Pacific	44.3	51.6	36.4	41.2
Latin America and Caribbean	26.4	12.6	22.1	9.7
Southern Asia	53.7	74.4	41.5	65.1
Sub-Saharan Africa	65.1	71	60.3	65.1

Source: Global Employment Trends for Women, ILO, 2009-data for 1998 and 2007

11

Socioeconomic analysis of beekeeping in Swaziland: A case study of the Manzini Region, Swaziland

Micah B. Masuku

Department of Agricultural Economics and Management, University of Swaziland, P. O. Luyengo, M205, Luyengo Swaziland.

Swaziland has substantial potential in beekeeping with her rich flora, proper ecological conditions and existence of colonies. However, the Swaziland beekeeping sector has not yet efficiently utilized the rich natural resources. The apiculture sector in Swaziland is still faced with challenges in respect to marketing and importation as a result of the quality of honey and competition from South African honey. The objectives of the study were to describe the socioeconomic characteristics of beekeeping farmers and determine the factors affecting honey production among smallholder beekeepers. Primary data was collected from 37 randomly selected respondents from a population of 63 beekeepers. The results revealed that 62.2% of the respondents were married, 32.4% were above the age of 55, and mostly 86.5% used the Swazi top-bar types of hives. The results further showed that honey production was explained by the farmer's experience and colony size, implying that an increase in the farmer's experience by 1% would result in 0.41% increase in the amount of honey produced, while a 1% increase in colony size would result in 0.57% increase in honey production. The study has shown that there are plenty opportunities to improve the livelihoods of smallholder farmers by engaging in beekeeping. In order for farmers to improve their honey production, they need to increase the colony size and also use langstroth beehives because of their high productivity.

Key words: Beekeeping, honey production, socioeconomic analysis.

INTRODUCTION

Agriculture and the economy in Swaziland

Agriculture is the backbone of Swaziland's economy and a major source of livelihood for rural households with about three quarters (70%) of the population relying on this sector for a living (Thompson, 2010). Some of the agricultural activities that take place in this country include sugarcane, citrus fruits, maize with other cereal crops and pulp production to name a few. The country has always benefited from the European Union (EU) markets which offered a higher price for sugar. However, such an arrangement was phased out and as a result the

price showed a decline. This development has caused the need to diversify the agricultural sector in order to enhance its contribution to the economic growth of the country. The identification of commodities with the opportunity for value adding is a priority for the country (Total Transformation Agribusiness (PTY) LTD, 2008). Thompson (2010), also state that agriculture contributes about 12.7% to the country's gross domestic product (GDP).

Swaziland is mostly covered by natural and man-made forests and the major parts of the country that are covered by forests are the highlands of Hhohho and

Shiselweni regions. Honey production has a positive effect to the vegetation, hence, the promotion of bee farming in the country has a significant potential.

Importance of beekeeping in rural development

Beekeeping also known as apiculture, is the act, science and or business of managing honey bees for the purpose of producing honey, beeswax and other bee products for personal consumption and industrial use. The most important component in the beekeeping industry is the bee as it is involved in the primary production of bee products. There are four well-known honeybee species in the world namely: *Apis mellifera, Apis dorsta, Apis cerana* and *Apis florae,* according to Admassu (2003). *A. mellifera* is native to Europe and Africa, while the rest are native to the Asian continent.

The honeybee *A. mellifera* is one of the most successful species in the animal kingdom judged by its ability to adapt to a wide climatic range. It is believed to have evolved in the tropics. It is highly productive and can adapt well in different climatic conditions. Although they are known as vicious and aggressive bees, they are good producers (Matavele, 2007). Beekeeping is an enterprise that offers great potential for development in Swaziland since it is easy and cheap to manage. For farmers to practise beekeeping they require little land and its quality is less important since the beehives are placed on trees (Oluwole, 1999). This enterprise serves as a means of empowering small-scale farmers who have low capital investments (Farinde et al., 2005).

According to Carruthers and Rodriguezi (1992), beekeeping provide local people with an economic incentive for preservation of natural habitat enhancing environmental quality thus, labour in rural areas can be utilized especially during dry seasons. Beekeeping is an activity that fits well with the concept of small-scale agricultural development. It is a labour-intensive undertaking, which can be easily integrated into larger agricultural or forestry projects. Bees not only aid in the pollination of some crops used in such projects, but also makes use of otherwise unused resources such as nectar and pollen. Previous studies indicates that the beekeeping activity provide benefits in terms of employment, pollination of crops and conservation of biodiversity (Didas, 2005); generates income through hive products (Jones, 2004) and renting bee colonies to pollinate crops (Gates, 2000). Ecological conditions and the floral composition, queen quality and resource management were found to be influencing profitability of beekeeping enterprises (Tucak et al., 2004; Cobey, 2001; Jong, 2000). Beekeeping potential was reported to be great in Swaziland given the economically valuable bee races, varied geography and rich floral resources in the country (Güler and Demir, 2005).

Beekeeping is of vital importance in starting and rebuilding of economic activities that would address socio-economic problems such as HIV and AIDS, poverty and unemployment. A range of products produced in beekeeping not only are rich in nutrients but also have medicinal properties, which people may benefit from. In arable farming, bees also improve crop yields through increased efficiency in pollination and also beekeeping diversifies agriculture as it can be integrated with other agricultural activities as well as agroforestry (Total Transformation Agribusiness (PTY) LTD, 2008).

In the context of agriculture-based major employment and economy of Swaziland, beekeeping has substantial contribution to income generation. In a family-based activity, it is very easy and less expensive to operate than any other income generating activity because a family, keeping 1-5 colonies does not require much land. Most of the time there will be no need to purchase raw materials as honey bees collect nectar and pollen from the available source of existing natural bee plants.

Swaziland has considerable potential in beekeeping with her rich flora, good ecological conditions and existence of colony. However, the beekeeping sector in Swaziland has not yet sufficiently utilized the rich natural resources. Beekeeping can play an important role in the urban and rural areas as small-scale farmers may produce products such as honey, beeswax, propolis to name a few, and selling them in order to generate income.

However, beekeepers encounter different challenges when in the course of the practice. The low yield of honey and other beekeeping products such as honey, beeswax, propolis may result from insufficient management practices and lack of adequate training. On the other hand, honey production is affected by climatic conditions and some bee diseases such as Varroa mites and the American Foulbrood.

Objectives of the study

The main objective of this study was to evaluate the problems that affect the economic performance of beekeeping farmers in Swaziland. The specific objectives were to describe the socio-economic characteristics of beekeeping farmers and determine the factors affecting honey production among smallholder beekeepers.

LITERATURE REVIEW

Status of honey production in Swaziland

Honey, which is one of the products of honeybees. It has been in use since time immemorial. Honey has been found to comprise mostly plant sugars that are readily absorbed by the human body. It is composed of water, protein, fat, carbohydrate, ash, calcium, phosphorus, iron, sodium, potassium, Vitamin A, thiamine, riboflavin and vitamin C (ascorbic acid). All these substances give honey it's nutritional and healing properties. The nutritional and healing properties of honey have been given much accolade through the ages. Honey contains a diversity of substances, which are indispensable to all living things (Olarinde et al., 2008).

Apart from honey, other bee products such as beeswax, bee pollen, propolis and royal jelly among others can be obtained from beekeeping, which are also enormous income generating products. Although, honey and other beekeeping products are very important (economically and socially), humans have not fully utilized the benefits of these products. Beekeeping is regarded as a vocation. It is yet to be practised as a paying occupation. It is obvious that the practice improves the ecology of an environment and helps in plant reproduction, which largely improves the living standard of the people and the nation's economy at large. Despite its numerous benefits and uses to humans and its importance in the society, very few people are engaged in bee keeping. Consequently, the few people who engage in beekeeping as a business are not only skeptical but are also not totally committed to it.

Another source of concern is that, because of the associated bee-keeping problems, especially the seemingly lack of technical know-how, only little or nothing is known about the level of technical efficiency of the few who practise it. This means that the sustainability of bee keeping for honey and the production of other products may not be ascertained. This stems from the fact that the ability to produce maximum output from a given set of inputs (technical efficiency), given the available bee-keeping technology has not been fully understood.

From the data that were obtained by WK Kellogg Foundation (2005) in a survey conducted in all four administrative regions of Swaziland, it was discovered that about 83 tonnes of honey were harvested of which approximately 30 tonnes were obtained from small-holder beekeepers. However, data were not readily available for some areas in the Hhohho, Manzini and Shiselweni regions but it was found that Hhohho had the highest yields of honey among the four regions with Lubombo lagging behind in production due to the fact that the region lacks a wide diversity of the vegetation that can sustain the bees throughout the year. Interviews indicated that these areas had large man-made trees and the natural vegetation receiving good rainfall as they are in the Highveld. The average harvest per hive was around 13 kg and as a result it was established that the beekeepers on average did not harvest more than two times in a year (WK Kellogg Foundation, 2005).

A study by Senger (2001) shows that in most high producing countries such as China, USA and Mexico, production rate is averaged between 50 to 150 kg per hive. However, the local data indicate that it is possible to achieve average production of 50 kg per hive in the Highveld. Improved hive management, selection and development of appropriate species are most of the critical issues that need to be addressed for honey production increment in the country. Currently, African bees are trapped in to catch boxes and transported to hives to produce honey. Vigorous efforts are needed to improve the technology to make the local industry more viable and competitive (WK Kellogg Foundation, 2005).

Factors affecting honey production

Honeybees have a lot to offer in terms of agricultural products and ecosystem services. However, bees are exposed to a number of threats such as climate change, reduced biodiversity, and invasive species that reduce their quality of health and longevity (UNEP, 2010). The cost of dealing with these problems is increasing for apiarists, thus making the beekeeping business less profitable. According to Pokhrel (2008), predators, parasites and diseases are some of other factors that affect beekeeping, thus reducing honey production. Since limitations of beekeeping may affect honey production in a way that these may feed on the honeybees, thus decreasing the population hence lowering production.

Parasites and diseases also affect beekeeping and this will eventually lower production due to the fact that honeybees will be engulfed by a lot of diseases, limiting the status of bees making honey. This is attributed to lack of adequate knowledge of management practices needed in beekeeping. Honeybees can also be affected when using plants for their nectar that had been treated with a high concentration of pesticides because the use of this treated nectar kills the bees and in that case lowers production in terms of output. Beekeepers therefore should control damage on vegetation planted close to the project area, by making use of less concentrated pesticides on such plants or crops (Pokhrel, 2008).

On the other hand, age can be a factor in beekeeping, during harvest times or hard operations you may find that only young adults are able to do all operations requiring man-power. Some literature depicts that only those individuals who are still at average ages of 20-40 years can be able to harvest honey from trees other than those above 50 years who are not able to do so. Gender is another factor that affects honey production in a country. Take for instance, a lot of women find it difficult harvesting their produce due to bees stings; and may be the division of labour that exist may limit the participation of women in beekeeping (Yahaya and Usman, 2008).

Lack of technical know-how can be another factor in honey production in the sense that beekeeping is mainly practised in rural areas. These areas have people who are less educated in agricultural practices due to the fact that they are unable to get funds for their education thus limiting the harvested honey yields (Yahaya and Usman, 2008).

According to Gamez et al. (2004), poor feeding

especially during winter affects honey production. When the colony is not well fed, it will leave the area at the same time affect the yield. Beekeepers therefore, introduce sugar syrup in their feeds at least 6 weeks prior to the onset of the first major nectar flow and this may encourage the production of bees that will be at the appropriate age for foraging by the time of the main nectar flow (Gamez et al., 2004). Further literature states that for honey to be increased, it is essential that there should be a well populated colony in areas where there is abundant nectarous flora.

METHODOLOGY

Research design

A descriptive cross-sectional research design was employed in the study with the aim of describing the farmers' characteristics and identifying factors influencing honey production.

Sampling procedure and data collection

The target population was 63 beekeepers in the Manzini region and this was based on a sample frame that was obtained from the Ministry of Agriculture, apiculture section. The study engaged 37 randomly selected beekeepers and random sampling technique was preferred because it is able to eliminate bias, both consciously and unconsciously. This helps in such a way that every member of the population has an equal probability of being chosen in the study (Key, 1997). The data were collected in December 2011 through the use of personal interviews with the aid of a structured questionnaire.

Data analysis

Descriptive and inferential statistics were used to analyze the data. These included mean, standard deviation and frequencies; whilst the inferential statistics included regression analyses to determine the factors affecting honey yields among small-scale beekeepers.

Analytical model

The analytical framework used in the study was based on the production function. The regression analysis was used to determine the relationship between beekeepers' socioeconomics characteristics and honey production. A Cobb-Douglas production function was used to determine the factors that influence honey production among beekeepers.
The regression model was specified as:

$$Y = \beta_0 X_1^{\beta_1} X_2^{\beta_2} X_3^{\beta_3} X_4^{\beta_4} X_5^{\beta_5} X_6^{\beta_6} e^u \tag{1}$$

Equation 1 was then linearized by taking the natural logarithm, which then yielded the following model:

$$Ln\, Y = \beta_0 + \beta_1 ln X_1 + \beta_2 ln X_2 + \beta_3 ln X_3 + \beta_4 ln X_4 + \beta_5 ln X_5 + \beta_6 ln X_6 + u \tag{2}$$

Where; Y = total honey yield (kg); X_1 = Experience of beekeeper (years); X_2 = labour measured in man hours; X_3 = size of the colony; X_4 = Age of beekeeper; X_5 = Gender of beekeeper; X_6 = Inputs

costs; β_0 = constant term; β_i = coefficients of the regression model; Ln = natural logarithm, and U = error term variable.

Table 1 presents the description of the variables used in the regression.

RESULTS AND DISCUSSION

Socio-economic characteristics of beekeepers

Table 2 presents the descriptive statistics of beekeepers. Most beekeepers were between the age of 15 to 40 years, which shows that beekeeping in the two areas studied (Ludzeludze and Mahlangatsha), was mostly done by the youth. Grown up farmers above the age of 40 years had low participation and this could be due to the fact that the operations that are done in beekeeping demand more man-power, which these old people do not have, so it becomes more difficult for them to conduct most of the beekeeping operations. The results in Table 2 show that about 54.1% of beekeepers had the age range of 15-40 years and 45.9% had more than 40 years old.

The results further show that there were 23 beekeepers that were married (62.2%). Fourteen (32.4%) of the farmers were above 50 years old. About 86.5% of the beekeepers used the Swazi top-bar hive, and they preferred it the most, whilst 13.5% of them preferred the Langstroth hive. Most farmers preferred the Swazi top-bar hive because it is cheap, easy to make and manage. According to the results most (56.8%) of the farmers had primary level of education, while 43.2% had secondary and high school levels of education. The number of beehives a farmer keeps influences the amount of honey produced. Farmers either use Swazi top-bar hive or langstroth hive. When asked about which one they prefer, the majority (86.5%) preferred Swazi top-bar because they are easy to harvest. However, they claimed that the langstroth allow for high yields.

As can be observed from Table 3, the amount of honey harvested by the beekeepers was between 10 to 100 kg per year (43.2%). Only one (2.7%) farmer had output above 700 kg. The results in Table 4 reveal that 78.8% of the variation in honey production is explained by the variables in the model. The results further revealed that honey production was positively and significantly influenced by the experience of the beekeepers in honey production ($p < 0.05$) and the size of the colony ($p < 0.05$).

The findings suggest that an increase in the beekeeper's experience by 1% would result in an improvement in honey production by 0.41%, while an increase by colony size by 1% would result in an improvement honey production by 0.57%. The more experienced farmers tend to have better management skills of bee farming. Experience helps farmers to master complex practices in bee keeping. The more bee colonies, the higher the production of honey. Hauser and Lensky (1994), also found significant influence of

Table 1. Description of variables used in the study.

Variable	Unit	Description	Apriori
Y	Kg	Honey output	+
X_1	Years	Experience of beekeeper	+ or -
X_2	Man-hours	Family labour	+
X_3	Number of bees group (hives)	Colony size	+
X_4	Years	Age of farmer	
X_5	Dummy(1= male; 0 = female)	Gender of beekeeper	+ or -
X_6	Emalangeni (E)	Inputs costs	

1US$ = E8.6

Table 2. Characteristics of beekeepers.

Item	Frequency	Percentage
Marital status		
Married	23	62.2
Single	14	37.8
Age		
15-20	9	24.3
21-30	5	13.5
31-40	6	16.2
41-50	5	13.5
Above 50	12	32.4
Education level		
Primary	21	56.8
Secondary	8	21.6
High School	8	21.6
Number of beehives		
1-5	18	48.6
6-15	15	40.5
Above 15	4	10.8
Preferred hive		
Swazi top-bar	32	86.5
Langstroth	5	13.5
Don't know	0	0.0

Table 3. Amount of honey harvested.

Harvest (kg)	Number of beekeepers	Percentage
10-100	16	43.2
101-200	8	21.6
201-300	3	8.1
301-400	2	5.4
401-500	1	2.7
501-600	4	10.8
601-700	2	5.4
Above 700	1	2.7

Table 4. Factors affecting honey production.

Variable	B	t- statistics	p- value
Constant	3.100	3.918**	0.000
Experience	0.410	2.548*	0.016
Family size (labour)	0.248	1.351	0.187
Age of beekeeper	0.054	0.268	0.790
Gender of beekeeper	0.058	0.199	0.843
Colony size	0.568	5.766**	0.000
Production costs	0.031	0.689	0.496

$**p<0.01$ and $*p<0.05$ respectively. $R^2 = 0.824$, Adjusted $R^2 = 0.788$, F- statistics $= 23.354*$

colony size on honey yield

CONCLUSION AND RECOMMENDATIONS

Although the involvement of small-scale beekeepers in beekeeping is still at an infant stage, the enterprise shows a great potential in improving the livelihoods of the farmers. The favourable natural environment and low disease incidence makes the farmers to be competitive in honey production. Most farmers in the study area use the local (Swazi) topbar hives and further enhance honey production by using langstroth because of their high productivity.

There are opportunities to improve the livelihoods of the smallholder farmers through beekeeping. Farmers need to gain experience in beekeeping in order to improve their honey production. This could be done through special trainings by government extension officers. Farmers also need to increase the colony size of their beekeeping enterprise. And use more of langstroth beehives because they are highly productive.

REFERENCES

Admassu A (2003). Botanical inventory and phenology in relation to foraging behaviour of the Cape honeybees (*Apis Mellifera Capensis*) at a site in the Eastern Cape, South Africa. Unpublished MSc Thesis, Rhodes University, South Africa.

Carruthers I, Rodriguez M (1992). Tools for Agriculture. Russell press, Nottingham, United Kingdom. P. 70.

Cobey S (2001). The quest for quality queens. Bee Cult. 29(6):18-20.

Didas R (2005). Beekeeping project in South Western Uganda. Bee World 86:69-70.

Farinde AJ, Soyebo KO, Oyedokan MO (2005). Improving farmers' attitude towards honey production experience in Oyo State, Nigeria. J. Hum. Ecol. 18(1):21-33.

Gamez S, Sait E, Banu T, Figen C (2004). The Economic Analysis of Beekeeping Enterprise in Sustainable Development: A Case Study of Turkey. Apiacta 38:342-351.

Gates J (2000). Apimondia. Bee Cult. 128(2):51.

Güler A Demir M (2005). Beekeeping potential in Turkey. Bee World, 86:114-119.

Hauser H, Lensky Y (1994). The effect of the honey bee (*Apismellifera* L.) queen age on worker population, swarming and honey yields in a subtropical climate. Apidologie 25:566-578.

Jong WD (2000). Micro-differences in local resource management: The case of honey in West Kalimantan, Indonesia. Hum. Ecol. 8:632-640.

Key JP (1997). Research Design in Educational Operation. Oklahoma State University. http://www.okstate.edu/ag/agedcm4h/academic/aged5980a/5980/ne wpage28.htm, accessed 18-12-2011.

Matavele R (2007). Situation Analysis of Beekeeping in Mozambique. Maputo, Mozambique.

Olarinde LO, Ajao AO, Okunola SO (2008). Determinants of Technical Efficiency in Bee-Keeping Farms In Oyo State, Nigeria: A Stochastic Production Frontier Approach. Res. J. Agric. Biol. Sci. 4(1):65-69.

Oluwole JS (1999). Completing farm children programme development through Agriculture Education in Nigeria. In Williams, S.B., Ogbimi, F.E., and Farinde A.J. (Eds) Farm Children and Agricultural Productivity in the 21st century, Book of proceedings. pp. 1-6.

Pokhrel S (2008). The ecological problems and possible solutions of beekeeping in hills and terai of Chitwan, Nepal. The Journal of agriculture and environment 19: 23-33 http://www.nepjol.info/index.php/AEJ/article/view/2113/1947. Accessed 18-12-2011.

Senger L (2001). China Republic of Honey, USDA, Foreign Agriculture Service Report.

Thompson CF (2010). Overview and performance on Agriculture. Swaziland Business Yearbook, Mbabane.

Total Transformation Agribusiness (PTY) LTD. (2008). Situation Analysis of Beekeeping Industry. http://www./apiservices.com/article/us/beekeeping regional analysis.pdf accessed 28-09-2010.

Tucak Z, Perispic M, Beslo D, Tucak I (2004). Influence of the beehive type on the quality of honey. Collegium Antropol. 28:463-467.

UNEP (2010). Environment for development: Indigenous knowledge in Africa. http://www.unep.org/ik/Pages.asp?id=Swaziland accessed 27-10-2010.

WK Kellogg Foundation (2005). Catalysing the development of honey industry in Swaziland.

Yahaya AT, Usman L (2008). Economic Analysis of Katsina State's Beekeeping Development Project, Savannah. J. Agric. 3:69-76.

12

Gender and small-farmer commercialisation: The case of two farming communities in Ghana

Ivy Drafor

Economics Department, Methodist University College Ghana, P. O. Box DC 940, Dansoman, Accra.

Different options of enhancing household financial status are explored by farmers in Ghana in order to cope with fast changing economic conditions. These include intensification of traditional crop production, diversification into new high value crops and off-farm activities. This paper examines small-farmer commercialisation (SFC) activities in the forest and transition zones of Ghana. Participatory appraisal methods including wealth ranking, livelihood analysis and interview of key informants and opinion leaders were used. The wealth ranking exercise resulted in the identification of three household categories as rich, intermediate and poor. Vegetable production was found to be an important commercialisation activity and pepper production was very successful in one subsidiary village in the forest zone, where the farmers formed a group for production and marketing of the produce. Adopters of SFC are motivated by profitability, regular flow of income from quick maturing crops, and important for women was the desire for financial independence and change in social status. A major barrier to participation in SFC is lack of credit as the adoption is both labour and capital intensive though large land holdings may not be required.

Key words: Women farmers and gender equality, farming systems, wealth ranking, small-scale farmer commercialization, participatory appraisal methods.
.....

INTRODUCTION

The starting point of structural transformation is broad-based smallholder-led agricultural growth and commercialisation, integrating traditional smallholder farmers into the exchange economy (Jayne et al, 2011; Heltberg and Tarp, 2002). Commercialisation of subsistence agriculture in developing countries has led to different levels of production and consumption changes for men and women (Adenegan et al., 2013). The impact of smallholder commercialisation on gender depends on the available resources and on who controls the income generated. According to Berhanu and Jaleta (2010), commercialisation entails market orientation and market participation, and enhances the links between the input and output sides of agricultural markets. Men and women in Ghana are faced with changing roles as a result of the transformation of agricultural enterprises from subsistence-based farming to market-oriented production systems and activities. The efforts of moving from subsistence-based production to more market oriented production is known as small-farmer commercialisation (SFC), the impact of which has not been rigorously ascertained.

Gender equality and the empowerment of women have been on the agenda for global development efforts for some time now. Indicators for this goal have focused on enrolment in school and status of women at all levels. Nota lot of attention has been devoted to exploring ways of empowering women in agriculture in general and in rural areas in particular. Fortunately, the impact of gender in improving the livelihoods of rural populations and people engaged in agriculture has recently been the focus of many global and continental institutions (IFAD, 2012; UNDP, 2012; WFP, 2012; FAO, 2011; World Bank, 2011; IFAD and AfDB, 2010). Studies on gender and agricultural commercialisation have focused on impacts of cash cropping on men and women and relations with nutrition and food security (von Braun and Kennedy, 1994; Webb, 1989). Not much work is available on what factors will make women adopt commercialisation activities. Little data exists in Ghana on men and women's agricultural commercialisation activities.

The aim of this paper is to assess the gender impacts of SFC in the forest and transition agro-ecological zones in Ghana, drawing experiences from the savannah zone. Each zone differs in population density, farming systems and livelihood experiences. The study identifies and examines small-farmer commercialisation activities, its pathways and constraints, and the motivation for SFC. It provides information for understanding how intra-household and inter-household gender relationships are affected by small-farmer commercialisation (SFC) in rural communities.

LITERATURE REVIEW

Given the interconnectedness of biological and social dimensions of human behaviour, gender should be seen to encompass both sex differences and social constructs that give rise to differences between men and women (Phillips, 2005). It is the central organizing principle of societies that governs the processes of production and reproduction, consumption and distribution (FAO, 1997). Gender analysis studies the different roles and responsibilities of women and men, the differences in women's and men's access to and control over resources, and their consequent constraints, needs and priorities. Incorporating gender analysis into the tools of participatory agricultural planning helps policy-makers and planners to understand how the structure of policies and programmes need to be designed to ensure that women benefit as well as men. Hunt (2004) added that gender analysis helps assess the impact of development activity on females and males, assess the differences in participation, and accrued benefits between men and women, towards sustainability and gender equality.

Globalization affects farmers around the world in different ways, based on their specific characteristics, the nature of their market networks and cropping patterns. Remoteness of a market reduces supply (Alene et al., 2008), and negatively affects farmer incomes. Market integration of producers of fruits and vegetables has been

shown to be higher than that of staple crop producers (Weinberger and Lumpkin, 2007). Inability of local agriculture to provide a reasonable standard of living pushes off farmers into low-paying jobs in towns (Jayne et al., 2011). As such, remaining in subsistence production with little market surplus that is sold in local markets limits the ability of smallholders to be better connected to the rest of the world.

Commercialisation is about increasing engagement with markets, increasing inputs and factors of production acquired from the market, using markets to hire labour, and borrowing funds to rent land, obtain technical advice and market information (Wiggins et al., 2011). It involves production of greater farm surpluses, expansion of participation in markets, and increases in farmer incomes and living standards (Jayne et al., 2011). Commercialisation of agricultural systems leads to greater market orientation of farm production (Pingali and Rosegrant, 1995; von Braun and Kennedy, 1994). Changes in product mix and input uses are determined largely by the market forces during the transition from subsistence production to market-oriented systems.

Smallholder farms are risk averse and do not make changes that could put them at financial risk or compromise their ability to ensure adequate supply of food for their household. Wiggins et al. (2011) noted that most examples of small farmers commercialising do not involve radical changes, but take place within existing farming systems, within existing land tenure systems, and are carried out by households using own labour.

Commercialisation leads to increases in income levels for small farmers. However, some researchers have expressed fears that agricultural commercialisation can weaken the role of women and their control over resources and income (Fischer and Qaim, 2012; Wiggins et al., 2011; Quisumbing et al. 1995; Quisumbing and Meinzen-Dick, 2001). According to Fischer and Qaim (2012), increasing degrees of commercialisation may worsen the role of women within farming households. Commercialisation is a major source of productivity growth in the future, yet, what is essential, as noted by Timmer (1997), is the need to deal with the risky environments facing farmers in order to speed up the commercialisation process.

METHODOLOGY

Study sites

The study was carried out in six rural communities in two important farming system zones in Ghana which represent a cross-section of SFC experiences across the country. Farmers in these areas produce a market surplus and the areas have strong trade links with the rest of the economy. At least some farm households in the area are actively involved in SFC or are in the process of adopting SFC activities. They are the transition zone (a major staple food supply zone in Ghana) and the forest zone (has farming systems that are important in terms of foreign exchange revenue generation for the country). The farming systems that characterize the

Figure 1. Map showing the three selected farming system zones.

transition zone are cereals, root and tubers, cotton, fishing, and livestock and those of the forest zone are tree crops (cocoa and oil palm), root and tuber crops, cereal and livestock.

One principal study village was first selected in each of the two farming systems. These are representative of the selected farming systems and have a growing incidence of commercialisation. They are Offuman for the transition zone and Bekwai for the forest zone. Two secondary villages were then selected in each farming system in the vicinity of the principal study village, which has relatively different production structures and market access. This helps to understand whether the SFC activities were also prevalent in smaller villages. Nyansuaka and Amoamo were the subsidiary villages in the forest zone, and Ampenkro and Adankranja were for the transition zone. The presence of a diversity of SFC activities was considered in the selection of communities.

The forest zone is located in the Ashanti Region of Ghana, in the Amansie East District with Bekwai as the district capital. Bekwai, which is about 40 km from Kumasi, the regional capital of the Ashanti Region, has a vibrant non-farm economy with significant marketing and trading activities. The site falls within the tropical rainforest with hilly topography and bimodal rainfall pattern. The transition zone is located in the Techiman District of the Brong-Ahafo Region of Ghana. It is the area between the forest zone in the south and the savannah zone in the north. Offuman is about 30 km from the district capital, Techiman, which has an international market patronised by traders from other parts of Ghana, and some West African countries including Togo, Burkina Faso, Mali and Cote d'Ivoire. The Techiman market goes on from Tuesday to Friday every week, unlike many markets that have a specific day of the week as market day. The presence of the market, coupled with improved road network to Offuman and to one of the subsidiary villages has resulted in vibrant market activities and trading in the community. Population density of the area is fairly low. A map of the study area showing the farming system zones is presented in Figure 1.

Analytical techniques

Participatory appraisal methods were used for case studies in

selected communities in the forest and transition zones of Ghana in order to capture changes that have occurred in their farming systems. Qualitative approaches were used coupled with in-depth interview of key informants to create a good database of the activities of the smallholders. The research methodology draws on rapid appraisal methods including wealth ranking, livelihoods analysis, income and expenditure matrices, benefit analysis flow chart, interview of key informants and opinion leaders, participant observation, and a review of secondary data. The combination of approaches helps to capture as much of the commercialisation activities in the communities as possible and reveal the challenges and barriers that limit their adoption of SFC.

Village entry approaches were used to prepare the communities ahead of actual visits for data collection. Community meetings were held in each of the principal and subsidiary villages, which were well attended by several households. Attendance at the community meetings in the selected villages ranged between 13 and 48 participants with female participation averaging about 40 percent of the total number. Women participated actively and were very outspoken in the two principal villages and Adankranja in the forest zone than in the other villages. It was observed that female participation improved whenever encouraged and also when the women were grouped separate from the men. Several days were spent holding meetings in each village.

The criteria for household classification were identified together with the community members for the wealth ranking exercise as no prior criteria for the classification was predetermined. Participants were given 100 cards to distribute according to wealth categories within the village. The criteria identified for household classification are farm size, asset ownership, livestock ownership, ability to educate children, type of housing, and adoption of improved production methods. Participants were also grouped by gender for income and expenditure matrix analysis.

For the income and expenditure analysis, the participants were divided into two groups based on gender and each group was given cards representing a specific amount of money, and was asked to distribute them among their main sources of income and expenditure. This exercise gave a clear indication of the patterns of expenditure of men and women as well as their income sources. The income generating activities were identified and documented in each village. The livelihood activities were characterised to identify areas and pathways of commercialisation. Some of the information was obtained from key informants such as relevant officials at the District Agricultural Development Unit (DADU), the District Assemblies and village leaders.

RESULTS AND DISCUSSION

Household characteristics by wealth

The wealth ranking exercise revealed three main wealth categories namely; those who are rich, those who are intermediate and those who are poor. These categories in the Akan language, which is widely spoken in the forest and transition zones of Ghana, are 'osikani' for rich, and 'dantemni' and 'ohiani' for intermediate and poor respectively. The household categories by wealth are similar in all the study communities. In the forest zone, the rich constituted 8% of the total households in the community, the intermediate households were 55% and poor households were 37%. In the transition zone, while only 5 percent are in the rich category, 71 percent of households are in the intermediate category and 24% are poor. Results from the household interviews show that the proportion of the households who are within the rich category ranges from 2% to 8% in the study areas, which is consistent with the finding from the focus group discussions. Majority of farmers are classified under the 'dantemni' (intermediate) category. Targeting development programmes at the intermediate and poor households can yield the best results for farmers in rural communities.

The wealth ranking exercise in Offuman, the principal village of the transition zone, showed that the rich had larger household size (more people living in the household) than the poor and the intermediate categories. Most of the households in the rich and intermediate categories have built their own houses but only 40 percent of the poor live in their own houses. The rich live in cement houses which are roofed with iron sheets. About 65% of those in the intermediate group have cement houses and 35% have brick houses roofed with iron sheets. All those who are considered as poor are in mud houses; 30% with thatch roofing and 70% had iron sheet roofing. Household size is not different in the forest zone, where the average household size is larger for rich households than for poorer households. According to the farmers, though there are very rich people who are part of their communities, they have migrated to live elsewhere. The rich and intermediate categories contribute significantly towards community development projects.

Farm size is related to wealth status. Average size of cultivated land is 170 hectares for rich households and 2 hectares for poor households. Production levels are also proportional to wealth status. Households with very small farm sizes are often food insecure as they also have low

incomes and limited range of economic activities. While the rich farmers are more diversified in both agricultural and non-agricultural activities, poorer households have farming as their only occupation and means of livelihood. Besides, richer households are able to adopt new technologies faster than poorer households.

The rich have more resources, are more educated, and have skills that enable them to produce on a large scale. There are differences in the level of education of household members among the categories. The poor and intermediate households are less educated, have limited skills, depend on traders who come to the village to sell their farm produce, and are often compelled to sell their produce early. The rich are able to move their produce to markets outside their local community to sell at competitive prices, with some engaged in trading and buying of farm produce from other farmers to sell in markets outside the village. The rich tend to have stronger market linkages and access to a wider range of information. To cope with livelihood difficulties, the poor resort to providing labour services on other farms for daily wage in order to provide food, pay school fees for children and meet other household needs.

Farming systems and small-farmer commercialisation activities

African smallholders have diverse sources of livelihood including crop and livestock farming and off-farm activities. In farming communities, commercialisation encompasses selling of a marketable surplus of traditional crops, diversification into the production of new crops, introduction of new income generating activities and post-harvest activities such as processing of farm produce. Livestock sales are undertaken in limited communities in the transition zone. Beyond keeping of few animals for household consumption, livestock production is not widespread in the forest zone. Different communities were found to have different production structures, potential for economic growth and value-added systems. Produce from food crops were consumed within the household and the surplus was sold for income. Where household members are engaged in non-farm activities or diversified agricultural production activities, they are able to finance the production of new crops and store farm produce to sell at a higher price at a later date.

Commercial production of vegetables (garden-eggs, tomatoes and pepper) was the most important pathway to commercialisation in the 6 villages visited (Table 1). Overall, about 31 percent of all cultivated land is devoted to vegetable production in the study area and 35 percent was to the production of root and tubers. Rich households can cultivate about 10 acres of vegetables while the intermediate households can cultivate about 5 acres of vegetables. Vegetable production was very

Table 1. Commercialisation pathways in two agroecological zones in Ghana.

Forest zone		
Asanso	**Adankranja**	**Denyasi**
Crops: Vegetables. Brought to village a few years ago from the Brong-Ahafo Region (Transition Zone). ***Non-agriculture:*** Trading in district and regional capital. Artisan work.	***Crops:*** Vegetables (pepper). Taro, cocoa, oil palm are also lucrative but limited to few people and few areas. ***Non-agriculture:*** Widespread small-scale trading in agricultural and non-agricultural products.	***Crops:*** Vegetables. Intensification of cocoa production. ***Non-agriculture:*** Trading

Transition zone		
Offuman	**Nyansuaka**	**Ampenkro**
Crops: Vegetables (tomatoes and garden-eggs). ***Non-agriculture:*** Trading in agricultural produce and ownership of stores ***Other:*** Keeping of livestock.	***Crops:*** Vegetables (very limited). Grows a lot of maize ***Other:*** Keeping of livestock	***Crops:*** Vegetables (tomatoes). Tomatoes processing factory being rehabilitated in a nearby town. ***Non-agriculture:*** Limited trading.

effective where the producers have formed a group for production and marketing. Only small amounts of vegetables are consumed at farm household level. Households consume a lot of cassava, plantain, maize and taro. Cocoa, oil palm and citrus are cultivated, but in limited quantities. As such, vegetables should be considered as cash crop.

Pepper production is very successful in Adankranja in the forest zone. A community member bought the seeds and began its production in 1983, a period when Ghana experienced extreme hardship and famine. After the first cultivation, he introduced four of his friends to it and all the four friends became wealthy through pepper cultivation. In the principal village, vegetable production was introduced from the transition zone (Brong–Ahafo Region). In these villages, the 1983 famine in Ghana led to a shift in the production of tree crops to the production of pepper in order to get quick money. Pepper production then expanded over the years.

The pepper farmers in Adankranja formed a group that had a membership of about 30 farmers. The cooperative enabled them obtain credit, which they paid up promptly. They were also able to access loans from the market women who bought the pepper. As a group, they negotiated for good and stable prices for their produce and agreed on a harvesting pattern whereby only a specific number of farmers harvested pepper at a time, to regulate the quantity available on the market at a given time.

The use of fertilizers and agro-chemicals started in 1988 due to low soil fertility and the incidence of pests and diseases. In the same year, the pepper farmers' cooperative bought a water pumping machine, which helped with dry season cultivation. Pepper cultivation gradually changed from small-scale farming to large-scale cultivation and new varieties were introduced with

time. However, the withdrawal of government subsidies which were on agricultural inputs through the Economic Recovery Programme (ERP) and Structural Adjustment Programme (SAP) resulted in very high cost of inputs and presents a constraint for adopting SFC.

Another example of SFC is maize. Farmers in Nyansuaka, a subsidiary village in the transition zone cultivate a lot of maize for sale. The driving force behind the cultivation of maize is its storability and contribution to household food security. It is consumed in large quantities throughout the year. Maize can be stored for a long time and sold during the lean season at a higher price. There is a high motivation for growing more maize as vegetables are perishable but are not processed. The farmers have constructed a maize storage unit where they store maize in bulk. Maize can be planted twice in a year and also brings quick income to farm households, and turns out to be the most profitable staple crop if it can be cultivated on a large scale and stored for a long period of time.

Ability to store storable farm produce makes it possible for farmers to sell them at a time when the price is favourable and when farmers are in need of money. Farmers who do not have money to pay off debts after the cropping season are compelled to sell their produce early. Rich households are more capable of storing farm produce than the intermediate and poor households. Obviously, the poor are compelled to sell immediately after harvest at prices that are usually dictated by the buyers. The farmers indicated that financial pressure, lack of alternative income generating activities and non-farm employment opportunities compel them to sell their produce early, which has implications for food security, investment and other financial obligations.

In addition to farming, there were a few off-season and non-farm activities such as firewood gathering, charcoal production and general trading, including moving of farm produce to sell outside the villages. Households in the farming system zones have limited post-harvest activities. Yam, cocoa and other tree crops were found to provide those engaged in their production with good income annually but the income is not frequent. Though taro cultivation is profitable, it does not present a general opportunity for many people as it only thrives well in valley bottom areas.

The availability of non-farm income was found not to be related to household typology. On average, 52 percent of households have non-farm income while 48 percent do not. Thirty-seven percent of poor households have non-farm income against 62 percent of intermediate households. Surprisingly, 67% of rich households have no non-farm income. It can therefore be said that wealth status is not determined by the extent of diversification into non-farm activities in the two farming system zones. The percentage of farmers in non-farm activity is, however, higher in areas that are characterised by a single farming season.

Generally, crop farming constitutes the major economic activity in most areas. However, focusing on traditional cropping activities makes the farmers vulnerable to economic and climatic shocks. Crop failure is on the increase due to land degradation, population growth, and climate change. Very few farmers are diversified, which reduces their production and financial risk. Differences in livelihood strategies lie in the differences in household resource endowment, institutional linkages, infrastructural development, and nearness to major marketing centre among others.

Motivation for SFC

Several factors motivated the farmers who adopted SFC in the study area. Regular flow of income, which comes from quick maturing crops like vegetables, and crops that have good yields with high demand and competitive pricing system are attractive to farmers. The need to come out of poverty was an important factor that motivated them to adopt SFC. Increase in income levels is therefore a major driving force. In addition, to women, economic independence is greatly desired either because they perceive that their husbands alone could not cope with the financial demands of the household or they are not in favour of requesting financial assistance from their husbands for every minor need. Women are attracted to high value crops which do not require large land holdings.

Vegetable production was therefore attractive to land poor farmers as it does not require large acres of land to adopt. It also does not hold the land for a long period of time. Belonging to an association is another major motivation as it is an effective means of obtaining credit and farmer information on inputs and prices.

The movement of households from one farming system

zone to settle at another led to the introduction of new crops in areas where they were not previously cultivated. An example is the introduction of beans and tomatoes production in the transition zone by settlers from Northern Ghana. The example of pepper in the forest zone by migrants from the Brong-Ahafo Region was mentioned earlier. The settler farmers explain the system of cultivating the new crop and farm households observe their cultivation and profitability. The profitability of a crop serves as an incentive for adoption or at least trial.

The level of profitability of the new crops, mostly high value crops which have good yields, is directly related to appropriate farm management practices. For example, vegetables are less resistant to harsh environmental conditions and require more care and attention. The attention includes frequent weeding, spraying against insects and diseases, fertilisation, and prompt harvesting. For those who adopt vegetable production, SFC has compelled them to adopt good farm management practices.

Farmers are aware that the production of non-staple or non-traditional crops can generate higher incomes. The reasons for adoption and the characteristics of adopters and non-adopters are presented in Table 2. Commercialisation has resulted in improved income levels that have enabled households to build houses, purchase pumping machines, some have purchased vehicles, cater for children, cater for themselves, and to improve household nutrition. Adopters of commercialisation had improved living standards than non-adopters.

Barriers to participation in commercialisation activities

The pathways of commercialisation often demands capital and labour as well as a thorough supervision of the process. Determination is necessary to adopt SFC. Access to credit and other means of financial support are necessary to enable farmers consider adopting commercialisation. Otherwise, community members who are resource poor are unable to participate. SFC requires large outlays of capital to purchase fertilizer and agro-chemicals, and to pay for labour services. Apart from credit, some farmers do not have fertile land on which to cultivate vegetables.

There is some degree of uncertainty in adopting vegetable production as output price is sometimes unfavourable. Farmers sell even when the price is very low because the produce is perishable and not stored or processed within the local setting. Farmers incur large losses when traders fail to come and buy the produce. Alternative marketing avenues need to be explored besides the role of the middleman.

Small-scale farmers are rather unwilling to purchase food items which they can grow themselves. This is

Table 2. Reasons for adoption and characteristics of adopters and non-adopters.

Adopters	Non-adopters
Characteristics	
• Have more income and own properties such as houses, television, and fridges. • Give better education to children. • Provide good and nutritious food for their family. • Good physical appearance (clothing). • Less borrowing	• Low income levels • Not able to educate children to higher levels. • Not able to provide good and nutritious food for the family. • Poor physical appearance (clothing). • Borrows money often.
Reasons for adoption and non adoption	
• The quest for better standard of living. • The need to get quick income to meet financial expenses, especially to pay for children's education. • In the case of vegetables, it is early maturing and can be harvested every week.	• Have another viable enterprise (taro, cocoa, oil palm, cassava and maize) • Adoption needs a lot of labour and capital. • High cost of chemicals and fertilizers. • Very intensive and difficult to undertake – requires hard work. • No interest in vegetable production. • Few fertile lands that can support such production activities.

particularly important considering their risk averse behavior. This confirms conclusions of a study by Drafor et al. (2013), which analysed the behavior of rural households in ensuring food security in lean seasons and showed that rural small-scale farmers will produce rather than purchase staples for household consumption under different policy scenarios. Consequently some community members in the farming system zones, especially the land poor, are hesitant to adopt SFC due to its implications for food security.

In communities where vegetable production is widespread, SFC is said to result in food shortages as vegetables are not consumed in large quantities and most of the fertile land is devoted to its production. Households involved in food production are key contributors to making commercialisation possible due to the complementary role they play in contributing to food security.

Gender impacts in agricultural commercialisation

The transformation of traditional farming economies into modernized small-scale farming has cultural implications, including important changes in indigenous patterns of gender relationships within the household and the community. The ability for women to move into commercial production requires resource availability, access to new technologies and market opportunities. Women often need to adopt strategies that allow them to bypass gender constraints to enable them have access to land, capital and other productive resources.

The key aspects of impact of SFC are increase in income, change in social status, economic and financial

independence, empowered decision-making position and gender equity. Some of these are particularly more important for women than men who usually play leadership and decision-making roles in society. Women adopters had better financial independence which improved their status in the household and community, especially when they control income generated from commercialisation activities.

Ability to control income from SFC activities depends on whether the activity was carried out as a household or at individual level. Most families farm together as a team, though there are individual farms. Many women also have their own farms. Access to and control of resources depends on who controls the income from economic activities in the household. Household members who have control over the income from SFC are able to rent land and hire labour, purchase fertilizers, agro-chemicals and farm equipment. As such, lack of control of income is directly linked with lack of access to productive resources. However, it was found that before some women could get access to a knapsack sprayer or a pump for work on their vegetable farm, they have to work for three days on the farm of the one who owns it. She is then allowed to have user access to these resources.

When both the man and the woman undertake commercialisation activities, they bring their resources together to educate their children and for the general welfare of the household. Children help on the farm after school and the entire household benefits. In the past, a division of labour existed, but everybody worked for the direct survival of the family – men, women and children. With the introduction of cash crops, women's responsibility to provide the required food crops increased, while men's main responsibility shifted to the production of cash crops, often with considerable labour contributions from women. An earlier study by Saito et al., (1994) showed that the introduction of cash crops resulted in the weakening of the traditional gender division of intra-household rights and obligations and farm women increasingly undertook tasks previously done by men.

There are changes in intra-household division of labour with the introduction of profitable commercialisation activities. In the study sites, women undertake the harvesting and marketing activities while the men carried out the land clearing, chemical application and some harvesting. The children do the planting and fertilizer application. In Nyansuaka and Ampenkro, women do most of the work on the farm after the men clear the land. With time, when more money is obtained from SFC activities, women and children work less on the farm in male-headed households since there is money to hire labour. When there is limited household income in the face of increasing farm size, women work more in the farm, which could affect the time left for them to undertake household activities. On the other hand, women in female-headed households (single women, the divorced, the separated and women with absentee husbands) work more on the farm with the introduction of SFC. Challenges in intra-household relationships stem from situations in which men complain of disrespectful behaviour from women whose income level have increased. Women also complain that some married men put pressure on the family when they adopt SFC by taking concubines.

Adoption of SFC is a gain to an entire village community. Inter-household relationships are strengthened through various forms of inter-dependence and collaboration. Non-adopters, including the youth, are employed to undertake various activities, for which they are paid either in kind or in cash. Borrowing from community members reduces as a result of financial independence of adopters. Adopters of SFC are major financial contributors towards community development, contributing more to enhance progress in the villages. This impact on community development is very important, especially with limited national development efforts in rural areas. Besides, SFC serves as motivation to stay in the villages and has resulted in reducing rural-urban migration.

CONCLUDING REMARKS

If we want agricultural growth to reduce poverty, it must be inclusive, leaving no real alternative to a smallholder-led agricultural development strategy (Jayne, et al. 2011). Interactions and interconnectedness of rich farmers and poor farmers can result in effective rural development and growth, without which many poor households can be left out completely. The outcome of small-farmer

commercialisation in two farming system zones reveals that entire communities benefit from SFC due to inter and intra-household relations.

There are a number of factors that motivate the adoption of small-farmer commercialisation in rural Ghana. Small farmers are attracted to activities that will bring quick and regular income, and which do not need large acres of land. Vegetables and maize satisfy these conditions. Farmers moving from one community to settle in another results in the introduction of new crops in the new communities, thus promoting small-farmer commercialisation. Membership of groups is also an advantage in benefiting from SFC activities in the farming system zones as it does not only encourage adoption of SFC, but also facilitates the process of obtaining credit and good prices. For maize however, production of a marketable surplus is key to improving income.

Women's entry into commercial agriculture is individual and therefore sustainable. Furthermore, the presence of SFC enhances gender equality and the empowerment of women in rural areas. When women have access to and control enterprises, resources and revenue from commercialisation activities, it enables them to achieve financial independence, increased social status and integrates them better into national and global markets. This process promotes the empowerment of woman in the agricultural sector.

Some of the advantages of adopting SFC can only be derived through the simultaneous adoption of improved farm and production management practices. SFC has compelled farmers to adopt better farm practices, which is unavoidable for vegetables as they are less resistant to harsh environmental conditions and require more care. Adoption of good agricultural practices can be increased if more farmers are given incentives to adopt SFC.

Small-farmer commercialisation improves the livelihood of rural households but requires access to productive resources and services. Access to credit and effective markets can serve as incentives for more women adopting SFC, leading to improved incomes, better social status, financial independence, and greater gender equality. SFC is generally capital intensive and many smallholders are unable to meet the high production costs from their own savings. It follows that rich households are more able to adopt SFC activities that require large capital outlays, followed by intermediate households. The role of credit and small starter packs are increasingly relevant for enhancing smallholder adoption of SFC. Poverty and the absence of alternative income sources in rural areas compel farmers to sell their produce early, limiting their ability to benefit from higher prices in lean seasons.

From the example of the pepper producers in the forest zone, market access, which addresses the role of middlemen that can diminish farm incomes, is a vital factor for successful commercialisation of agriculture. Consistent with Weinberger and Lumpkin (2007), market integration of vegetable producers is higher than that of staple crop producers. A revisit of the system of marketing agricultural products across the country with specific policies that protect the interest and income of small-scale farmers is an urgent need. Effective marketing systems and alternative avenues for value addition for vegetables should be explored due to their perishable nature.

Conflict of Interests

The author(s) have not declared any conflict of interests.

ACKNOWLEDGEMENT

The author would like to thank the FAO for providing the funding for this research.

REFERENCES

Adenegan KO, Adams O, Nwauwa LEO (2013). Gender Impacts of Small-Scale Farm Households on Agricultural Commercialisation in Oyo State, Nigeria. Br. J. Econ. Manage. Trade 3(1):1-11.

Alene DA, Manyong VM, Omanya G, Mignouna HD, Bokanga M, Odhiambo G (2008). Smallholder market participation under transactions costs: Maize supply and fertilizer demand in Kenya. Food Pol. 333:18-328.

Berhanu G, Jaleta M (2010). Commercialization of smallholders: Does market orientation translate into market participation? Improving Productivity and Market Success (IPMS) of Ethiopian farmers project Working Nairobi, Kenya, ILRI. P. 22.

Drafor I, Kunze D, Sarpong D (2013). Food Security: How Rural Ghanaian Households Respond to Food Shortages in Lean Season. Int. J. Agric. Manage. 2(4):199-206.

Fischer E, Qaim M (2012). Gender, agricultural commercialization, and collective action in Kenya. Food Sec. 4(3):441-453.

FAO (2011). The State Of Food And Agriculture 2010-11: Women In Agriculture Closing the Gender Gap for Development. Rome: FAO.

FAO (1997). Gender: The key to Sustainability and Food Security. SD Dimensions/ Women and Population/Special. www.fao.org/WAICENT/FAOINFO/SUSTDEV/Wpdirect/WPdoe001.htm.

Heltberg R, Tarp F (2002). Agricultural supply response and poverty in Mozambique. Food Pol. 27:103-124.

Hunt J (2004). Introduction to gender analysis concepts and steps. Dev. Bull. 64:100-106. Accessed on August 26, 2013. Available at: http://www.vasculitisfoundation.org/wp-content/uploads/2012/11/development_studies_network_intro_to_gender_analysis.pdf.

IFAD (2012). Grants: An overview. Accessed on March 18, 2013. Available at: http://www.ifad.org/operations/grants/index.htm.

IFAD and AfDB (2010). Towards purposeful partnerships in African agriculture: A joint evaluation of the agriculture and rural development policies and operations in Africa of the African Development Bank and the International Fund for Agricultural Development. Accessed on December 20, 2012. Available at: http://www.ifad.org/evaluation/jointevaluation/docs/africa/africa.pdf.

Jayne TS, Haggblade S, Minot N, Rashid S (2011). Agricultural Commercialization, Rural Transformation and Poverty Reduction: What have We Learned about How to Achieve This? Synthesis report prepared for the African Agricultural Markets Programme Policy Symposium, Alliance for Commodity Trade in Eastern and Southern Africa April 20-22, 2011, Kigali, Rwanda.

Phillips SP (2005). Defining and measuring gender: A social determinant of health whose time has come. Int. J. Equity in Health, 4(11):1-4.

Pingali PL, Rosegrant MW (1995). Agricultural commercialization and diversification: processes and policies. Food Pol. 20(3):171-185,

Quisumbing AR, Meinzen-Dick RS (2001). Empowering Women To Achieve Food Security: Overview. FOCUS 6 • Policy Brief 12. International Food Policy Research Institute, Washington DC.

Quisumbing AR, Brown LR, Feldstein HS, Haddad L, Pena C (1995). Women: The Key to Food Security. Food Policy Statement No. 21, Aug. 1995. International Food Policy Research Institute, Washington, DC.

Saito AK, Mekonnen H, Spurling D (1994). Raising the Productivity of Women Farmers in Sub-Saharan Africa. World Bank Discussion Papers. Africa Technical Department Series, P. 230. Washington DC.

Timmer P (1997). Farmers and Markets: The Political Economy of New Paradigms. Am. J. Agric. Econ. 79(2):621-627

United Nations Development Programme (UNDP) (2012). Africa Human Development Report 2012: Towards a Food Secure Future. UNDP Report.

von Braun J, Kennedy E (Eds.) (1994). Agricultural commercialization, economic development and nutrition. Johns Hopkins University Press for the International Food Policy Research Institute, Baltimore and London.

Webb P (1989). Intrahousehold Decisionmaking and Resource Control: The Effects of Rice Commercialization in West Africa. International Food Policy Research Institute, Washington DC.

Weinberger K, Lumpkin TA (2007). Diversification into horticulture and poverty reduction: A research agenda. World Dev. 35(8):1464-1480.

WFP (2012). Purchase for Progress: Gender. Accessed on December 28, 2012. Available at: http://documents.wfp.org/stellent/groups/public/documents/reports/wfp252603.pdf.

Wiggins S, Argwings-Kodhek G, Leavy J, Poulton C (2011). Small farm commercialisation in Africa: Reviewing the issues. Future Agricultures Consortium, Research Paper.

World Bank (2011). World Development Report – 2012 – Gender Equality and Development, Washington D.C.: World Bank

13

Determinants of soybean market participation by smallholder farmers in Zimbabwe

Byron Zamasiya[1], Nelson Mango [1] Kefasi Nyikahadzoi [2] and Shephard Siziba [3]

[1]International Centre for Tropical Agriculture, (CIAT), P.O. Box MP228 Mt Pleasant, Harare, Zimbabwe.
[2]Centre for Applied Social Sciences, University of Zimbabwe, P. O. Box MP167 Mt Pleasant, Harare, Zimbabwe.
[3]Department of Agricultural Economics and Extension, University of Zimbabwe, P. O. Box MP167, Mt Pleasant, Harare, Zimbabwe.

This article examines the determinants of soybean market participation by smallholder farmers in Zimbabwe, with a view to identifying key policy entry points for increasing farmer incomes. Market linkages have been identified as key to the successful integration of grain legumes into the smallholder farming systems of southern Africa. Data for this article is derived from a baseline household survey in Guruve district of Zimbabwe. Using a sample of 187 smallholder farmers, we employed the Heckman's Probit model with sample selection to firstly, identify the factors affecting a farmer's decision to participate in soybean markets and secondly, evaluate the factors that affect the intensity of a farmer's participation. Study findings show that the use of inoculants and improved soybean seed varieties are significantly correlated with participating in soybean markets. Results also show that ownership of radios has a positive effect on the household's decision to participate in the soybean market. Further results show that male-headed households are less likely than female-headed households to participate in soybean markets because legumes are seen as women's crops in Zimbabwe. We conclude that in order to leverage smallholder farmers' market participation in soybean markets, it is important to improve access to inoculants and improved soybean seed varieties and improving access to market information. We recommend that authorities could improve access to market information to improve farmers' decision making on soybeans market participation.

Key words: Soybean, market participation, determinants, smallholder farmers, Zimbabwe.

INTRODUCTION

Market linkages have been identified as key to the successful integration of grain legumes into the smallholder farming systems of southern Africa (Chianu et al., 2009). Soybean (Glycine max) is a commodity with relatively higher prices and that has shown great potential to sustain production in smallholder farming systems due to its multiplicity of use. Soybean can be used as cash crop, as food and also as means of improving soil fertility through Biological Nitrogen Fixation (BNF). The net income benefits derived from soybean production depend on the extent to which farmers participate in output markets. According to IFAD (2003), market participation can be an effective route for rural smallholder farmers to move out of abject poverty and increase income. Studies show that market participation by smallholder farmers in developing countries is very low (Barret, 2008). This scenario has slowed down agriculture driven economic growth and exacerbated poverty levels. As such farmers cannot benefit from the welfare gains and income growth associated with market participation. However, for agriculture to meaningfully contribute to economic growth, smallholder farmers have to commercialize their farming activities to produce marketable surpluses (Jagwe et al., 2010). The issue of why most smallholder farmers who happen to make the larger proportion of the poor in developing countries self select themselves out of the remunerative markets remains largely unanswered. It is therefore necessary to identify the key determinants of soybean market participation by smallholder farmers in order to be able to identify key entry points and interventions that can increase household income.

The trade theory posits that if households participate in markets by selling surplus of what they produce on a comparative advantage, they are set to benefit not only from the direct welfare gains but also from opportunities that emerge from economies of large-scale production (Siziba et al., 2011; Barrett, 2008).

Indeed, they will also benefit from technological change effects from the improved flow of ideas from trade-based interactions (Barrett, 2008). Consequently, there will be improved factor productivity. Despite the stream of benefits that are inherent with market participation, evidence from studies in southern Africa shows that smallholder farmers' participation in agricultural output markets is low due to high market transaction costs, information asymmetries, institutional constraints among other constraints. Barret (2008) argues that inducing market participation through trade and price based market interventions does not provide the sufficient conditions to induce improved participation. In addition to these policies, households need to have access to productive assets, adequate private and public investment, institutional and physical infrastructure to access remunerative markets (Siziba et al., 2011; Barret and Swallow, 2006). As noted by Barret (2008) such smallholder farmers with access to production, private and public sector goods, properly functioning institutions and well developed physical infrastructure actively participate in markets contrary to their counterparts.

However, the general trend in most southern African countries is that most agricultural produce is lost soon after production largely because of poor post harvest handling and failure to access the formal markets (Phiri and Otieno, 2008). This trend is attributed to several factors and barriers in agricultural commodity marketing that discourage smallholder farmers from participating in formal markets. These factors range from household characteristics for instance low education levels, labor shortages, inadequate government services, high transaction costs and lack of physical infrastructure (Siziba et al., 2011, Jagwe et al., 2010; Pingali et al., 2005). In response to these challenges, most

governments in Sub Saharan Africa implemented marke liberalization policies in the 1980s and 1990s which sought to open new market led economic growth opportunities (Barrett, 2008). It involved the abolition of commodity boards, introduction of free markets and encouragement of private sector participation. According to Jayne and Jones (1997), although the overall aim of the liberalization was to improve the functioning and effectiveness of markets, it produced mixed results. In some cases, there was actual retreat to subsistence agriculture while in others there was increased market participation in more remunerative markets, technological progress and improvements in institutions and physical infrastructure.

This study sets to establish factors affecting soybean market participation and the level of marketed surplus among smallholder farmers. The results of this study are essential in contributing to the existing body of knowledge on soybean market participation which is scant locally as most previous research concentrated on biophysical aspects of soybean production. Therefore, understanding smallholder marketing of soybean is vital for increased participation which may lead to increased farmer incomes, improved soil fertility and ultimately reduced poverty. Information from this study will be useful to agricultural policy makers to create or amend existing policies in an effort to develop the soybeans production and markets as well as motivate producers to access soybean commodity markets.

Smallholder soybean production in Zimbabwe

Historically, soybean production in Zimbabwe was highly mechanised and carried out by commercial farmers in high rainfall areas (Estehuizen, 2011). The commercial farmers had easy access to inputs, financial capital, irrigation services and well developed marketing channels (Madanzi et al., 2012). The output from commercial farmers accounted for 95% while smallholder farmers contributed only 5% of national soybean output (Estehuizen, 2011). Smallholder farmers used unimproved retained seeds and did not have access to Bradyrhizobium inoculant and this contributed to yields as low as 0.6 t ha-1 compared to 3 to 4 t ha^{-1} in the commercial sector (Mabika and Mariga, 1996). The smallholder farmers lacked general knowledge on good agronomic practices. Shumba-Munyulwa (1996) noted that agronomic research on soybean production was confined to the commercial sector and extension in smallholder farming sectors was limited. This implies that the recommendations from such agronomic studies could not be applied to smallholder farming.

In 1996, the government formed the National Soybean Task Force (NSTF) whose mandate was to help increase the participation of smallholder farmers in soybean production and marketing (Madanzi et al., 2012). In

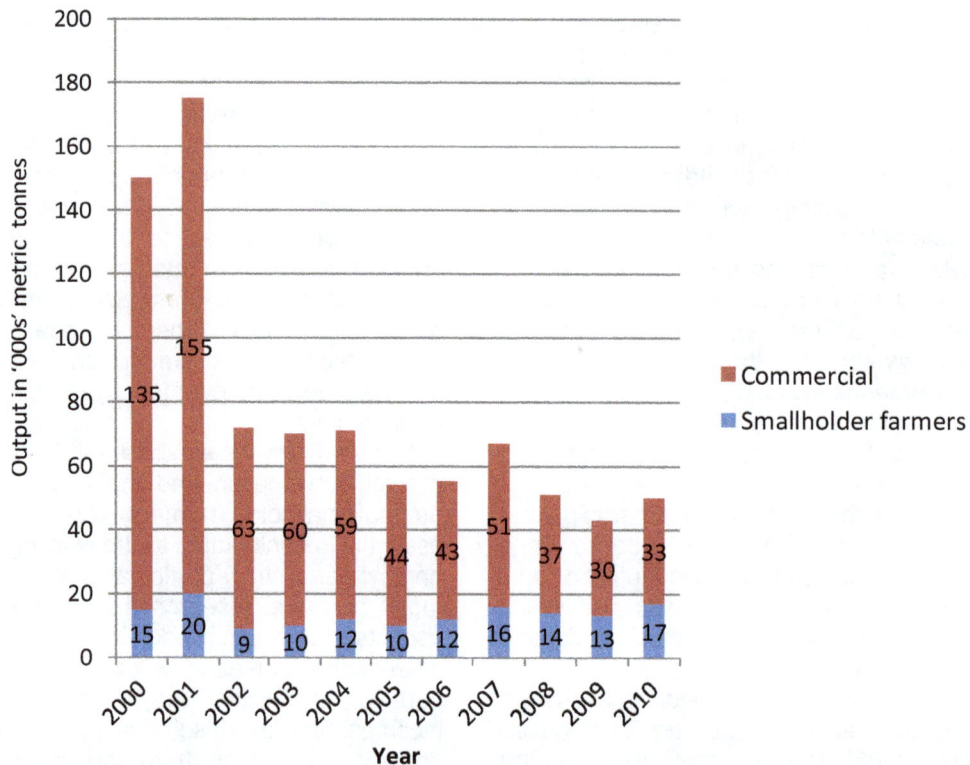

Figure 1. Soybean production in Zimbabwe in "000' metric tonnes. Source: Technoserve, 2011.

particular the programme provided agricultural extension, access to cheap inputs and linkages to markets to smallholder farmers. When the programme started, it enrolled 55 smallholder farmers but by end of 2006 the programme had reached a total of 55,000 smallholder farmers who produced 40,000 t per annum (Chianu et al., 2009). Complimentary efforts have been done by Africare and the N2Africa Project in Zimbabwe who are assisting the smallholder farmers with agronomic knowledge on soybean production in addition to market linkages. Despite these efforts, soybean producing smallholder farmers face challenges such as access to cheap inputs and rhizobium (Madanzi et al., 2012). Although the Rhizobium is produced by Zimbabwe's Soil Productivity and Research Laboratory (SPRL) at a break-even price of $3.20 and distributed through Agricultural Technical and Extension services (AGRITEX) at a retail price of $5.00, some farmers claim that they access the inoculant at more than double the cost (Woomer et al., 2013). The seed houses are not producing sufficient quantities of soybean seed for the market as the smallholder farmers do not purchase the improved seed.

Despite the government's efforts in distributing land from the commercial farmers to landless peasants, Zimbabwe is still facing huge deficits in soybean production with demand far outstripping current production levels. Zimbabwe's annual demand for soybean is 125,000 metric tonnes while production has

been fluctuating far below the equilibrium quantity (Varia, 2011). At present, the demand deficits have been filled by imported soybeans from South Africa, Zambia and Malawi. Zimbabwe is only producing 30% of its national demand of 125,000 metric tonnes and capacity utilization at the major soybean processors is only 16% (Technoserve, 2011). The huge demand deficit in soybean production offers an opportunity for smallholder farmers to produce large quantities of soybeans, participate in markets and improve household income. Since soybean is renowned for its high propensity to fix nitrogen, intensive market participation by smallholder farmers would also improve soil fertility and yields for subsequent crops such as maize if farmed on the same land in rotation. However, despite this market opportunity particularly from the booming livestock and poultry industries where soybean is used to produce animal feed, smallholder farmers are producing very low quantities of soybean for sale and market participation is very low as shown in Figure 1.

Figure 1 shows the contribution of smallholder farmers to national output has remained very low between 2002 and 2010. The observed trends in soybean production, presents an opportunity for smallholder farmers to exploit the market by increasing production of soybeans, as well as participating in its supply chain for income generation. However despite the income generation potential of soybean for smallholder farmers and the huge supply

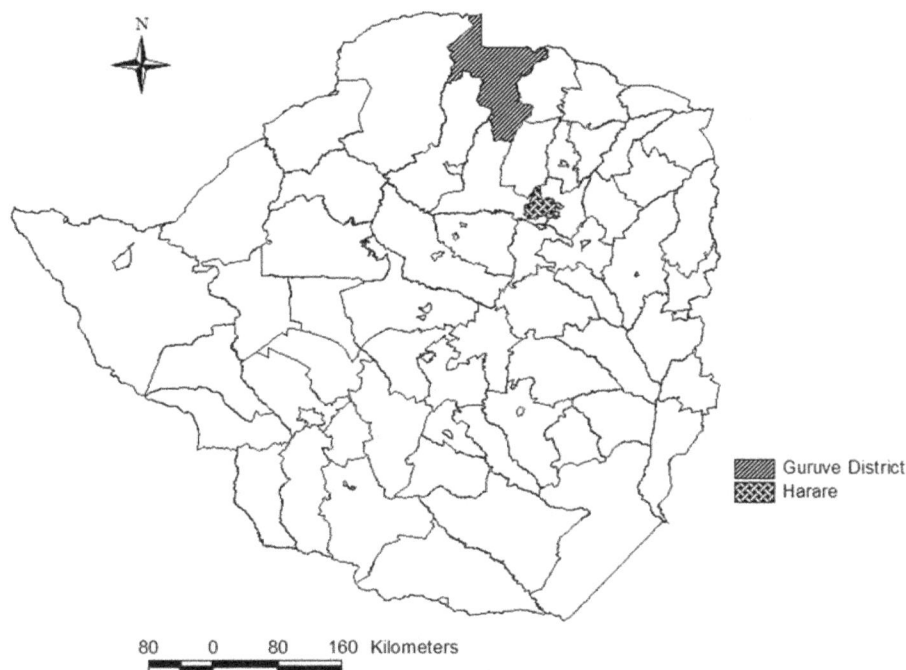

Figure 2. Map of study site: Zimbabwe's Guruve district.

deficit in Zimbabwe, research on soybean has largely focused on biophysical aspects such as yield enhancement, production practices and nutrient use efficiency. There is a lack of information on soybean market participation by smallholder farmers and in particular the factors that influence the level of marketable surplus. Smallholder farmers' market participation is equally important if the full benefits from soybean production are to be realized. The studies on factors affecting smallholder market participation have not been fully exploited especially for soybeans. Most studies (Siziba et al., 2011; Okoyo et al., 2010; Jagwe et al., 2010) conducted on factors influencing smallholder market participation have concentrated on staple crops, that is, maize, cassava and bananas.

Since staple crop markets are very different from soybean markets, recommendations from such studies may not be applicable to soybean markets. Thus this study is an attempt to fill the knowledge gap on soybean market participation by smallholder farmers. To the best of our knowledge, this is the first such study in Zimbabwe, which seeks to identify factors influencing soybean market participation and the intensity of market participation by smallholder farmers.

THE STUDY APPROACH

Study site

This study was conducted in Guruve district, which is in linked to

Mashonaland west province of Zimbabwe (Figure 2). The district is linked to the main legume market, Harare, by a 151km tarred road. Although most of Guruve district lies in natural farming region IV, which is a semi-arid and marginal zone, the study sites lie in natural farming region II. The annual average rainfall is 600 mm while the annual average temperature is 26.5°C. This natural farming region is an agro-ecologically high potential zone suitable for growing soybeans, maize and common beans. The altitude range is 800 to1500 m above sea level. The main livelihood activity is farming with maize being the dominant cereal crop while soybeans and common beans constitute the main legume cash crops.

Sampling and data analysis

This study uses cross sectional household data from the baseline survey collected using a questionnaire with semi structured and structured questions. A sample of 187 of actual greater than 128, an apriori power analysis computed using G Power. It therefore means that the sample provides acceptable statistical power (that is 0.80) for moderate correlation r = 0.30, at two tailed 0.05 level of significant (Franzel et al., 2007). Random sampling was used to select the wards and the households for interviewing from the lists that were provided by resident agricultural extension officers. In the first place, 10 households per ward were randomly selected from six wards where the project is being implemented while the 127 all came from a counterfactual site.

A counter factual site is a site similar to the intervention (treatment) in agroecological and market conditions but did not receive a treatment (Binam et al., 2011). The 127 sampled households in the counterfactual site were randomly sampled from 6 wards that did not participate in the project. The sampling approach followed by the project was meant to allow the use of propensity score matching approach in impact assessment. Data collection for this study was done in October 2011 through face- to-face administration of questionnaires. The survey collected information on household composition and characteristics, crop production, household market participation, access to infrastructure, household incomes, ownership of land and non land assets, livestock ownership and access to agricultural inputs on credit.

The analytical approaches

The data was entered, cleaned and then analyzed using STATA Version 11.2. The study uses the Heckman's model with sample selection to identify the factors that affect smallholder farmers' decision to participate in soybean markets and then to evaluate the factors that affect intensity of soybean market participation. This model is adopted on the basis that it models the market participation decision as a two step process that involves (1) the household deciding on whether or not to participate in the soybean market (2) the level of market participation. The factors influencing the farmers' decision to participate are estimated using the Probit model (selection equation) while the level of participation is estimated using the Ordinary Least Squares approach (Outcome equation). Goetz (1992) and Huang et al. (1991) noted that the use of Heckman's model with sample selection allows the interpretation of results by distinguishing between factors that affect the farmer's decision to participate in the market and those that affect the level of market participation.

According to Greene (2003), in instances where observed characteristics only occur in subsets, incidental truncation occurs. As such, this study uses this model as it corrects for sample selection bias and incidental truncation. The selection bias arises due to the observation of sales from a subset of households who participated in the soybean markets. The empirical analysis in this study is premised on three constructs namely household characteristics, information and assets. In this study, the econometric analysis is based on these constructs to reflect the effect of transaction costs on farmer's decision to participate in the market and also the level of market participation. Variables hypothesized to explain smallholder farmers' soybean market participation and level of participation were identified based on theoretical framework and on past empirical work on market participation under transaction costs (Goetz, 1992; Holloway et al., 2000; Key et al., 2000; Alene et al.,2008; Jagwe et al., 2010; Siziba et al., 2011).

This study builds on earlier studies on smallholder market participation under transaction costs by applying this to smallholder market participation in soybean markets. Based on these constructs as in Jagwe et al. (2010), in this study household head's gender, head's age, head's age squared and household size are used as proxies for household characteristics. Livestock wealth or resource endowment is represented by number of cattle owned while information is represented by contact with extension, household head education, distance to nearest market, ownership of radio and ownership of a mobile phone. These constructs are used in the analysis to reflect the influence of transaction costs on the farmer's decision to participate in a soybean market and to estimate the significant factors that influence the level of market participation.

The outcome regression

The outcome model is conditional on market participation and it is estimated using the Ordinary Least Squares (OLS). In the OLS equation, the dependent variable is amount of soybeans sold (continuous variable). In this paper we hypothesized that gender of household head, age of household head, size of the household, farming experience; ownership of cattle and distance to the market

affect the intensity of a household's participation in the soybean market—following Jagwe et al. (2010).

Selection equation

In the selection equation, that is the Probit model, the dependent variable is a dichotomous variable 'participation in soybean market (represented as 1 when a household participates in the market and 0 otherwise'). The independent variables that condition the participation of smallholder farmers as adapted from literature are gender of household head, age of household head, size of the household, farming experience; ownership of cattle, ownership of radio, ownership of cellphone, access to extension, use of rhizobial inoculants and use of improved soybean seed varieties (Table 1). Age may influence market participation through various channels such as experience, access to resources and risk preferences. The expected direction of the effect of age is thus ambiguous. The gender of a household head is likely to reveal the differences in market orientation between male and female household heads. Cunningham et al. (2008) argues that male household heads sell their produce when prices are high while female household heads keep their produce for household food self sufficiency. We thus expect the sign to be positive meaning that male-headed households are more likely to participate in soybean markets as compared to their female counterparts.

Alene et al. (2008), posit that the household size is an indicator of the amount of family labor that is available for production activities.

It also explains the consumption levels for a household. We thus expect the sign to be positive when a household's labor resources are efficient that is they produce far more output than what they require for household consumption. In such a case, there is high marketable surplus. However, if the sign is negative it is an indicator of household labor inefficiency that is, a larger household produces far less than what it needs for household consumption and thus less marketable surplus. According to Omiti et al. (2009), the distance to the market negatively influences both the household's decision to participate in the market and the amount sold (intensity of participation). The further the distance to the market, the higher the transport costs and the lower the net benefit to the household. Key et al. (2000) note that farmers who stay in remote areas have low input use that is, they normally substitute high value commercial varieties with locally easily obtainable varieties.

Consequently, this input substitution has adverse effects on productivity, market participation and marketable surplus. We thus expect a negative relationship between distance to market and likelihood to participate in marketing. This implies that the higher the distance to the nearest selling points, the lower the likelihood of a household to participate in markets. However, Fafchamps and Hill (2005) observed that wealthy farmers can sell their produce to distant markets as they can afford the high transport costs compared to the poor farmers. This then implies that we expect the resource constrained farmers to participate in local markets while the resource endowed farmers participate in distant markets.

Most economists argue that relative prices form critical incentives to induce market participation and increase the amount of marketable surplus (Alene et al., 2008; Fafchamps and Hill, 2005). Smallholder farmers in Zimbabwe access market information on prices of inputs and output through contact with extension agents, radios and phoning the buyers using cell phones. Knowledge of input prices enables farmers to make informed decisions on input use intensity and also the area to commit to soybeans. We argue that access to price information positively influences the farmers' decision to participate in soybean markets while the lack of it acts as a disincentive. We therefore expect a positive relationship

Table 1. Description of covariates used in the regression models.

Variable	Description	Measurement	Expected sign
Household characteristics			
Age	Age of household head	Number of years	+
Age squared	Age of household squared		+
Gender	Gender of household head	0=female; 1=male	
Household size	Number of people in a household	Number	+
Farming experience	Number of years household head has been farming as a household	Number of years	+
Information			
Distance to market	Average distance from household's home to nearest point of sale	Km	-
Household head's education	Education level of household head	0=no secondary education 1=has secondary education	+/-
Access to extension	Access to agricultural extension for crop production advice	0=no access	+/-
Own cellphone	Ownership of a cellphone	0=does not own 1=owns a cellphone	+/-
Own radio	ownership of radio	0=does not own 1=owns a radio	+/-
Assets			
Number of Cattle Owned	Number of cattle owned	Ratio	+

between a household's decision to participate in the soybean market and its access to market information, ownership of a radio and or cellphone. By accessing extension agents, farmers get advice on good agronomic practices, improved technologies and market prices. We therefore expect the sign to be positive when farmers have access to extension agents and negative otherwise. According to Zingore et al. (2007), ownership of cattle is a major determinant of the timeliness of agronomic operations. We assume that the resource-endowed farmers may use their livestock for traction to till larger pieces of land and for transportation to the market. According to Alene et al. (2008) and Zingore et al. (2007) cattle ownership has a wealth effect, in that those households who own animals are more likely to use fertilizers than those without. The resource endowed households are also more likely to have cash resources to finance basal fertilizer purchases, inoculants and improved soybeans germplasm (Zingore et al., 2007). Varia (2011), notes that resource constrained smallholder farmers lack access to finance, give less priority to their non staple crops and use poor agronomic practices. The combined effect of these factors is very low yields and low market participation compared to the commercial farmers who have higher use of herbicides and fertilizers. We thus expect a positive relationship between wealth (resource endowment) and intensity of market participation as such households are more likely to have higher marketable surplus. According to Alene et al. (2008), access to agricultural extension services enhances market participation and marketable surplus as agents provide technical assistance and information on improved varieties and technologies.

Extension agents are the information exchange platform between research and farmers; they decode information from researchers into a format understandable by farmers and also provide feedback to the researchers. These results were also observed by Siziba et al. (2011), who noted that access to extension services reduces farmers risk perceptions and thus improve market participation. We thus expect a positive relationship between access to extension services and market participation in soybean markets.

RESULTS AND DISCUSSION

Sample characterization

The household survey results in Table 2 show that only 28.88% (54 out of 187 farmers) of the sampled households participated in the soybean market. The average marketable surplus for households that participated in the soybean market is 211.26 kg. These results are consistent with findings by Ojiem et al. (2007) and Giller et al. (2006) who note that soybean output is very low in smallholder farming communities largely because farmers apportion at most 5% of their land to legumes and do not fertilize them leading to low yields. The low levels of marketable surplus could also be a

Table 2. Description of sample household and socioeconomic characteristics.

Parameter	Market participants	Non market participants	p-values
Sample n (prop)	54 (28.88)	133 (71.12)	
Head age (years)	43.76(12.96)	50.43(16.60)	0.0089
Household size	5.33(3.16)	5.18(2.77)	0.7433
Head education (% prop with secondary)	59.26(0.50)	46.62(0.50)	0.1184
Farming experience (no. of years)	15.13(11.94)	20.42(15.51)	0.0257
Gender (%prop of male)	75.93(0.43)	79.7(0.40)	0.5709
Own mobile phone (%prop)	68.52(0.47)	63.91(0.48)	0.5511
Own Radio (% prop)	68.52(0.47)	52.63(0.50)	0.0469
Number of cattle owned	2.35(3.46)	2.35(3.61)	0.0027

result of low input usage and the substitution of commercial high value varieties with low yielding locally available varieties. The results show that the average household head for market participating households (43.76) is significantly lower with a standard deviation of 19.96 than that of non-participating households (50.43) that has a standard deviation of 16.60 and this is significant at 1% level of significance. The probability of younger farmers to participate in soybean market is higher than that of older farmers. The results from the survey show that amongst the market participating households, 75.93% are male headed while 79.70% of the non-market participating households are male headed. Since the p-value is 0.5709, there is thus no statistically significant difference between the two groups of soybean farmers.

Results for the average household sizes show that the mean household size for market participants is 5.33 with a standard deviation of 3.16 while that for non-market participants is 5.18 with a standard deviation of 2.77. Although the household sizes were slightly lower than the national average household size of six, the p-value of 0.7433 indicates that there were no significant differences in household sizes between the market participating and non-market participating farmers. In terms of farming experience, there were statistically significant differences observed between soybean market participating households and the non-market participants at 5% level of significance. Households that participated in the soybean market on average had 15 years of farming experience compared to their counterparts with over 20 years. The 2 sided t test results show that the difference in farming experience is statistically significant at 5% level of significance. This implies that the probability of less experienced to market soybean is very high.

The results also show that 68.51% (standard deviation 0.47) households who participated in the soybean market owned radios while 52.63% (standard deviation 0.50) amongst non-market participants owned radios. Since the p-value is 0.0469, we observed significant differences between the two groups at 5% level of significance. This

means that ownership of radios is common among market participating households than non-market participating households. As such, owning a radio increases the probability of marketing soybeans.

Although, we estimated that 68.5% of the soybean market participating households owned cellphones with a standard deviation of 0.47 compared to 65% with a standard deviation of 0.48 for non-participating households; the p-value of 0.5511 shows that there were no statistically significant differences in the proportions. This suggests that cellphone ownership is not a determinant of soybean market participation among the smallholder farmers.

Econometric results

The results from the econometric analysis for the market participation (Probit Model results) and intensity of market participation (OLS regression model) are presented here. Intensity of market participation is estimated conditional on the smallholder farmers' market participation decision.

Factors affecting soybean market participation

Table 3 presents the OLS results for intensity of market participation and the Probit model results for smallholder farmers' decision to participate in the soybean market. The OLS regression model estimates the factors affecting the intensity of participation in a soybean market while the Probit model estimates the determinants of the dichotomous soybean market participation variable.

Selection model results (Probit model results)

The results in Table 3 show that for the Probit model, gender of household head, ownership of a radio, access to agricultural extension services, use of inoculants and use of improved soybean seeds affect the farmers

Table 3. OLS and Probit Estimates for soybean market participation and intensity of participation.

Dependent variable	Probit (selection model) (soybean bean market participation		OLS (outcome) (Amount of soybean sold)	
	β	p-value	β	p-value
Gender	-0.847	0.004***	6.249	0.345
Head age	-0.063	0.172	0.210	0.536
Head age squared	0.000	0.488	0.004	0.490
Household size	0.045	0.511	-0.403	0.256
Farming experience	-0.001	0.939	-0.367	0.183
Ownership of cattle	0.003	0.283	0.023	0.094*
Distance to market	-	-	3.921	0.014**
Own radio	0.672	0.0060***		
Own cellphone	0.003	0.992		
Access to extension	0.4185	0.086*		
Used Inoculants	0.894	0.016**		
Use improved seed varieties	0.684	0.041**		

*** Significant at 1% level; ** significant at 5% level; * significant at 10% level.

decision to participate in the soybean market as a seller. The gender of the household head negatively influences the likelihood of smallholder farmers' participation in the soybean output market, that is male headed households are less likely to participate in soybean markets than female headed households. The probable explanation is that in Guruve district as in other parts of Zimbabwe, most legumes are culturally viewed as women's crops. These results are consistent with the findings of Alene et al. (2008) for Kenya but contrary to the findings of Cunningham et al. (2008) in a study on gender differences in marketing styles in western Oklahoma. Ownership of a radio, which represents access to a communication asset positively and significantly, influences a smallholder farmer's likelihood of participating in the soybean market. It represents access to formal sources of market information that increases the likelihood of market participation. In Zimbabwe, radio stations frequently air broadcasts on rainfall patterns, crop varieties and input and out prices. Access to this information lowers the transaction costs and road accessibility to the market. According to Siziba et al. (2011) access to such information reduces smallholder farmers risk perceptions and improves the likelihood of participating in the soybean market. These results are consistent with the findings of Siziba et al. (2011) on cereal market participation in southern Africa. Access to agricultural extension agents positively influences the likelihood of participating in soybean markets. The results demonstrate the importance of improved technology and support services in promoting soybean market participation. The likely explanation for this is that agricultural extension workers are the bridge between research programmes and farmers. They

provide information on good agronomic practices, production technologies, soybean varieties and market information. This interaction is likely to improve productivity, marketable surplus and enhance a smallholder farmer's likelihood of participating in a market. These results are consistent with the findings of Alene et al. (2008).

The use of rhizobial inoculants in the production of soybeans by smallholder farmers in Guruve district is significantly positive and increases likelihood of participating in the soybean market. The likely explanation for this is that rhizobial inoculants increase average yield and total soybean production with lower costs than using inorganic fertilizers (Chanaseni and Kongngoen, 1992). Thus the results show that smallholder farmers who used rhizobial inoculants for soybeans had a higher likelihood of participating in soybean markets than their counterparts. Similarly, the use of improved soybean seed varieties has a significantly positive influence on soybean market participation by smallholder farmers. The likely explanation is that improved seed varieties (germplasm) have high yield potential and are disease and pest resistant thus improve productivity and marketable surplus (Technoserve, 2011).

OLS regression model results

The results for the OLS regression model are shown in Table 3. Livestock wealth (cattle owned) and average distance to the market explained the intensity (amount of soybean sold) of smallholder farmers' participation in soybean market. Number of cattle owned had a positive and significant influence on the intensity of market participation conditional on market participation. The probable explanation is that resource endowed households have more cattle which they can use for traction and transportation, a development which reduces production and market related transaction costs. The resource endowed households are likely to have finances from which they are able to hire labor, purchase inoculants, buy improved soybean germplasm and thus can grow soybeans on bigger pieces of land compared to the resource constrained smallholder farmers. Furthermore, households who own cattle are more likely to use good agronomic practices to produce their soybean. Resultantly, this will increase yield and marketable surplus. These results are consistent with the results of Alene et al. (2008). Zingore et al. (2007) noted that resource endowed farmers had higher yields in their fields compared to resource constrained farmers.

Distance to the market positively and significantly influences the intensity of soybean market participation by smallholder farmers. This means that as distance to the market increases, the amount of soybean sold by smallholder farmers also increases. These results are in contrast to findings from studies on staple crops in which distance negatively influences smallholder farmers' intensity of market participation (Siziba et al., 2011; Alene et al., 2008, Makhura et al., 2001; Key et al., 2000). A common finding in all these studies is that as distance from the market increases, variable transport costs increase and this discourages resource constrained smallholder farmers from selling high volumes. However, a possible explanation for the Zimbabwean case is that, local buyers offer very low prices compared to well established distant buyers. This is so because established soybean buyers are based in Harare, which lies over 151 km from the study sites. As such most farmers are set to benefit from price differentials between local prices and prices in distant markets.

Conclusion

This article did set out to identify through empirical evidence the determinants of soybean market participation and further evaluate the factors that affect intensity of market participation by smallholder farmers in Guruve district of Zimbabwe. This study used cross sectional household data of 187 randomly selected smallholder farmers in Guruve district in Zimbabwe. Econometric analysis was done using the Heckman model with sample selection, which corrects for selection bias at market participation decision by smallholder farmers. Choice of covariates for the OLS and Probit was guided by economic theory, literature and in some cases intuition. Descriptive results from the survey show that only 28.88% of the survey households participate in soybean market. The market participating households

averagely sold 211.26 kg of soybean. Most of the market participating households owned communication equipment such as radios (68.52%) and had bigger land sizes (3.52 ha) compared to the non-participating households. The econometric analysis results from this study show that for the OLS model, livestock wealth or resource endowment and distance to the market have positive influence on marketed surplus. However, for the Probit model, only gender negatively influences the smallholder farmers' decision to participate in soybean market while household ownership of a radio, access to agricultural extension, use of rhizobial inoculants and use of improved soybean varieties have a positive influence on household's likelihood to participate in the soybean market.

Based on these findings from the analysis of the factors affecting soybean market participation by smallholder farmers in Guruve district, we recommend that policy makers can improve farmer to extension worker ratio as this will improve access to technical information and support services on improved technologies such as use of inoculants, biological nitrogen fixation and knowledge on improved soybean seed varieties. Furthermore, policy makers could improve the dissemination of market information as it is currently available through radio broadcasts. Access to market information would improve farmers' knowledge of markets and aid in decision making on market participation as well as the level of marketed surplus. This will lead to increased productivity, high marketable surplus and enhances the likelihood of participating in the soybean market.

REFERENCES

Alene AD, Manyong VM, Omanya G,Mignouna HD, Bokanga M, Odhiambo G (2008). Smallholder market participation under transactions costs: Maize supply and fertilizer demand in Kenya. Food Pol. 33(4):318-28.

Barrett CB (2008). Smallholder market participation: Concepts and evidence from eastern and southern Africa. Food Policy. 33:299-317.

Barrett CB, Swallow BM (2006). Fractal Poverty Traps, World Development. 34(1):1-15.

Binam JM, Abdoulaye T, Olarinde L, Kamara A, Adekunle A (2011): Assessing the Potential Impact of Integrated Agricultural Research for Development (IAR4D) on Adoption of Improved Cereal-Legume Crop Varieties in the Sudan Savannah Zone of Nigeria. J. Agric. Food Info. 12(2):177-198.

Chanaseni C, Kongngoen S (1992). Extension programs to promote rhizobial inoculants for soybean and groundnut in Thailand. Can. J. Microbiol. 38:594-597.

Chianu JN, Ohiokpehai O, Vanlauwe B, Adesina A, De Groote H, Sanginga N (2009). Promoting a Versatile but yet Minor Crop: Soybean in the Farming Systems of Kenya. J. Sustain. Dev. Afr. 10(4):1-21.

Cunningham LT, Brown BW , Anderson KB, Tostao E (2008).Gender differences in marketing styles.Agric. Econ. 38(1):1-7.

Estehuizen D (2011). Zimbabwe GAIN Annual Report. Global Agricultural Information Network. USDA Foreign Exchange Service. pp. 1-12.

Fafchamps M, Hill RV (2005). Selling at the Farmgate or Traveling to Market. Am. J. Agric. Econ. 87(3):717-734.

Franzel F, Redfelder E, Lang AG, Buchner A (2007). G*Power: A flexible Statistical power analyses program for social, behavioural and biomedical sciences. Behav. Res. Methods 39(2):175-191.

Giller KE, Rowe E, de Ridder N, van Keulen H (2006). Resource use dynamics and interactions in the tropics: Scaling up in space and time. Agric. Syst. 88:8-27.

Goetz SJ (1992). A selectivity model of household food marketing behavior in Sub-Saharan Africa. Am. J. Agric. Econ. 74(2):444-452.

Greene W H (2003). Econometric Analysis, Fifth edition. Pearson Education International, USA.

Holloway G, Nicholson C, Delgado C, Ehui S, Staal S (2000). Agroindustrialization through institutional innovation: transactions costs, cooperatives and milk-market development in the east African highlands. Agric. Econ. 23:279-288.

Huang C, Raunikar R, Misra S (1991). The application and economic interpretation of selectivity models. Am. J. Agric. Econ. 73(2):496-501.

IFAD (2003). Promoting Market Access for the Rural Poor in Order to Achieve the Millennium Development Goals. Discussion Paper. Rome.

Jagwe J, Machethe C, Ouma E (2010). Transaction costs and smallholder farmers' participation in banana markets in the Great Lakes Region of Burundi, Rwanda and the Democratic Republic of Congo. Afr. J. Agric. Res. 6(1):1-16.

Jayne TS, Jones S (1997). Food marketing and pricing policy in eastern and southern Africa: a survey. W orld Dev. 25(9):1505-1527.

Key N, Sadoulet E, de Janvry A (2000).Transactions costs and agricultural household supply response. Am. J. Agric. Econ.82(1):245-259.

Mabika V, Mariga IK (1996). An overview of Soyabean Research in smallholder farming sector of Zimbabwe. In:Soyabean in the smallholder cropping systems in Zimbabwe. Mpepereki, S., Giller K.E and Makonese, F. (Eds). Government Printers Zimbabwe. Pp.12-17.

Madanzi T, Chiduza C, Kageler SJR, Muziri T (2012). Effects of different plant populations on yield of different soybean (Glycine max (L) Merrill) varieties in a smallholder sector in Zimbabwe. J. Agron.11(1):9-16.

Makhura MN, Kirsten J, Delgado C (2001). Transaction costs and smallholder participation in the maize market in the Northern Province of South Africa. Seventh Eastern and Southern Africa Regional maize conference, 11-15th February. P. 463.

Ojiem JO, Vanlauwe B, de Ridder N, Giller KE (2007). Niche-based assessment of contributions oflegumes to the nitrogen economy of W estern Kenya smallholder farms. Plant Soil 292:119-135.

Okoyo BC, Onyenweaku CE, Ukoha OO (2010). An Ordered Probit Model Analysis ofTransaction Costs and Market Participation by Small-Holder Cassava Farmers in South-Eastern Nigeria.MPRA Munich Personal RePEc Archive MPRA Paper No. 26114, posted 22. October 2010. http://mpra.ub.uni-muenchen.de/26114/

Omiti JM (2009). Factors affecting the intensity of market participation by smallholder farmers: A case study of rural and peri-urban areas of Kenya. Afr. J. Agric. Resour. Econ. 3(1):57-82.

Phiri NA, Otieno G (2008). Report on managing pests of stored maize in Kenya, Malawi and Tanzania. Nairobi: The MDG Centre, East and Southern Africa.

Pingali P, Khwaja Y, Meijer M (2005). Commercializing small farmers: Reducingtransaction costs. FAO/ESA W orking Paper No. 05-08. FAO, Rome, Italy.

Shumba-Munyulwa D (1996). Soyabean in semi arid Zimbbawe: Production and Utilisation. Proceedings of the preparatory workshop on research into promiscouos nodulating soyabeans. February 8-9, University of Zimbabwe, Harare. pp. 18-21.

Siziba S, Nyikahadzoi K, Diagne A, Fatunbi AO, Adekunle AA (2011). Determinants of cereal market participation by sub-Saharan Africa smallholder farmer. Learning Publics J. Agric. Environ. Stud. 2(1):180-193.

Technoserve (2011). Southern Africa Regional Soybean Roadmap, Zimbabwe value chain analysis. Technoserve.

Varia N (2011). Technical Report: Soybean Value Chain. USAID Southern Africa Trade Hub.

Woomer PL, Balume I, W afullah, Zamasiya B, Mahamadi D, Chataika B (2013). Report on Market Analysis of inoculant production and use. Putting Nitrogen Fixation to work for smallholder farmers in Africa. www.N2Africa.org.

Zingore S, Murwira HK, Delve RJ, Giller KE (2007). Influence of nutrient management strategies on variability of soil fertility, crop yields and nutrient balances on smallholder farms in Zimbabwe. Agric. Ecosyst. Environ. 119(1-2):112-126.

14

Financial constraints and entrepreneurial activity choice among clients of micro finance institutions in Jimma area

Misginaw Tamirat

Department of Agricultural Economics, Jimma University College of Agriculture and Veterinary Medicine, P. O. Box 307, Jimma Ethiopia.

The research is intended to assess the small holder entrepreneurs' enterprise choices under financial constraint. Adapting economic model of household-production interactions, results from a survey of 140 smallholders was used on multinomial logit regression techniques. The paper makes the case that the access to finance has limited effect on the choice of entrepreneurial activity than individual differences did. It was also found that majority of the problems the entrepreneurs faced have no significant association with access to credit rather with macroeconomic and institutional factors. There has also been strong association of human capital, physical and social capital with entrepreneurial activity choice, implying enhancement of smallholder's entrepreneurship need to take into account other socio-economic factors besides the access to credit. The activity analysis has also showed that there is an out-flock of entrepreneurs from agriculture to non agricultural sector which would have a critical implication on the country's endeavor to food security. Generally, the study reveals sets of key variables relevant to the smallholders' entrepreneurial activity choice, and provides an evaluation of intensity of the effects of the variables. The paper concludes by bringing these critical insights to bear on possibilities for designing microfinance programs that would help flourish smallholder entrepreneurship which would gear towards realization of the country's long run development plan.

Key words: Entrepreneurship, activity choice, micro finance.

INTRODUCTION

Background and justification

Sub-Saharan Africa as a whole remains the world's most technologically backward, food-insecure and politically instable region with a considerable part of the population remain undernourished. However, recently, countries like Ethiopia have been growing at a relatively fast rate, which in turn has led to improvements in several areas such as trade, mobilization of government revenue, infrastructure development, and the provision of social services (UNCTAD, 2012). Nonetheless, sustainability of the economic progress and diversification of potential sectors requires technical progress tailored to the country's varied agro-ecologies, development of supporting institutions and moreover boosting entrepreneurial skill of the smallholders[1]. It has been suggested as way to break the poverty trap is to encourage petty entrepreneurship among the poor, in order to foster production surpluses and hence economic progress in the region (Khalid, 2003; Naude, 2010).

Currently, global development is entering a phase where entrepreneurship will increasingly contribute to economic development by facilitating self employment, income distribution and competition. Entrepreneurship in this context pertains to the actions of a risk taker, a creative venture in to a new business or the one who revives on existing business. The rapid ascent of emerging markets has sparked a renewed interest in understanding the role entrepreneurs play in shaping the transformation of developing countries, and what determines smallholders' entrepreneurship (Andre´ van et al., 2005; Antoinette, 2009). There is ample evidence that entrepreneurship is a key factor for economic development by carrying out innovation specifically in the flourishing of small businesses (Levine 1997; Naude, 2010).

In contrast to the old 'top down' development, the current approach which emerged over the past decade is the development 'from below'. This approach assumes that development is based on stimulating local entrepreneurial talent and subsequent growth of indigenous companies. Despite several interventions by the government and various development practitioners to improve the livelihoods of smallholders, in Ethiopia, the issue of small scale entrepreneurship development remains a key challenge (Khalid, 2003). This partly related with the fact that most policy makers as well as researchers treat entrepreneurs as a homogeneous group of actors that are uniformly affected by economic conditions or policy interventions. This view misses very fundamental differences among the types of entrepreneurs (particularly smallholders) who choose to be engaged-in varieties of activities (businesses), which affect the economy in various ways.

Virtually all of the literatures on factors facilitating entrepreneurship development noted that financial constraints are one of the biggest concerns impacting potential entrepreneurs around the world (Khalid, 2003; Beck et al., 2009; Popescu and Crenicean, 2012). Studies have shown that the relevance of credit to entrepreneurial activity choice depends on the individual level differences than macroeconomic conditions or access to finance, that is, the attitudes, skills and actions of smallholder producers (Sanyang and Huang, 2010; Popescu and Crenicean, 2012). Impact studies of microfinance institutions on development have concentrated on assessing the effects of credit programs on borrowers' as individuals, and as members of their household and enterprises wellbeing, largely overlooking the effects of financial access on choice of entrepreneurial activities.

Though, there is solid empirical evidence that improved access to credit spurs enterprise growth; little is known about what type of enterprises are preferred by smallholders and what factors influence entrepreneurial activity choices of borrowers.

Reviews of literatures on microfinance and economic growth display several dimensions of financial constraints but few were concerned with the association of access to finance and entrepreneurial activity choice. Moreover, there is hardly any work on the relationship between entrepreneurial activity choice and loan utilization among smallholders. Therefore, this paper characterizes the entrepreneurial behavior of smallholders based on their access to microfinance among the clients of Eshet, Harbu and OCSCO (Oromia Credit and Saving Share Company) in South western Ethiopia.

RESEARCH METHODOLOGY

Description of the study areas

The study was conducted in Jimma zone, which is one of the 13 administrative zones in Oromia Regional State. Jimma zone is one of the major coffee growing zones in the country; currently the total area of land covered by coffee in the zone is about 105,140 ha, which includes small-scale farmers' holdings as well as state and private owned plantations. The Zone accounts for a total of 21% of the export share of the country and 43% of the export share of the Oromia Region (Anwar, 2010). The survey considered smallholder households that are the clients of Eshet, Harbu, OCSCO (Oromia Credit and Saving Share Company) microfinance institutions in *Seka, Yebu* and *Agaro* districts.

Data and data sources

A community based cross-sectional study design was employed based on the framework of household production model. The data for the study were collected from secondary and primary sources. The secondary data were collected from documentations of the financial institutions surveyed, and District and Zone Finance and Economic development office. In the survey both formal and informal methods were employed to collect the required information from clients of the microfinance, and key informants. Self administered semi structured questionnaire and individual interviews using the pre-tested questionnaire were made to generate the household level data.

Sample size and method of sampling

A multi-stage mixed sampling procedure was adapted for selecting the sample of borrower, in which a two stage purposive sampling (to select the districts and the FA[2]s) followed by random sampling techniques (to select the households) was used. The sample districts were selected based on secondary information with the help of knowledgeable people about the area and information from the microfinance institutions. Three major FAs from each district were then identified based on distribution of the microfinance institutions, and accessibility. From total of these nine FAs,

---

[1] the dominant social/economic group in Ethiopia

[2] Farmers' Association

proportional to the population (clients of the microfinance institutions), 140 households were selected for the study.

Methods of data analysis

Descriptive analysis and econometric analysis were used for analyzing the data.

Descriptive statistics analysis

Descriptive and inferential statistics were applied in documentation of the basic characteristics of the sampled clients along with the portrayal of entrepreneurial activity in the area. This employed use of descriptive statistics. The study also tested variables individually whether they had an effect on entrepreneurial utilization of credit using the Chi-square, F- test and t-tests.

Econometric analysis

In household production, model households are basic economic units making a number of decisions in their day to day life. To analyze factors that determine household's choice of entrepreneurial activity, the multinomial logit model was used. Multinomial logistic regression is used to analyze relationships between a non-metric dependent variable and metric or dichotomous independent variables. Based on Liao (1994), when a single dependent variable takes on three or more discrete and/or when their natural ordering is not clear then the responses are usually called multinomial responses.

The multinomial version of these models has logit and probit specifications. But the multinomial logit model is preferred, not only because of its computational ease but also it is based on basic economic theory of utility maximization (Liao, 1994). The model is derived from random utility function (McFadden, 1973). In random utility model it is assumed that individuals maximize their utility by choosing one of the alternatives available to them. In this case, it is assumed that the borrower maximizes his/her utility by choosing one among the available mutual exclusive alternative to invest their return from microfinance institutions.

Specification of multinomial logit model

The specification of multinomial logit probability model is given below: First, let j denotes a given discrete business alternative for the borrower, which takes the value from 0 to 2 whereby; j = 0) represent household's choice to support [3]their job before the loan; j = 1 represents households who diversified their businesses; j = 2 represents household who begin new business. Then, choosing the j = 0 as standard regime and assuming that the sum total of probabilities of all the three entrepreneurial alternatives must be unity. Using the unordered random utility model specification used in Wooldridge (2002), the model for the ith respondent faced with j choice presented as follows: Suppose that the utility of choice j is:

$$U_{ij} = X_{ij} + \varepsilon_{ij} \qquad (1)$$

In general, for an outcome variable with J categories let the jth business strategy that the ith household chooses to maximize its utility could take the value 1 if the ith household chooses j^{th} entrepreneurial alternative and 0, otherwise. The probability that a household with characteristics X chooses business option (entrepreneurial alternative) j, P_{ij} is modeled as:

$$p_{ij} = \frac{\exp(\beta_j' X_i)}{\sum_{j=0}^{J} \exp(\beta_j' X_i)}, \quad J = 1, 2, 3 \quad i = 1, 2 \ldots n \qquad (2)$$

Where: P_{ij} = probability representing the ith respondent's chance of choosing entrepreneurial option j, X_i = Predictors of response probabilities, β_j = Covariate effects specific to j^{th} response category with the first category as the reference.

$\beta_1 \ldots \beta_J$ are m vectors of unknown regression parameters (each of which is different, even though X_j is constant across alternatives). By setting the last set of coefficients to null (that is, $\beta_J = 0$), the coefficients β_i represent the effects of the X variables on the probability of choosing the Jth alternative over the reference category. In fitting such a model, J −1 sets of regression coefficients are estimated.

RESULTS AND DISCUSSION

Demographic and socio-economic characteristics

Age and gender of the respondents

Tables 1 to 5 present the summary statistics of several key household variables. The results of the household survey show that the mean age of the respondents was 39 years with average 9 years of working experience in the main occupational activities. The mean age for male respondents was 39 (n = 97) and female's was 41 (n = 43). With regard to nature of the business (whether respondents have changed their main occupational activity because of the credit) and its relation to age, respondents were grouped into two categories, where households opted to expand their main occupation (59%) are found to be greater than those undertook new activity (41%). As displayed in Table 1, the mean age 42 years for the former category is significantly greater than the later (35 years), which may imply that older people are either reluctant to take on new businesses.

As shown in Table 2, from a total of 140 respondents, 69% were male and 31% were female. Majority of male respondents (76%) have changed their main occupation as a result of the loan, while 59% of female were found to change their occupation, though no significant association was observed between gender and change of entrepreneurial activity. Female's lower propensity to change business (as compared to the male counterpart) is more likely due to their lack of access to information, or lack of appropriate incentives to act on the information as well as restricted decision power on some basic resources.

Educational status of respondents

Entrepreneurship is a high risk investment, and as such only non-risk-averse individuals are likely to begin new

---

[3] This may include loan use to expand the enterprise or to compensate the loss in the previous period and continue on the same business

Table 1. Demographic and economic characteristics of the households by nature of business.

Parameter	Nature of the business		Average	t-value
	New venture mean (SD)	Expansion mean (SD)	Mean (SD)	
Age	35(9)	42(10)	38(11)	3.6***
Family size (Adult equivalent)	3.4(2.1)	4.2(2.3)	3.8(2.2)	1.98**
Formal education level/grade	7.28(3.4)	7(3.8)	7(3.6)	0.55
Land size (Hectares)	1.3(1.4)	1.6(1.8)	1.5(1.6)	1.06

***, ** statistically significant at 1%, 5% significance levels, SD = standard deviation.

Table 2. Gender and pre-loan economic activity of respondents by the nature of business.

Description		Nature of the business		Chi-square value
		New venture (N %)	Expansion (N %)	
Main occupation	Crop dominated livestock (I)	9	26	13.3***
	Pity trade dominated livestock (II)	15	12	
	Pity trade dominated crop (III)	10	14	
	Crop dominated pity trade (IV)	7	5	0.46
Gender	Female	11	20	
	Male	30	39	

***, statistically significant at 1% significance levels.

Table 3. Respondent's characteristics and post loan entrepreneurial activity.

Parameter	Entrepreneurial activity			Average	F - value
	Agriculture mean (SD)	Pity trade mean (SD)	Off-farm mean (SD)	Mean (SD)	
Age	43(10)	37(10.2)	35.6(9)	38(11)	7.3***
Family size (Adult equivalent)	4.5(2.5)	3.5(1.9)	3.4(1.2)	3.8(2.2)	3.6**
Formal education level/grade	7.4(3.9)	7.5(3.8)	6.2(3.5)	7(3.6)	1.5
Land size (Hectares)	2.3(2.3)	0.85(0.45)	0.88(0.46)	1.5(1.6)	8.1***

***, **, statistically significant at 1%, 5% significance levels respectively, Adult equivalent = AE, Hectares = ha.

Table 4. Relationship between nature of the business and the major problems faced.

Major problems	Nature of the business		Total %	Chi-square value
	New venture N (N %)	Expansion N (N %)		
Inadequate training/inception	11(8)	26(19)	27	
Poor follow up/support	6(4)	22(16)***[a]	20	13.254***
Limited marketing support	30(21)***[b]	21(15)	36	
Liquidity constraint	11(8)	13(9)	17	

*, **,* statistically significant at 1%, 5%, 10% significance levels.

venture (Miner and Raju, 2004). Education influences the selection to become an entrepreneur through various mechanisms. Primarily, human capital influences occupational choice and performance patterns within occupations. Mean educational attainment of household heads was 7 years of schooling, and 77% of respondents were found literate (Table 1). The survey found that women without formal education out-number men in the

Table 5. Relationship between nature of the business and Loan cycle.

Loan cycle	Nature of the business		Total N%	Chi-square value
	New venture (N%)	Expansion (N%)		
1st	17	14	31	
2nd	4	14***a	18	18.31***
3rd	6	20***a	26	
4th	15***b	10	25	

***, statistically significant at 1%, significance level, ***a there are significantly higher numbers of respondents expanded their business within 2 to 4 cycle than the other category, **b those who have embarked on new business in their 1st and last season are significantly higher than the ones did not.

same category. However, women's average formal education (7.41 years) is greater than that of male (6.21 years). This possibly is because of the fact that micro finances target on improvement of disadvantaged social classes; however males with higher education level mostly have further prospects elsewhere. In the independent sample t-test analysis, the average year of formal schooling for the ones changed their main occupation (7.3 years) was greater than (6.9 years) the respondents did not change their occupation.

Farm size and land tenurial status

The farm size was expressed in terms of amount of land actually cultivated in any farming season. the result showed that 16% (n = 23) respondents did not have title to land, among the respondents having land use right (83%) worked on pieces of land less than two hectares, only twelve (one percent) worked on more than two hectares of land. As illustrated in Table 1, the average farm size in the sampled households is 1.7 ha.

The evidence on the relationship between land size and change of business activity because of the loan demonstrates (Table 1), the respondents that changed their main occupation have less average land holding (1.3 ha) than the respondents opt to expand their existing business (1.6 ha). However, no or weak statistically significant correlation has been observed between land holding and the nature of the entrepreneurial activity undertaken.

Main occupational activities

The result of the survey on main occupational activities prior to the loan recognized four main activities. These are crop dominated livestock production, petty trade dominated crop production, petty trade dominated livestock production and crop production dominated petty trade. However, the activities are not mutually exclusive, for some of respondents simultaneously engaged in two or more occupations in varying degrees. As shown on

Table 2, out of 140 household heads, 41% reported crop dominated livestock as their main occupational activity and 26% exercised petty trade dominated livestock, 24% engaged in petty trade-dominated crop farming and while 9% crop farming-dominated petty trade as their main source of income. As evident from Table 2, due to the loan, 40% of crop dominated livestock, 53% of petty trade dominated livestock production, 36% of petty trade dominated crop farming, and 31% of the crop dominated petty trade have changed their main businesses. The analysis of response on change of business entails that only less than a third of the respondents preferred to be engaged on agricultural production, while the remainders resorted to pity trade with varying degree of intensity. The out flock of entrepreneurs from agriculture to pity trade has a remarkable implication on the country's endeavor to food security and curbing food price hicks even though it widens the economic pillars. The activity shift from agriculture to non agriculture sector increases the general consumers while decreasing the number of food producers at least in the short run which may end up hiking up food prices.

The main occupations the respondents engaged-in after the loan are displayed in Table 2, where a significant association has been observed on age, family size and landholding of the respondents with the entrepreneurial activity engaged-in because of the loan. As it is noticeable in Table 3, household heads in the agriculture businesses (average of 43 years) are significantly older than that of petty trade (37 years) and off-farm activities (35 years) implying the scarcity of productive labor and product in agriculture sector. Similar to age of the respondents, the family size (4.5 AE) and agricultural land size (2.3 ha) in agricultural activity of the respondents was significantly higher than the ones in petty trade with 3.5 AE and 0.85 ha, and that of off-farm activity was 3.4 AE and 0.88 ha, respectively.

The main product the micro finances offer in the survey areas is a group-liability loan, followed by saving. Groups are formed by average of 6 to 10 members who agree to mutually guarantee the reimbursement of their loans. The loan size increases by 50% as the client progresses from one loan cycle to the next. The loan amount ranges from

1,000 Br to 5,000 Br per member depending on loan purpose and length of client ship. The entrepreneurs take on the following three major activities after the loan:

i. Petty trade (local food and drinks processors, cart transport, small hotels and tearooms and other retail activities) accounted 44%;
ii. Off-farm loan (handicraft, cattle fattening, cereal vending) 32%;
iii. Agricultural (purchase oxen, dairy) 24%.

As displayed in Table 4, respondents' rank the major problems encountered in the business decision process, where limited marketing support (36%) was the dominant problem followed by inadequate training during inception of the business (27%). There was a significant association between the problem faced and the entrepreneurial activity adapted. The number of respondents that reported limited 'market support' as a dominant problems are significantly larger under new business option than the ones expanded their older job, whereas respondents who ranked 'poor follow up' as a major problem are significantly greater in the group expanded their older business than the ones embarked on new ventures. This implies that majority of the problems faced by the entrepreneurs are not directly related to finance, but rather arise as a result of weak institutional support and linkages. This may mean for financial institutions to integrate their training and monitoring with extension and marketing services of concerned stakeholders.

Significant association was observed between lengths of participation in the lending program (loan cycles) and the nature of entrepreneurial activity tailored. Table 5 reveals that new clients have changed their business more proportionally than the relatively established clients. The evidence from Table 5 shows that majority of respondents (17%) that get on new ventures managed to change their activity on the first loan cycle. Possibly it is related with lack of entrepreneurial skill (poor training during inception and follow up). Focus group participants indicated that most of the clients are doubtful whether their business would be able to pay the debt, mainly in the first season. However, according to key informant's discussion, based on the performance form the first season, in the second and third years of their client ship, borrowers look for entrepreneurial solutions for their businesses.

Determinants of clients' choice of entrepreneurial activities

Prior to conducting the analysis multicolliniarity among the explanatory variables was checked so that the parameter estimates will not be seriously affected by the existence of multicolliniarity among variables. The

variables were tested for hetroskedastisity and the test rejected for all variables, the null that there is a significant difference among the variables in the same group variances. Besides, practicality of the multinomial logit model depends on the independence of the alternatives (Liao, 1994).

In order to check the independence of the alternatives Haussmann's specification test of independence was undertaken. The test did not reject the null hypothesis of independence of the included business options suggesting there is no evidence against the specification. Also, because of the Haussmann's endogeneity test 'income of the household head' is left out of the model for it is endogenous with occupational choice. Finally, as shown in Table 6, the estimated model fitted the data reasonably well; the likelihood ratio tests indicated that the slope coefficients were significantly different from zero at less than 1% significance level.

DISCUSSION OF RESULTS

Age

The age of the household head measured in years is a continuous variable implying experience in his/her main activity. It was found to be significantly and positively related to diversification option but negatively related to business expansion option. This positive sign entails that, keeping all other variables constant, the likelihood of diversifying the business at hand increases as the age of household head increases, as compared to expanding/sustaining the business. Whereas the negative sign in the expansion column imply that an increase in age is negatively related with the probability of expanding businesses as compared to staying in the same business (Table 6).

This is principally, at older ages the physical ability of the household head decreases to manage the available business let alone to expand it; however households diversify their enterprises to sustain that level of income which may support livelihood of the family. Furthermore, in relation to an increase in age, social responsibility shares the time otherwise would have been used for the main occupation. Additionally, individuals who have stayed for long in a business may establish a goodwill or social capital (regular client) in the business which they are less willing to change because of loss of their regular clients and fear of institutional arrangement in the new business.

An increase in the age of the household head by a year increases the odds of choosing the diversification option increases by 16% and the likelihood of expanding the business at hand decreases by 1.2%. Corresponding to this, Sinha (1992) also elucidated that older people are risk averse and choose to widen their means of guaranty. This result is also consistent with standard job-shopping

Table 6. Multinomial logit estimates of determinants of clients' entrepreneurial options (Marginal effects).

Explanatory variables	Expansion $\frac{\partial y}{\partial x}$ (Z value)	Diversify $\frac{\partial y}{\partial x}$ (Z value)	NewBizz $\frac{\partial y}{\partial x}$ (Z value)
AGE (age number of years)	-0.010(-1.19)	0.016(2.03)**	-0.005(-0.51)*
EDULEV (education in years of formal schooling)	-0.048(1.24)	0.01(2.19)**	-0.05(1.2)*
FAMSZ (family size in Adult equivalence)	0.121(0.78) ***	.000(2.76)***	-0.121(-2.69)
TLU (livestock In TLU)	-0.001(-0.27)	-0.002(-0.41)	0.003(0.64)
Land (farmland in hectare)	-0.000(-0.91)	0.000(2.8)	-0.000(-1.19)
CRIS (expected risk in probability)	-0.049(-1.34)	0.03(0.79)	0.019(-0.80)
Social expenditures in Birr	0.094(-1.69)	-0.149(-2.95)**	0.055(0.52)
Marketing information (Yes/no)	-0.105(-2.54)	0.100(2.34) **	0.005(0.67)
Market price (in Birr)	0.012(0.78)	-0.000(-2.12)	-0.011(-2.21)
Multiple sourcing (Yes/no)	0.09(0. 90)	-0.171(-2.01)	0.08(0.80)**

***, **,*significant at 1%, 5%, 10% respectively, $\partial y/\partial x$ = marginal effects. Number of observations = 140, Wald chi^2 (24) = 60.05, Prob > chi^2 = 0.0001, Log pseudo likelihood = -119.790, Pseudo R^2 = 0.1971.

models such as Johnson (1978) and Miller (1984) which predict that younger workers will try riskier occupations first, and their argument that the probability of switching into new ventures is roughly independent of age and total market experience.

Education

Formal years of schooling of household head (a discrete variable) was found to be negatively and significantly correlated with the new venture option, and positively with the diversification option. The negative sign points out that as education level of the household head increases the possibility that the household chooses to engage in new business contracts as compared to expanding the old business. This is possibly because as level of education increases, households analyze the risks associated and interpret the available information in a more productive way than lower education level.

Hence, smallholder households with a better knowledge did seen to engage in new businesses (lose their guaranty) unless the information they get persuades them to do (opportunity cost of adapting the strategy is lower). More importantly, poor households are known to distribute risks over portfolios of asset (Siegel and Alwang, 1999). Thus, as the level of education increases by a year of schooling, the probability that the household will choose to engage in new business falls by 5% while, the probability to diversify the business increases by 1% as compared to the reference category. Similarly, Van der Sluis and Van Praag (2008) studied the relationship between education on entry into and performance in entrepreneurship in developing countries; the relationship between schooling and performance is unambiguously positive.

More education increases the outside opportunities and

drive potentially successful entrepreneurs to other occupations where the marginal value of additional education is higher than for entrepreneurship. This result is consistent with the view that men with better education level are more likely to switch into new ventures if they have better assets. However, micro finances are meant to serve the disadvantaged social groups that lack basic resources. Therefore, besides the education level, the wealth status of the household determines entrepreneurial activity choice.

Family size

Family size measured by adult equivalence was found to have a positive and significant relation with the new business and diversification options. The positive relationship between economically active labor force and choosing the new business position entails that keeping all other variables constant, the probability of being engaged in new ventures increases as the economically active family size increases.

The marginal effect of an increase in amount of labor by one adult equivalent increases the likelihood of opting for new venture by 7.2%. It is possibly because changes in family composition and in the roles as well as relations of family members have implications for the emergence of new business opportunities (entrepreneurial skill), opportunity recognition, business start-up decisions, and the resource mobilization process (Aldrich and Jennifer, 2003).

In other way, in view of the fact that the household head need to support all the members, he/she looks for opportunities to diversify and secure livelihood. Therefore, an increase in family size by an adult equivalence increases the probability to diversify by infinitesimally smaller percentage. The percentage is so smaller because the unit of measurement, adult equivalence, gives higher weight to the more economically active labour (and so less for the non productive family member).

Land size

The variable measures the size of productive land holding in hectares. It represents household's physical asset holding and influences the nature of the activity the household may undertake. The analysis exposed that, as the size of land holding increase, the households' likelihood to expand the available business increases. In other way, land shortage is positively associated with the likelihood of new business option. Families without land usually rely on their livelihood income from working as hired labor or non-farm activities. This enhances the opportunity to come across and learn the nature of different businesses, which may boost the likelihood of engaging on new ventures whenever limitations are alleviated. The marginal effect of an increase in a hectare of land increases the probability of expanding the existing business by less than one percent. As Vollrath (2007) discussed, land inequality can be an important factor influencing the propensity to become an entrepreneur through different channels mainly for land can be used as collateral for bank loans, especially in cases of loans needed to start a new firm or to enlarge an existing one.

Multiple sourcing

It is a dummy variable having value of 1, if the household has multiple sources of credit and 0 otherwise. The variable, whether the household has borrowed from multiple sources or not, correlates positively and significantly with new business option if the household had more than one source. The positive relationship shows that, other variables fixed the odds in favour of choosing to set up new business increases, if the borrower had multiple sources as compared to expanding the existing business. Roughly, it means that if the household has single source of credit, the possibility of expanding his business increases. This may seems to correspond with Crépon et al. (2011), argument that money is fungible and credit is only loosely monitored, and one might have expected that the loans would help those who desired to start something new. However, particularly in this case, it is the inadequacy of amount of loan sighted as the reason for the positive relation between multiple sourcing and the new business option. As the household has multiple sources in reference to single source, the probability that the household would opt to start a new business increases by 8% as compared to the expansion option.

Besides, in rural areas, micro finances are encouraged

to finance existing activities, which had a track of records. This was to make sure that repayment rates would be high. The close attention paid to repayment rates, which may lead to certain conservatism by credit officers, and may reduce the extent to which microcredit indeed leads to starting new, profitable activities (Beck et al., 2009; Field et al., 2011).

Collect and utilize market information

The variable was used as a dummy variable taking value 1 for collecting and using market information and 0, otherwise. Having access to market information is positively and significantly related to diversifying options. Household heads that collect and make use of market information are encouraged to diversify their business as compared to expanding the existing business. The positive relationship indicates that keeping all other variables constant, the likelihood of choosing to diversify the existing business increases as the household collect and utilize market information than not by 10%, as in reference to the expansion option.

Recent research indicated that frequent interaction with customers (the use of formal procedures for collecting and utilizing market information) has a positive impact on new product performance, which in turn should impact new venture performance (Parry and Song, 2010). It implies that having access to market information enlightens the household about the market prices, and demand, if they found the market to be disgusting, they refrain from diversifying the business.

Expenditure on social purposes

This variable is a continuous variable measured in terms of amount of money expend on social rationale. The variable is used in the model to include expenditures like marriage and circumcision expenditures, funeral and other religious or traditional ceremonies. The model result shows these expenditures correlate positively and significantly with continuing on the established business option. The positive signs entail that the increase in the likelihood of continuing in the established business increases as households engaged more on social commitments (Table 6). On the other hand, the variable was found to have negative and significant relation with the new business option, indicating that expenditures on social issues curtain the possible amount of money that would otherwise be used for new venture establishment.

Participation in social commitments increase the social capital of the household head may be to the extent that it serves as a trade mark for his business. Therefore, it was evident that the increase in the amount of social purposes increases the likelihood of staying on the same business, increases by 93%. Pertaining to the new business option, since the social purpose and its return decreases the amount of time, money and marketable surplus, the likelihood of choosing new business option decreases by 15% in contrast to the reference category.

CONCLUSION AND RECOMMENDATION

The results indicate the presence of a fundamental set of reasons for business start-up, expansion and or diversification in addition to a mere access to finance. It appears that, like the economic theories, rather than the access to credit, the amount of loan is important in determining entrepreneurial activity decision. Variables employed have been found to have different effects on the choice of entrepreneurial activity options both in the trend and magnitude. However it has been difficult to single out the effects of other source of income besides loan (remittances, windfall gains *etc.*). This study also found that access to credit increased a move to non-farm activities and this have an ambiguous consequence in the long run development of agriculture and hence food security. Microfinance institutions give trainings and monitor the activities of their clients; however, there is lack of distinction of problems faced by different entrepreneurs, some being beginners fail in marketing their products, others lack basic skill of operation. The significant relationship between problems faced and entrepreneurial activity choice supports this conclusion. In addition to the external factors like credit, entrepreneur's own characteristics and interaction of the factors affect the choice of the entrepreneurial activity.

Therefore, if micro finance is to boost entrepreneurship, it should be on the enterprises having long last impact on the country's development and social welfare. Further, micro finance programs should be aligned with the country's strategic plan in such a way that it can expand and strengthen sectors that have higher multiplier effect. Also, it is advisable for microfinance institution's endeavor to enhance the quality of their advisory services by focusing on specific problems the entrepreneur faced rather than giving general training perceiving entrepreneurs as homogenous.

Financial institutions and entrepreneurship development organizations need to establish specialized units to provide the framework and strategy necessary in designing and delivering effective credit policies as well as programs for attracting and enlightening members of the small business sector. Finally, the results of this paper have important implications for microfinance institutions and other stakeholders making general efforts to support entrepreneurial activities of smallholders. Future studies should investigate the relationship between enterprise choice and credit service, employing larger sample sizes, wider variables, encompassing wider and different geographical, cultural and economic aspects.

REFERENCES

Aldrich HE, Cliff EJ (2003). The Pervasive Effects of Family on Entrepreneurship: Toward A Family Embeddedness Perspective. J. Bus. Venturing 18(5):573-596.

Andre' van S, Martin C, Roy T (2005). The Effect of Entrepreneurial Activity on National Economic Growth. Small Bus. Econ. 24:311-321.

Antoinette S (2009). The Divide between Subsistence and Transformational Entrepreneurship NBER Innovation Policy and the Economy. May, 29 2009, MIT.

Anwar A (2010). Assessment of Coffee Quality and Its Related Problems in Jimma Zone of Oromia Regional State. Msc Thesis February 2010. Jimma University.

Beck T, Asli D, Honohan P (2009). Access to Financial Services: Measurement, Impact, and Policies. World Bank Res. Obser. 24(1):119-145.

Crépon B, Devoto F, Duflo E, Parienté W (2011). Impact of Microcredit in Rural Areas of Morocco: Evidence from a Randomized Evaluation. March 31, 2011. Working Paper.

Field E, Pande R, Papp J, Rigol N (2011). Term structure of debt and entrepreneurship: experimental evidence from microfinance. Harvard Working Paper. P. 41.

Johnson W (1978). "A Theory of Job Shopping." Quart. J. Entrep. 92:261-278.

Khalid M (2003). Access to Formal and Quasi-Formal Credit by Smallholder Farmers and Artisanal Fishermen: A Case of Zanzibar. Ministry of Agriculture, Natural Resources, Environment and Cooperation, Zanzibar, Tanzania Research on Poverty Alleviation. Research Report No. 03.6. Mkuki na Nyota. Dar Es Salaam, Tanzania. P. 35.

Levine R (1997). Financial development and Economic growth. Views and Agenda. J. Econ. Lit. 35(2):688-726.

Liao F (1994). Interpreting Probability Models Logits, Probits and Other Generalized Linear Models. Sage University Paper Series on Quantitative Applications in the Social Science, Series No. 07-101. Thousand Oaks', CA; SAGE. P. 87.

McFadden D (1973). Conditional Logit Analysis of Qualitative Choice Behavior. In Frontiers in Econometrics, New York: Academic Press. pp. 105-142.

Miller A (1984). Job Matching and Occupational Choice. J. Polit. Econ. 92(6):1086-1120.

Miner J, Raju N, (2004). Risk Propensity Differences between Managers and Entrepreneurs and Between Low- and High-Growth Entrepreneurs: A reply in a more conservative vein. J. Appl. Psychol. 89(1):3-13.

Naude W (2010). 'Entrepreneurship, developing countries and development economies: New Approaches and insights', Small Business Economics: An Entrepreneurship J. 34(1).

Parry E, Song M (2010). "Market Information Acquisition, Utilization, and New Venture Performance," J. Prod. Innov. Manage. 27(7):1112-1126.

Popescu M, Crenicean L (2012). Considerations Regarding Improving Business Competiveness from an Entrepreneurial Perspective. Int. J. Acad. Res. Account. Finan. Manage. Sci. 2(1):9-15.

Sanyang SE, Huang W-C (2010). Entrepreneurship and economic development: the Empretec showcase. Int. Entrepreneurship Manage. J. 6(3):317-329.

Sinha T (1992). "Are Older People More Risk Averse?" Bond University School of Business. Gold Coast, Queensland 4229. Discussions Paper P. 32:17.

UNCTAD (2012). Economic Development in Africa. Report 2012. Structural Transformation and Sustainable Development in Africa. UNCTAD/ALDC/AFRICA/2012. United Nations Publication. ISSN 1990–5114.

Van der Sluis J, van Praag M (2008). "Education and entrepreneurship selection and performance: A review of the empirical literature", J. Econ. Surveys 22(5):795-841.

Vollrath D (2007). Land Distribution and International Agricultural Productivity. Am. J. Agric. Econ. 89(1):202-221.

Wooldridge JM (2002). Econometric Analysis of Cross Section and Panel Data. The MIT press.

15

Factors influencing rural household food insecurity: The case of Babile district, East Hararghe Zone, Ethiopia

Tilksew Getahun Bimerew and Fekadu Beyene

Department of Rural Development and Agricultural Extension, Haramaya University, Haremaya, Ethiopia.

The paper examines the status and factors affecting food insecurity of rural household in Babile Ethiopia. A two-stage random sampling procedure was used to select 150 sample households from four kebeles. Both primary and secondary methods of data collection were used. Descriptive statistics and binary logit model were used as methods of data analysis. Binary logit model identified five out of ten variables included in the model as significant factors of rural household food insecurity. Size of cultivated land, educational status of the household head, annual farm income, use of improved variety, and insect and pest infestation problem were found significant factors influencing household food insecurity. The results of econometric analysis made it clear that these factors were the major determinants of household food insecurity in the study area.

Key words: Food, factors, binary logit model, rural households, Babile district, Ethiopia.

INTRODUCTION

Food security has become a crucial agendum all over the world because food is a very fundamental human right that transcends cultural, political background, and religious beliefs. In addition, the right to food is acknowledged in universal declaration of human rights as well as the international covenant on economic, social and cultural rights (ICESCR) which bring consequences to the state to ensure right to food which consists of obligation to respect, protect and fulfill (Hadiprayitno, 2010). Despite progress witnessed in reducing poverty in some parts of the world over the past couple of decades, dealing with persistent rural poverty has continued to constitute the economic development agenda of sub-

Saharan Africa (IFAD, 2010). The region is the most vulnerable region to food security, in which about half of its population in food insecurity (Shapouri et al., 2009). The region is highly dependent on food import and food assistance.

Ethiopia remains one of the poorest countries in the world with human development index ranking 157 out of 169 countries reported (UNDP, 2010). With US$ 350, the country's per capita income is much lower than the sub-Saharan Africa average of US$ 1,077 in the year 2009 (World Bank, 2011). Despite the effort from the Ethiopian government and farmers' community, Ethiopia remains highly vulnerable to severe and chronic food insecurity in

a large extent (CSAE, 2010).

According to Ministry of Agriculture (2012), Ethiopia has experienced high economic growth in recent years which was 11%, however despite this, significant poverty and chronic food insecurity remains in the country. It was estimated that about 38.7% of households were food insecure. Most of these food insecure households are subsistence farmers, and vulnerable to weather fluctuations. High population growth has also contributed to decline in farm sizes, and environmental degradation remains a problem. Dramatic variations in rainfall and repeated environmental shocks further contribute to poverty and food insecurity.

Based on the joint government and humanitarian partners' requirement document released, about 3.2 million people required food assistance in the first half of 2012. The highest needs were identified in Somali and Oromia regions where 34% of the total population of each region is estimated to be in need. The net food requirement is reported to be around 158,000 metric tons (USAID, 2012).

Consider the agro-ecological zone and farming system of Babile district, there are high spatial variations of food insecurity. This might leads to raise a fundamental question about how this variation occurred among household living in the area. Besides, factors influencing household food insecurity in the area are not yet known and documented before. This indicates that there exist information gap on the factors influencing rural household food insecurity to implement different food security programs. The main objectives of the paper were to identify status of household food insecurity, and to examine factors influencing rural households' food insecurity in the area.

Assessing factors influencing rural household food insecurity is very crucial as it provides information regardless of food insecurity status of the household level that helps the policy makers for effective implementation of food security programs. Besides, the output of this research may help development practitioners and policy makers to acquire better knowledge to carry out development interventions at the right time and the right place in rural areas to decrease vulnerability to food insecurity. In addition to this, the study may help to know and document the factors influencing household food insecurity in the area.

Food security is defined in different ways by international organizations and researchers. Food security is a situation that exists when all people, at all times have physical, social and economic access to sufficient, safe and nutritious food that meets their dietary needs and food preferences for an active and healthy life (FAO, 2002). Food insecurity exists when this condition is not met. Similarly, Caraher and Coveney (2004) defined as, food poverty and food insecurity signify the inability to consume an adequate quality or sufficient quantity of food in socially acceptable ways, or the uncertainty that

one will be able to do so. According to Andersen (2009), food security is used to describe whether a country has access to sufficient food to meet citizen's dietary energy requirements. Some experts used the term national food security to refer to self-sufficiency, means that the country has the ability to produce the food demanded by its population. Thus, food security is a multidisciplinary concept which includes economic, political, demographic, social (discriminatory food access), cultural (eating habits) and technical aspects. Making food security a reality therefore also implies to take into consideration the role of non food factors.

The international human rights approach then has critical potential to highlight food insecurity as symptoms of a system which fails both to ensure individuals and households have adequate income, and to ensure that what is available to purchase or consume, at affordable cost (that is, physically and economically accessible for all), is appropriate for health. There is a clear interdependence and indivisibility between the right to food and the right to health, as articulated throughout United Nations general comment 14 on the right to the highest attainable standard of health. This embraces a wide range of socio-economic factors promoting conditions under which people can lead a healthy life, as well as the underlying determinants of health including food and nutrition (CESCR, 2000).

Food security is commonly conceptualized as resting on three pillars: availability, access, and utilization. As Webb et al (2006) noted, these concepts are inherently hierarchical, with food availability is necessary but not sufficient to ensure access, which is in turn necessary but not sufficient for effective utilization. Availability reflects the supply side of the food security concept. In order for all people to have sufficient food, there must be adequate availability. But adequate supplies do not ensure universal access to sufficient, safe and nutritious food, nor do they ensure that the food to which people has access is used to its full potential to advance human health and well-being (Webb et al., 2006). Food availability solely does not assure access to food and enough calories do not necessarily guarantee a healthy and nutritional diet (Andersen, 2009).

Hence, the second pillar of the food security concept is access. Access is most closely related to social science concepts of individual or household well-being: what is the range of food choices open to the person(s)? It reflects the demand side of food security, especially as manifest in the role food preferences plays in the definition of food security. This is meant to capture cultural limitations on what foods are consistent with a population's prevailing values. Two people from different traditions with access to exactly the same diet might not consider themselves equally food secure given variation in religiously or culturally determined food tastes. Inter and intra household distributional questions also influence access (Webb et al., 2006). According to

Stamoulis and Zezza (2003), food access is access by individuals to adequate resources (entitlements) to acquire appropriate foods for a nutritious diet.

The third pillar of food security is food utilization. Utilization reflects concerns about whether individuals and households make good use of their food access. Do they acquire nutritionally essential foods that they can afford or do they forgo nutrient intake in favor of consumption of an inadequately varied diet, of non-food goods and services, or of investment in their future livelihoods? Are the foods they purchase safe and properly prepared, under sanitary conditions, so as to enjoy their full nutritional value? Do individuals have adequate access to preventive and curative health care so as to be free of diseases that can limit their ability to absorb and metabolize essential nutrients? In particular, over the past generation, widespread concerns have arisen about micronutrient deficiencies associated with inadequate intake of essential minerals such as iodine, iron or zinc, and vitamins, in particular A and D (Webb et al., 2006).

Some agencies, such as the United Nations Food and Agriculture Organization (FAO), consider stability to be a fourth dimension of food security. Stability captures the susceptibility of individuals to food security due to interruptions in access, availability or utilization. Certain individuals within communities or households may be more vulnerable to instability and are at greater risk of food insecurity. This matter for targeting of interventions and the design of safety nets intended to safeguard food security for vulnerable subpopulations (Christopher and Erin, 2009).

According to Renzaho and Mellor (2010), food security should be based on four inter-related pillars of food availability, food access, food utilization and asset creation. Asset creation is concerned with putting in place structures and systems that sustain a household's or individuals' ability to overcome sudden shocks which threaten their access to food including economic and climatic crises. Their conception of food security is not highly different from the general food security concept. They, for instance, explain that food availability is about the amount of food which is available through domestic production or import, including from food aid. Furthermore, Renzaho and Mellor explain that access to food means distribution nutritious food which can be accessed by all household members.

Renzaho and Mellor (2010), explain that food utilization comprises of physical utilization and biological utilization. Physical utilization is concerned with household's entitlement on physical means that can be used to utilize food, whereas biological utilization involved the ability of human body to absorb the nutrients from the food effectively. Therefore, food security is highly related with public health matters such as access to clean water, housing condition and sanitation. The last pillar is asset creation according to Renzaho and Mellor (2010) which is

concerned with creating an enabling environment that able to protect individuals from a sudden shock that harms their access to food. It is built through certain structures and system that comprises of five different capital assets: human, natural, financial, social and physical. Examples of these capital assets for instance roads, water supplies, schools, food production, food processing and packaging, food marketing or market regulation, income transfer, affordable credits, trust, reciprocity, and social networks. In line with this concept, Braun (2009) stated that ensuring food security does not only require appropriate agricultural management and utilization of natural resources and eco-systems, but good governance and sustainable political system. This is obvious since food secures life and because the mission of national security is to secure society and defend its existence. This implies that food also an essential element of national security (Fullbrook, 2010). In addition, Fullbrook states that to secure food supply, it must be universally viewed not only as a commodity but as a security good. Food must be put as a priority above other activities and its positions must be recognized as an inviolable foundation of human existence and security.

MATERIALS AND METHODS

Location of the study district

The East Hararghe zone has 17 districts from which Babile is the one. It is located 35 km away from the city of Harar and about 555 km East of Addis Ababa. It lies between 8°, 9'- 9°, 23' N latitude and 42°, 15'- 42°, 53' E longitude. It shares its border with Gursum from the North, Fedis from the West, Harari National Regional State from the North West, and Somali National Regional State in the East, South, and South West (DARDO, 2011) (Figure 1).

Sampling techniques and methods of data collection

A two-stage random sampling procedure was used to select 150 sample rural households. Firstly, 4 kebeles were randomly selected from 21 kebeles of arid and semi-arid agro-ecological zones of the district. Secondly, based on probability proportional to size technique 150 sample rural households were randomly selected from the corresponding 4 kebeles of both arid and semi-arid agro-ecological zones. Both secondary and primary data collection methods were employed. The primary data required for this study was collected from sample respondents using structured questionnaire; data like Caloric intake and factors affecting food insecurity were the major once. Data collection was started after pretest was conducted and modifications were made.

Methods of data analysis

Measuring food insecurity status

The major food types used are sorghum, maize, ground nut and sweet potato. Animal products, fruits and vegetables are rarely consumed by rural households in this area. The common ways of acquiring food were own-farm production and purchase from markets. Other ways of acquiring food include gifts, food loans and

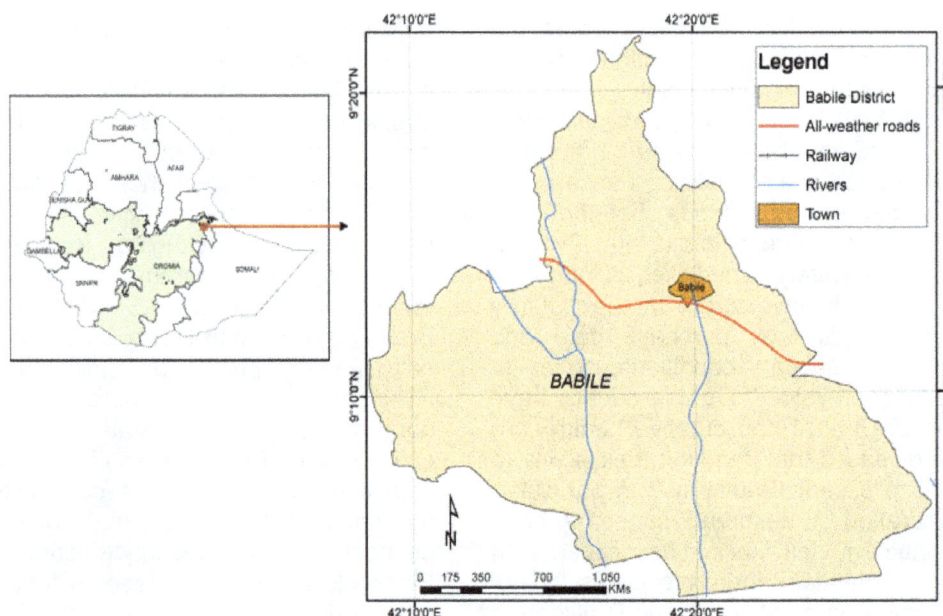

Figure 1. Map of Babile district, East Hararghe zone, Ethiopia.

Table 1. Conversion factor to calculate adult equivalent (AE).

Sex	Age	Adult equivalent (AE)
Boys	<13	0.4-0.80
Girls	<13	0.4-0.88
Male youth	13-18	1.0-1.20
Female youth	13-18	1.0
Male	19-59	1.0
Female	19-59	0.88
Old Male	>59	1.0
Old Female	>59	0.80

Source: Gassmann and Behrendt (2006).

food aid from governmental or nongovernmental Organizations.
Data on a household's caloric acquisition per adult equivalent per day were obtained on available food consumption from purchase and stock for two periods (before and after harvest) to the households. This is because measuring food insecurity status at the household level by direct surveys of dietary intake in a single period doesn't take in to account the down ward risks that rural households might face. The down ward risk might be resulted in the level of, and changes in, socioeconomic and demographic variables such as real wage rates, employment, production, price ratios and migration, etc. Thus, to taken in to consider these downward risks that rural households might face, collecting the amount food that rural households consumed in two periods (that is, before harvest season as first period for seven days and after harvest season as second period for 7 days) and calculating average calorie intake per adult equivalent of each sample households in both period is better way of measuring food insecurity status.

The information was obtained from the household member that is knowledgeable in the preparation and consumption of the commonly used instead of kilogram and/or liter were converted in to

a standard metric system and to do that conversion factor were calculated between metric units and local units. Firstly, the amount of food consumed was converted in to calorie for the periods of one (before harvest season in the month October for the seven days) and period of two (after harvest season in the month of January for the seven days) with the aid of standard nutrient composition table, then divide the calorie intakes of each sampled household in to seven in order to obtain daily calorie intake of each selected households for both periods. Secondly, the household's daily calorie intakes per adult equivalent (calorie per AE per day) for both periods were calculated by dividing the daily caloric intakes by the family size after adjusting for adult equivalent using the consumption factors for age-sex categories. Thirdly, the average households' daily calorie intake per adult equivalent was calculated. In order to calculate the average household's daily calorie intake per adult equivalent (calorie per AE per day) for two periods, the sum of each household's calorie intakes per adult equivalent (calorie per AE per day) for the two periods were divided by two. The calculation of AE for food consumption takes into account the household through recall. The local units that rural households

Table 2. Descriptive statistics of dummy variables.

Variable	Description	Food insecure	%	Food secured	%	Chi-square (χ^2)
		Food Insecurity status				
Eduhhh	Illiterate	66	44	38	25.3	6.376**
	Literate	19	12.7	27	18	
Improvvari	User	18	12	42	28	29***
	Non user	67	44.7	23	15.3	
Pestinfes	Yes	62	41.3	22	14.7	22.84***
	No	23	15.3	43	28.7	
Off/Nonfarm	Yes	30	20	40	26.7	10.20***
	No	55	36.7	25	16.7	
Irrigatscheme	Yes	2	1.4	5	3.3	2.36*
	No	83	55.3	60	40	

*, ** and*** significant at less than10, 5 and 1%, respectively (Source: Own computation result, 2012).

Table 3. Descriptive statistics of continuous variables.

Variable	Food insecure HHS Mean	Standard deviation	Food secure HHs Mean	Standard deviation	t-value
Age	36.25	7.51	39.09	7.80	2.32**
Famesize	5.29	1.63	4.80	1.52	-1.82*
Sizecult	1.17	0.67	1.45	0.88	-2.13**
Totfarin	4,474	2,978	6,965	4,504	4.06***
Hhexpend	6,822	3,337	7,972	3,162	2.14**

*, ** and*** significant at less than10, 5 and 1%, respectively; (Source: Own computation result, 2012).

age and sex of the household members, as described by Gassmann and Behrendt (2006) (Table 1).

To identify food insecure households and analyze the contributing factors of food insecurity an international minimum calorie requirement was used as cutoff point between food insecure and secure households. Thus, households whose average daily per capita intake higher than or equal to 2200 Kcal per adult equivalent per day (recommended per capita daily calorie intake), were considered as food secure where as those whose average consumption is below 2200 kcal per AE per day were considered as food insecure households.

Analytical models

The food insecure status of sample households was determined using descriptive statistics. Factors influencing household food insecurity were analyzed using Descriptive statistics and Binary Logit Model (Tables 2 and 3). The results of significant variables using descriptive statistics are follows:

Following Gujarati (1995); Aldrich and Nelson (1984); Hosmer and

Lemeshow (1989); the functional form of logistic model is specified as follows:

$$P_i = E(Y=f/x) = 1$$

$$P_i = E(y = 1/x) = \frac{1}{1+e^{-(B_0+B_1X_1)}}$$

(1)

For ease of exposition, we write (1) as:

$$P_i = \frac{1}{1+e^{-zi}}$$

(2)

The probability that a given household is food insecure is expressed by (2), while the probability for food secure is:-

$$1-P_i = \frac{1}{1+e^{zi}}$$

(3)

Therefore, we can write as:

$$\frac{P_i}{1-P_i} = \frac{1+e^{Z_i}}{1+e^{-Z_i}} \quad (4)$$

Now $\left(\frac{P_i}{1-P_i}\right)$ is simply the odds ratio in favor of food insecurity. The ratio of the probability that a household will be food insecure to the probability of that it will be food secure. Finally, taking of the natural log of equation (4) we obtain:

$$Li=\ln\left[\frac{P_i}{1-P_i}\right]=Z_i=\beta_0+\beta_1X_1+\beta_2X_2+....+\beta_nX_n \quad (5)$$

Where P_i = is the probability of the household to be food insecure; $1-P_i$ = is the probability of the household to be food secure; Z_i = is a function of n explanatory variables (x) which is also expressed as:

$$Z_i = \beta_0+\beta_1X_1+\beta_2X_2+.....\beta_nX_n \quad (6)$$

β_0 = is an intercept; $\beta_1,\beta_2,....\beta_n$ are slopes of the equation in the model; Li = is log of the odds ratio, which is not only linear in X_i but also linear in the parameters; X_i is vector of relevant household characteristics If the disturbance term (U_i) is introduced, the logit model becomes:

$$Z_i = \beta_0+\beta_1X_1+\beta_2X_2+.....\beta_nX_n+U_i \quad (7)$$

The dependant variable in this study is food insecurity which is dichotomous dependent variable in the model taking value of 1 if a household is food insecure and 0 otherwise.

Explanatory variables

Family size (FAMESIZE): This refers to the total number of family members of the household in adult equivalent (AE). It was expected that family size and household food insecurity associated negatively.

Age of household head: It was measured in number of years. Rural households devote most of their time or base their livelihoods on agriculture. The older the households head the better he/she has social network as well as the more experience on farming and weather forecasting. Thus, it was hypothesized that household head age has negatively related to household food insecurity.

Educational status of the household Head (EDUSTATUS): This is a dummy independent variable taking the value 1 if the household head is literate, 0 otherwise. It was expected that education status of the household head will have negative association with household food insecurity.

Size of cultivated land (SIZECULT): This is a continuous variable representing the total landholding of the household measured in

hectares. It was expected that size of cultivated land will have negative association with household food insecurity.

Access to improved variety (IMPRVAR): This is a dummy independent variable taking the value 1 if the household uses improved variety, 0 otherwise. It was expected that access to improved variety negatively associated with household food insecurity.

Off-farm/Nonfarm income (OFFNONFI): This is a dummy independent variable taking the value 1 if the household participate in off/none farm income sources, 0 otherwise. Participation in non/off-farm activities was expected to be negatively associated with household food insecurity.

Annual farm income (TOTFARIN): Farm income can be defined as the total annual income earned from farm produces i.e. livestock and crop production in Birr. It was hypothesized that farm income and food insecurity status of a household will have negative association.

Annual household expenditure (HHEXPEND): The proportion of income spent on food expenditure matters the status of household food insecurity. The proportion of income spent on food expenditure matters the status of household food insecurity. It was hypothesized that proportion of food expenditure and food insecurity are related negatively.

Insect and pest infestation (PESTINFEST): Insect and pest infestations are important biological factors restraining crop production and causes of food deficit in the study area. In light of this, it was hypothesized that insect and pest infestations will have positive association with food insecurity status of the households.

Use of irrigation scheme (IRRIGSCME): is a dummy variable in the model taking value of 1 if the household uses irrigation and 0, otherwise. It was expected that use of irrigation scheme and household food insecurity are negatively related.

RESULTS AND DISCUSSION

Descriptive results

Food Insecurity status of the households

Using 2200 kcal per AE per day as a benchmark to classify food insecure and secured sample households, 85 sample households were found to be unable to meet the minimum subsistence requirement and 65 sample households met the minimum subsistence requirement. In other words, 57 and 43% of the sample households were food insecure and food secure, respectively.

Econometric results

The econometric results of hypothesized variables were presented using binary logit model. This model was used to identify potential explanatory variables affecting household food insecurity through maximum likelihood (ML) estimates. Before running the analysis, it was necessary to check for the existence of multicollinearity among continuous variables and verify the degree of

Table 4. Maximum likelihood estimates of binary logit model.

Variable	Coefficient	Z	Significance
Constant	.3973121	0.25	0.801
Age	.0099781	0.29	0.775
Famsize	-.2173352	-1.23	0.219
Educstatu	-1.091609	-2.05**	0.040
Sizecultilan	-.7317101	-1.95*	0.051
Annfarmin	-.0001918	-2.71***	0.007
Annexpend	.0000834	1.02	0.306
Offnon	-.16031	-0.32	0.747
Imrvari	-.4256152	-0.64	0.521
Irrgschme	-2.466635	-2.10**	0.035
Pesinfest	1.495444	3.20***	0.001

Log likelihood = -48.743881; Number of Observation (N) =150; Log likelihood ratio value: $\left(\chi^2_{\ df}=18\right)$ = 86.10 *** Pseudo R^2 = 0.8229; *, ** and*** significant at less than 10%, 5% and 1%, respectively; (Source: model output, 2012).

association among dummy variables. Variance inflation factor and contingency coefficient were computed to detect multicollinearity for continuous variables and high degree of association for dummy variables respectively.

It is possible to conclude that there were no multicollinearity and association problems between set of continuous and dummy variables as the respective coefficients were very low. This shows that for all continuous explanatory variables the VIF was less than 10 (Table 6). For dummy explanatory variables CC was less than 0.75, which revealed the absence of a severe multicollinearity problem among potential explanatory variables (Table 7).

Factors affecting household food insecurity

With the exception of Linear Probability Model, estimation of binary choice models usually makes use of the method of maximum likelihood (Table 4).

Explanation of significant explanatory variables

Size of cultivated land: Production or output can increased either by intensification or by using higher size of cultivated land. As the cultivated land size increases, the likelihood that the holder gets more output is high. Size of cultivated land negatively and significantly affected the household food insecurity at less than ten percent probability level. The negative sign of size of cultivated land indicates that the size of cultivated land increases, the likelihood of the household to be food insecure will decline. This result coincides with the findings of (Frehiwot, 2007).

Annual farm income: Availability of farm income helps

the farmers to purchase agricultural inputs like fertilizers and improved varieties. Therefore, the more rural households use improved technologies, the higher the probability to increase production and productivity, and consequently achievement of food security. The result of the regression analysis indicates that annual farm income negatively and significantly influences household food insecurity at less than one percent probability level. The negative sign of annual farm income indicates that annual farm income increases the likelihood of the household to be food insecure will decrease. Similar study was reported by (Belayneh, 2005).

Irrigation scheme: It was hypothesized that use of irrigation scheme negatively associated with the household food insecurity. The result of the regression analysis supports this hypothesis. Use of irrigation scheme negatively and significantly affected the household food insecurity at less than five percent probability level. The negative sign of use of irrigation scheme indicates that when the households continue in use of irrigation scheme, the likelihood of the household to be food insecure will decrease.

Insect and pest infestation: Pests are one of the constraints of food security in the rural society (Ehrlich, 1991). It was hypothesized that insect and pest infestation have a positive association with household food insecurity. The result of the regression analysis supports this hypothesis. The result of the analysis indicates that insect and pest infestation problem positively and significantly affected the household food insecurity at less than one percent probability level. The positive sign of insect and pest infestation indicates that insect and pest infestation problem persists in the area, the likelihood of the household to be food insecure will

Table 5. Marginal effect of significant explanatory variables.

Variable	Change in the probability of food insecurity	Z	P>|z|
Educatio	-.2603695	-2.07	0.039
Annfarmin	-.0000448	-2.69	0.007
Irrgschme	-.5158117	-3.39	0.001
Pesinfest	.3435646	3.40	0.001
Sizecultil	-.1710421	-1.98	0.048

Change in the probability of food insecurity is calculated at the mean values of Xs.

Table 6. Variance inflation factor test for continuous variables.

Factor	Variable	1/VIF
Age	1.19	0.84
Family size	1.47	0.68
Dependency ratio	1.17	0.85
Farm Income	1.14	0.87
HH expenditure	1.37	0.73
Size of cultivated land	1.05	0.95
TLU	1.12	0.89
Asset possession	1.13	0.88
Mean VIF	1.21	

increase.

Educational status of the household head: Education may help rural people to be easily equipped with new ideas, thinking, and technology that help them to change their negative attitude in to positive once. The result of the regression analysis indicates that educational status of the household head negatively and significantly influences the household food insecurity at less than five percent probability level. The negative sign of educational status indicates that as rural households' continue in upgrading their educational status, the likelihood of the household to be food insecure will decrease. This result coincides with the findings of (Frehiwot, 2007).

Marginal effect of significant explanatory variables

In binary logit model, the changes in probabilities (slopes) can be computed, though not constant, and are termed as marginal effects or the change in log-odds ratio for a unit change in a covariate. In this study the changes in probabilities (slopes) computed by using marginal effects (Table 5).

Size of cultivated land: The marginal change in the size of cultivated land influenced negatively to the probability of food insecurity. The computed result indicates that if the size of cultivated land increases by one hectare, then

decreases by 0. 171 when all other variables held at their mean values. With increasing population land size per household member will not increase. So when land size /person decreases, the food insecurity increases.

Annual farm income: The marginal change in annual farm income influenced negatively to the probability of food insecurity. The computed result indicates that if the annual farm income of the households increases by 1000 unit, then the probability of the households to be food insecure decrease by 0. 0448 when all other variables held at their mean values.

Use of irrigation scheme: The marginal change in use of irrigation scheme influenced negatively to the probability of food insecurity. The computed result indicates that if the sample households keep using irrigation scheme, then the probability of the household to be food insecure decreases by 0.516 when all other variables held at their mean values.

Insect and pest infestation: The marginal change in insect and pest infestation problem influenced positively to the probability of food insecurity. The computed result indicates that if insect and pest infestation problem persists in the area, then the probability of the household to be food insecure increases by 0.346 when all other the probability of the households to be food insecure variables held at their mean values.

Table 7. Contingency coefficient test for dummy variables.

Variable	EDUS	IRSE	IRRIC	EXTE	CRE	OFFN	PEST
Edus	1.000						
Irsee	-0.165	1.000					
Irric	0.010	0.012	1.000				
Exte	-0.089	0.238	0.076	1.000			
Cred	0.081	-0.028	0.089	0.076	1.000		
Offn	-0.102	0.300	-0.016	0.150	0.066	1.000	
Pesti	0.022	-0.180	-0.122	-0.045	-0.028	-0.059	1.000

Educational status of the household head: The marginal change in educational status of the households influenced negatively to the probability of food insecurity. Educational status of the household favor the probability of the household to be food secure. The computed result indicates that if the sample households keep in upgrading their educational status, then the probability of the households to be food insecure decreases by 0.260 when all other variables held at their mean values.

Conclusions

The finding of the study indicates that 57% of samplehouseholds were unable to meet the minimum average daily calorie intake per adult equivalent. These food insecure households couldn't obtain the required average daily minimum calorie requirement from their production, purchase, or stock they had. Moreover, their participation in off/nonfarm activity, utilization of irrigation scheme, utilization of improved variety, and their educational status couldn't take out of them from food insecurity status. In addition to this, the existence of insect and pest infestation was significant in the district that inhibits their effort to be out of food insecurity.

The results of econometric analysis for the factors of household food insecurity have shown that the direction and influence of various factors on household food insecurity has varied. Educational status of the household head, annual farm income, use of irrigation scheme, and size of cultivated land associated negatively. Whereas, insect and pest infestation demonstrates positive and significant association with household food insecurity. Finally, the results of econometric analysis made it clear that these factors were the major factors of household food insecurity.

RECOMMENDATIONS

1. Size of cultivated land and household food insecurity associated negatively. However, population increases beyond the carrying capacity of land which fastens the vulnerability of rural households towards food insecurity.

Therefore, measures such as appropriate land use, improved technologies and proper extension services should be in place to raise existing land productivity.
2. As annual household farm income and food insecurity are associated negatively on the model result, searching and providing productive technical skill that make farmers competitive on the current farming system and generate income should be sought and promoted. Farm income-food insecurity relationship leads to propose high value of cropping pattern.
3. It was found that insect and pest infestation and household food insecurity associated positively. Thus, provision and awareness creation about different biological and chemical conservation measures should be provided so as to reduce the problem. Therefore, governmental and nongovernmental organizations that are working in the area should give due attention to reduce the problem.
4. The result of the analysis indicates that the use of irrigation scheme and household food insecurity associated negatively. Therefore, the agricultural and rural development office and nongovernmental organization that are working in the area should encourage, facilitate and strengthen the farmers to use small scale ground water irrigation activities so as to increase food production and reduce food insecurity.
5. Educational status of the household head in relation to food insecurity confirms that negative and significant. Therefore, farmer training centers should give due attention in strengthening the already provided training to the farmers to change their attitude and upgrade their production potential. In addition to this, strengthening informal education and vocational or skill training should be promoted.

Conflict of Interests

The author(s) have not declared any conflict of interests.

ACKNOWLEDGMENTS

The author would like to extend his unshared thanks to

the almighty God. His special thanks go to all my family members.

REFERENCES

Aldrich J, Nelson FD (1984). Linear Probability, Logit and Probit Models: Quantitative applications in the Social Science: Sera Miller McCun Sage pub Inc., University of Minnesota and Iowa.

Andersen PP (2009). Food security: definition and measurement. Food Security (2009) 1:5-7.

Belayneh B (2005). Analysis of Food Insecurity Causes: The case of Rural Mete Woreda, Eastern Ethiopia, School of Graduate studies, Haramaya University

Braun JV (2009). Addressing the food crisis: governance, market functioning, and investment in public goods. Food Security1:9-15.

Caraher M, Coveney J (2004). Public Health Nutrition and Food Policy. Public Health Nutri. 7:591-598.

CESCR (Committee on Economic, Social and Cultural Rights) (2000). General Comment 14: The right to the highest attainable standard of health. UN Doc. E/C.12/2000/4.

Christopher BB, Erin CL (2009). Chapter for the International Studies Compendium Project, Robert, Cornell Univ. CSAE (Centre for Studies of African Economies) (2010). Department of Economics Oxford University Oxford, United Kingdom.

DARDO (2011). District Agricultural and Rural Development Office.

FAO (2002). The State of Food Insecurity in the World 2001. Rome: FAO, pp. 4-7

Frehiwot F (2007). Food insecurity and its determinants in Amhara region, School of Graduate Studies, Addis Ababa University.

Fullbrook D (2010). Food as security. Food Security. 2:5-10.

Gassmann F, Behrendt C (2006). Cash Benefit in Low Income Countries: Simulating the Effects on Poverty Reduction for Senegal and Tanzania. Discussion Paper No. 15, Issues in Social Protection. Gebeva, Switzerland: International Labor Office.

Gujarati DN (1995). Basic Econometrics. Second Edition. MacGraw-Hill, New York.

Hadiprayitno II (2010). Food security and human rights in Indonesia. Dev. Pract. 20(1):122-130.

Hosmer D, Lemeshew S (1989). Applied Logistic Regression. A Wiley-Inter- Science Publication, New York.

International Fund for Agricultural Development (IFAD) (2010). 'Rural Poverty Report 2011: New realities, new challenges: new opportunities for tomorrow's generation'. Rome: IFAD.

Ministry of Agriculture (2012). Productive Safety Net APL III Additional Financing. Report No: 63924.

Renzaho AMN, Mellor D (2010). Food security measurement in cultural pluralism: Missing point or conceptual misunderstanding? Nutrition 26:1-9.

Shapouri S, Rosen S, Meade B, Gale F (2009). Food security assessment 2008–2009. Washington, DC: Economic Research Service, USDA.

Stamoulis K, Zezza A (2003). A conceptual framework for national agricultural, rural development and food security strategies and policies, ESA Working Paper No 03-17, FAO Agricultural and Development Economics Division, retrieved 5 March 2010

UNDP (2010). 'Human Development Report 2010: The Real Wealth of Nations: Pathways to Human Development', New York: Palgrave Macmillan.

USAID (2012). Ethiopia Food Security Outlook.

Webb P, Coates J, Frongillo EA, Rogers BL, Swingdale A, Bilinsky P (2006). Measuring Household Food Insecurity: Why it's So Important and Yet So Difficult to Do. J. Nutri. 136(S1):1404S-1408S.

World Bank (2011). 'Ethiopia: Country Brief', the World Bank Group.

16

Cost-minimizing food budgets in Ghana

Francis Addeah Darko, Benjamin Allen, John Mazunda, Rafiullah Rahimzai and
Craig Dobbins

Department of Agricultural Economics, Purdue University, 403 West State Street,
Krannert Bldg.West Lafayette, IN 47907, Indiana, United States.

Attaining the daily required nutritional recommendations is a major challenge in Ghana where the average person earns about $1.89 per day. A linear programming diet model is used to determine the cheapest basket of food items that satisfy the recommended daily nutritional requirements of the average Ghanaian. Initial findings show that an average Ghanaian requires $0.36 per day to meet his daily nutritional needs. This would be met with a food basket made up of sorghum, yam, cassava, coconut and milk. With this food basket and the estimated food expenditure, the average person in Ghana would save about 40% of his/her daily food expenditure. Sensitivity analyses are also performed to test the robustness of the findings.

Key words: Developing countries, nutrition, minimum costs, linear programming.

INTRODUCTION

Economic livelihood is largely dependent on nutrition. Good nutrition promotes growth and development in children, and improves work capacity and productivity of adults directly or indirectly through its concomitant reduced vulnerability to diseases (Smith and Haddad, 2000). That notwithstanding malnutrition, underweight and overweight, remains a major problem in Sub-Saharan Africa, South Asia and other developing countries. In 1995, about one-third of children aged five and below living in developing countries were under-nourished (Smith and Haddad, 2000). Between 2000 and 2002, 96% of the 852 million undernourished people in the world were living in developing countries (Muller and Krawinkel, 2005). Ziraba et al. (2009) also reports that about 30% of urban dwellers and 10% of rural dwellers in Africa are either overweight or obese.

Like other developing countries, malnutrition is very prevalent in Ghana. About 14.3% of children are in Ghana are underweight (CIA, statistics, 2012). A study conducted by DHS-Africa over the period of 2003 to 2007 reported that about 18% of Ghanaian women are obese. A similar study by Ziraba et al. (2009) observed that about 35% of urban dwellers and 15% of rural dwellers in Ghana are either overweight or obese. Anthropometric measures of nutritional status show a strong nutritional pattern, with malnutrition roughly increasing from the southern to the northern part of the country (Alderman, 1990). Given the importance of nutrition to economic livelihood, it is important that the malnutrition problem in Ghana is addressed. The components of food basket in Ghana differ from one ecological zone to another. Whilst roots, tubers and plantains dominate the food basket in the forest and rural coastal zones, grains (millet, sorghum and maize) dominate the food basket of households living in the savannah zone. The difference in the components of zonal food baskets follows a pattern – the average Ghanaian only spends on food items that are not produced in his/her zone of residence. For instance, forest and rural coastal zone dwellers produce mainly roots and tubers and virtually nothing of grains. Because they do not produce grains, grains sell at a relatively higher price in the forest and rural coastal zone compared to the savanna zone where they are largely produced.

Problem statement

The average Ghanaian earns $1.89 per day (World Bank, 2008), and depending on the area of residence, spends 61 to 76% of this income on food. In view of this, although the national food balance sheet (available at FAO statistical database, FAOSTAT) indicates that sufficient food is available, many people resort to eating mainly what they produce, or at least what is produced in their ecological zone of residence, in order to curtail food expenditures. This behavior tends to affect the nutritional status of Ghanaians as the locally produced food items in a particular ecological zone do not frequently make nutritionally excellent diets. Against this background, this study seeks to identify a combination of food items that can be purchased at a minimum cost and at the same time, meet the nutritional requirements of the average Ghanaian.

LITERATURE REVIEW

The question of obtaining the Recommended Dietary Allowance (RDAs) at the lowest possible cost has been a matter of concern for quite a long time now. Garille and Gass (2001) illustrates that economic literature credits Stigler's (1945) diet problem for its role in present linear programming applications. Stigler's interest was to find how much of his chosen 77 foods would be consumed by a man weighing 154 pounds so that his intake of nine nutrients would be at least equal to the recommended dietary intake (as suggested by the National Research Council) while maintaining minimum costs. Stigler's RDAs of interest were calories, protein, calcium, iron, vitamin A, niacin, riboflavin, thiamine and vitamin C.

Stigler (1945) argues that no one before his study had attempted to determine the minimum cost of obtaining the amount of calories, protein, minerals, and vitamins which other studies have accepted as adequate or optimum. He argues that a minimum cost of an adequate diet is governed by the nutritive values and the costs of food eligible for inclusion. He reasons that the other conventional diets cost so much because dieticians take account of the palatability of foods, variety of diet, prestige of various foods and other cultural aspects of consumption. Only natural foods were included in his diets since vitamin pills do not contain all the nutrients needed for a man's good health. In his solution, Stigler identified nine food items that minimized costs while providing the required nutrients. His diet consisted of

varying amounts of wheat flour, cornmeal, evaporated milk, peanut butter, lard, beef liver, potatoes, spinach and dried navy beans. Garille and Gass (2001) points out the inadequacies of Stigler's minimal subsistence diet in terms of palatability, variety, and overall adequacy. In her article describing the evolution of the diet model into a more acceptable menu-planning approach, Lancaster (1992) observes that "the solution to the least cost diet is the equivalent of the human dog biscuit." The combination of food items may not be desirable for consumption but nutrition and costs are controlled. Darmon et al. (2002) illustrates that Linear Programming is important as it can be utilized to help explain observational studies by modeling underlying structures of food choice, independent of social or cultural factors or the declaration bias inherent to dietary surveys.

MODEL FORMULATION AND DESCRIPTION

A linear programming model is used to find the cheapest combination of food items that satisfies the most important daily nutritional requirements of a 22 year old Ghanaian male. Linear programming models optimize (minimize or maximize) linear functions of a set of decision variables, while respecting multiple linear constraints (Ferguson et al., 2003). A first step in developing a linear programming model for diet optimization is to establish a clear question that has to be addressed by the model. An objective function that best answers the question must then be formulated and expressed as a linear function of the decision variables. Finally, nutritional and palatability constraints must be identified to govern the diet optimization process (Breind et al., 2003).

According to FAO statistical database (FAOSTAT) the following food items are the most significant in the food balance sheet of Ghana: cassava, yams, plantains, maize, tomato, rice, oranges, sorghum, coconuts, milk, poultry, cattle meat, and pig meat. The model is specified as follows:

$$Minimize\ Z = \sum_j C_j X_j$$

Subject to:

$$B_i \le \sum_j A_{ij} X_{ij}$$

$$\sum_j S_{ij} X_{ij} \le K_i$$

$$X_j \ge 0$$

$$\sum_j X_j \le 3Kg$$

The objective of the model is to minimize food expenditure, Z (in US$). X_j is the quantity (in kg) of food item, j; A_{ij} denotes the amount of nutrient i, in one kilogram of food item, j; C_j is the cost of a kilogram of food item, j; and B_i and K_i denote largest and smallest acceptable quantity of nutrient, i respectively. Constraints in the model include the maximum amount of daily food consumption and the minimum and maximum nutritional requirements: energy, protein, carbohydrates, vitamin A, vitamin C,

Table 1. Price and nutritional content of one kilogram of food items.

Food Item	Price (US $)	Energy (Kcal)	Protein (g)	Carbohydrate (g)	Vitamin A (mcg)	Vitamin C (mg)	Iron (mg)	Calcium (mg)
Cassava	0.12	1600	13.6	380.6	10	206	2.7	160
Yam	0.35	900	20.1	207.1	9610	196	6.9	380
Plantains	0.46	1160	7.9	311.5	450	109	5.8	20
Maize	0.29	970	33.4	217.1	0	62	5.5	20
Tomato	0.72	180	8.8	39.2	420	127	2.7	100
Rice	0.62	1120	23.2	235.1	0	0	5.3	100
Oranges	0.24	470	9.4	117.5	110	532	1	400
Sorghum	0.35	3390	113	746.3	0	0	44	280
Coconut	0.15	3540	33.3	152.3	0	33	24.3	140
Milk	0.28	610	31.5	48	460	0	0.3	1130
Poultry meat	2.84	2720	113.9	0	740	15	15.7	1380
Cattle meat	2.54	2540	171.7	0	0	0	19.4	180
Porcine meat	2.56	1280	210.6	0	20	0	8.7	130
Constraints		≥2400	≥56	≥130	900≤X≤3000	90≤X≤2000	8≤X≤45	1000≤X≤2500

iron, and calcium. According to Anderson and Earle (1993), where only minimum requirements are set, there is a tendency for a linear programming application to have solutions showing a gross imbalance of some nutrients. The Food and Agriculture Organization (FAO) and the World Health Organization (WHO) model ensured that we prevented the problem of nutritional imbalance that is common in linear programming.

Data

The significant food items in the food balance sheet of Ghana and their producer prices were obtained from FAOSTAT. The producer prices were converted into U.S. dollars from the Ghanaian currency (Cedis). The minimum energy requirement (2400 calories) was obtained from US Department of Health and Human Services (HHS). The average age (22 years) was obtained from 2000 population census of Ghana (Ghana Districts, 2009) and the corresponding nutrition requirements from the US Department of Agriculture (USDA). Three kilogram was used as the maximum amount of food that an average Ghanaian can consume.

The daily minimum and maximum amounts of each required nutrient (Table 1) were obtained from HHS and the Dietary Reference Intake respectively. The nutritional value of each food item was obtained from the USDA nutrient data laboratory (USDA National Nutrient Database for Standard Reference).

RESULTS

According to the model, $0.36 is the minimum per day amount that the average Ghanaian needs to spend on food to meet the necessary nutritional requirements. In order to achieve this minimum food expenditure, Table 2 (the 'level' column) shows that, the food basket must consist of 0.348 kg of cassava, 0.058 kg of yams, 0.173 kg of sorghum, 0.212k g of coconuts and 0.747 kg of milk.

Table 2 (the 'marginal' column) further shows that although other food items like plantains, maize, tomatoes,

have documented the harmful effects that may arise from excess consumption of some nutrients (FAO/WHO, 2003). Vitamin D and excess calcium are associated with kidney stone formation. High levels of vitamin A are associated with hair loss, bone pain and dry skin. Specifying upper and lower limits for each nutrient in our rice and oranges as well as some meat products (poultry, beef and pork) are widely produced and consumed in Ghana, their inclusion in the food basket will increase the minimum food expenditure. While consumption of a kilogram of plantains will increase food expenditure by $0.393, consumption of tomatoes will increase it by $0.645 per kilogram of tomatoes. For maize and rice, a kilogram consumption of each will increase food expenditure by $0.185 and $0.541, respectively. Quite insignificant is the $0.03 that a kilogram of orange consumption adds to food expenditures. A kilogram consumption of any meat product: beef, pork or poultry would increase food expenditure by at least $2. Needless to say, the consumption of any of the above food items will increase food expenditure because they are relatively more expensive than the components of the food basket suggested by the model.

As shown in Table 3, all the maximum constraints and the carbohydrate and iron minimum constraints are not binding in the model. The binding constraints in the model are minimum energy, protein, vitamins A and C and calcium requirements. According to Table 3, although these constraints are binding, a unit increase in the right hand side of any of them will not increase food expenditure significantly. A unit increase in the minimum energy, protein and calcium constraints will increase food expenditure by only $0.000010750, $0.002 and $0.00016504, respectively. A unit increase in the right hand side of the minimum vitamins A and C, on the other hand, will increase the food expenditure by $0.000019765 and $0.0002121, respectively. Tables 4 and 5 present the results of the sensitivity analysis. In

Table 2. Optimal values of food items.

Food Item	Lower	Level	Upper	Marginal
Cassava	0	0.345	+ ∞	0
Yam	0	0.058	+ ∞	0
Plantains	0	0	+ ∞	0.393
Maize	0	0	+ ∞	0.185
Tomato	0	0	+ ∞	0.645
Rice	0	0	+ ∞	0.541
Oranges	0	0	+ ∞	0.03
Sorghum	0	0.173	+ ∞	0
Coconut	0	0.212	+ ∞	0
Milk	0	0.747	+ ∞	0
Poultry meat	0	0	+ ∞	2.296
Cattle meat	0	0	+ ∞	2.076
Porcine meat	0	0	+ ∞	2.029

+ ∞, Positive infinity.

Table 3. Optimal values of nutrient requirement.

Nutrient	Lower	Level	Upper	Marginal
Energy	2400	2400	+ ∞	1.08E-05
Protein	56	56	+ ∞	0.002
Carbohydrate	130	341.47	+ ∞	0
Vitamin A	900	900	+ ∞	1.98E-05
Vitamin C	90	90	+ ∞	2.12E-04
Iron	8	14.307	+ ∞	0
Calcium	1000	1000	+ ∞	1.65E-04

+ ∞, Positive infinity.

general there were small ranges for the optimal solution when considering the prices of the food items presented in Table 1. It would take a minimal altering of the price of a food item to cause the composition of the optimal food basket to change. A broader range was however observed for the nutrient requirements.

For all the binding constraints, assuming non-degeneracy, any change in their right hand side values ('current' column of Table 4), regardless of how small, will result in a change of the model's solution. The components of the food basket, and the marginal values of the food items, will however not change within the lower and upper limits ('lower' and 'upper' columns of Table 4) of each of the binding constraints.

If the price of a component of the food basket increases (decreases) by a small amount, the components of the food basket, and the solution, will remain unchanged while food expenditure will increase (decrease). If an increase in price is outside the lower and upper limits ('lower' and 'upper' columns of Table 5), the amount of that food item in the food basket will increase. However if the decrease is outside the range, the value of the food item will decrease. If the decrease is large enough the

value will become zero (the components of the food basket will change)

Demand curves were developed for rice, cassava, and milk. As observed in Figure 1, these curves show the relative consumption of each food item that the model would advise given a change in price of the food item and keeping all other values constant. The graphs indicate that the quantity of milk suggested by the model is more sensitive to price changes than quantities of cassava and rice.

The model suggests a daily consumption of 0.747 kg of milk. With reference to the Ghanaian setting however, this figure is too high to be acceptable. We therefore constrained the maximum amount of milk that can be consumed to 0.5 kg. This introduced oranges into, and cassava out, of the original food basket suggested by the model, and increased food expenditure by about 27.6%. It also increased the total per day amount of food consumption by about 19%; and increased the marginal values of the binding constraints and the food items that are not in the food basket. Further, the price of milk obtained from FAOSTAT seems quite low, so we increased it by 50% while maintaining the maximum milk consumption constraint. This affected the model in the same way as the maximum milk consumption alone did, except that the food expenditure increased by only 8.33% more than the original one suggested by the model. While maintaining the 50% increase in the price of milk, we decreased the maximum milk consumption constraint to 0.25 kg. The effect of this on the model is similar to that of the 50% milk price increase plus the 0.5 kg maximum milk consumption constraint, except that the original food expenditure increased by about 65% and the amount of each food item, particularly oranges, in the food basket increased significantly. When the price of milk was increased by 50% without the maximum milk consumption constraint, the components of the original food basket suggested by the model did not change, but the food expenditure decreased by about 29%; and the marginal values of the constraint and the food items that are not in the food basket changed. Next, we decided to do away with milk in the model. When this was done, the original food expenditure increased by about 84% and orange was introduced into the original food basket. The daily consumption of oranges suggested by the model when milk was taken of the model is about 2.2 kg, which is significantly higher than the other food items in the food basket. After this change any further manipulation of the model resulted in unrealistic objective function values.

DISCUSSION AND CONCLUSION

The model simulation for the least cost diet determined that an average Ghanaian can spend $0.36 per day on his nutritional requirements. The optimal solution is low because the average Ghanaian produces his own food and buys only the food items that he does not produce.

Table 4. Nutrient constraint sensitivity.

Equation name	Lower	Current	Upper
Minimum energy requirement	1900	2400	3935
Minimum protein requirement	42.67	56	72.28
Minimum carbohydrate requirement	-∞	130	341.5
Minimum vitamin A requirement	353.4	900	3000
Minimum vitamin C requirement	23.56	90	173.8
Minimum iron requirement	- ∞	8	14.31
Minimum calcium requirement	217.9	1000	1583
Maximum energy requirement	2400	1000000	+ ∞
Maximum protein requirement	56	1000000	+ ∞
Maximum carbohydrate requirement	341.5	1000000	+ ∞
Maximum vitamin A requirement	900	3000	+ ∞
Maximum vitamin C requirement	90	2000	+ ∞
Maximum iron requirement	14.31	45	+ ∞
Maximum calcium requirement	1000	2500	+ ∞
Maximum food consumption	1.538	3	+ ∞

"+ ∞, Positive infinity; - ∞, negative infinity.

Table 5. Food item sensitivity analysis.

Food Item	Lower	Current	Upper
Cassava	0.07924	0.1197	0.1306
Yam	0.1637	0.3515	3.958
Plantain	0.06649	0.4599	+ ∞
Maize	0.106	0.2907	+ ∞
Tomato	0.07451	0.72	+ ∞
Rice	0.08348	0.6241	+ ∞
Orange	0.2083	0.2386	+ ∞
Sorghum	0.1675	0.3502	0.4458
Coconut	0.1222	0.147	0.231
Milk	0.104	0.2767	0.6115
Poultry meat	0.5445	2.841	+ ∞
Bovine meat	0.4636	2.539	+ ∞
Porcine meat	0.5343	2.564	+ ∞

+ ∞, Positive infinity.

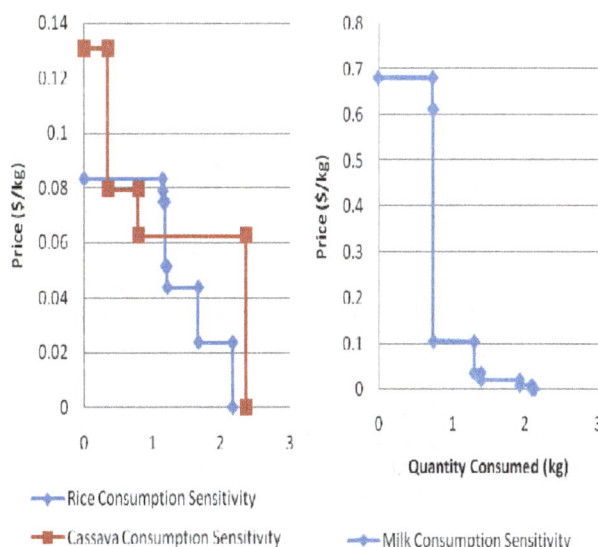

Figure 1. Sensitivity of rice, cassava, and milk to price variation.

The $0.36 represents 19% of an average Ghanaian's daily income of $1.89 (World Bank, 2008), and is significantly smaller than the 60 to 70% of daily income that is spent on food in developing countries. Therefore, if the findings of this model are adhered to, the average Ghanaian will be able to curtail food expenditure by about 40% and have enough of his income (about 80%) left for other financial obligations. Thus, all things being equal, human livelihood and poverty can be improved in Ghana if the results of this study are adhered to.

To satisfy his/her nutritional requirement while minimizing costs, an average Ghanaian should consume sorghum, yams, cassava, coconuts and milk. Cassava, yam, sorghum and coconuts are produced in many parts of Ghana, and thus readily available and frequently consumed by many people - both rich and poor. Although milk is produced in high quantities, unlike the other food items, because it is relatively more expensive, milk is not a common food item in the food basket of poor households especially. In order to encourage the consumption of milk by poor households, milk has to be reasonably priced. This can be done through a milk production subsidy program and/or a milk price control program. Notwithstanding the difference in the components of food basket across the various zones of the country, the food basket recommended by this study is adoptable. The difference in the components of the food basket across zones is more of a matter of availability and food expenditure than it is of culture since the staple food of many cultural groups are made from about the same food items. But for milk, the food items recommended by this study are widely available in all parts of the country, and per the results of this study, the food items can be obtain at a minimum cost.

The minimum constraints on energy, protein, vitamin A, vitamin C and calcium are binding in the model. If the recommendations of this model are adhered to, the marginal values of the binding minimum constraints show that the nutrient intake of a Ghanaian can be increased without any significant effect on food expenditure. Assuming non-degeneracy for all the binding constraints, any change in their right hand side values will result in a change of the model's solution. The components of the food basket will however not change within the lower and upper limits of each of the binding constraints. Note that the marginal values of the food items will also remain constant within the specified range. If the price of a component of the food basket increases (decreases) by a small amount, the components of the food basket, and the solution, will remain unchanged while food expenditure increases (decreases). If an increase in price is outside the lower and upper limits the amount of that food item in the food basket will change.

Stigler (1945), Smith (1959), Foytik (1981) and Colavita D'Orsi (1990) used models similar to the one used in this study to formulate normative diets for poor households, and for young and elderly population groups. These studies (including ours), implicitly assume that the nutritional requirements are independent, all nutrients are equally important, and that all sources of nutrients are equivalent. These assumptions are however not very realistic in that nutritional requirements are not independent, and it is common knowledge that some nutrients are more important than others. In view of this, some contemporary studies (Conforti and D'Amicis, 2000; Briend et al., 2003) include, or advocate the inclusion of additional constraints derived from observed habits and behaviors to limit the amount of some food items in the solution. Conforti and D'Amicis (2000), for instance, constrained the amount of each food item that enters the solution to equal a given proportion of the food group (for example, meat, vegetable etc.) to which the food item belongs; and the various food group are in turn constrained to equal specific proportions of the total food in the solution.

Although the underlying assumptions of this study are not very realistic, given the many cultural groups in Ghana, and the fact that different cultural groups consume food items and food groups in different proportions, it will be impractical to impose constraints like those of Conforti and D'Amicis (2000). The results of the study are therefore considered to be for Ghana in general and not for any specific cultural group. Future

studies can consider the food consumption patterns of specific cultural groups, and impose relevant constraints. Despite the general nature of the results, important implications regarding malnutrition in Ghana can be drawn, because like the results of any linear programming diet model, it provides a stylized description (not a normative tool) of the diet of the average Ghanaian. Policy makers who desire to diminish the impacts of malnutrition can focus efforts on price stabilization and subsidization of a few key foods determined by the model. Milk is especially important contributing more than half of the cost minimizing consumption bundle.

An important limitation of the study is our use of producer prices, instead of retail prices, of the food items, and the implications that this would have on the results. Food and Agriculture Organization database had only producer prices. Meanwhile, producer prices do not necessarily present any problem where individuals consume their own produce as it would have cost them exactly the (producer) price to buy it. Problems however arise when an individual has to buy a food item that is not widely grown in his/her region. If such a food item comes from a different region then it has to be purchased at the retail price. Taking this into consideration the food budget and the composition of our food basket would change.

ACKNOWLEDGEMENT

Presentation of this research was made possible, in part, through support provided by the Bill and Belinda Gates Foundation, and the Bureau of Economic Growth, Agriculture and Trade, U.S. Agency for International Development through the BASIS Assets and Market Access Collaborative Research Support Program. The opinions expressed herein are those of the authors and do not necessarily reflect the views of the sponsoring agency

REFERENCES

Alderman H (1990). Nutritional status in Ghana and its determinants. Social Dimensions of Adjustments in sub-Saharan Africa. World Bank Working Paper nr 3. World Bank, Washington.

Anderson AM, Earle MD (1993). Diet Planning in the Third World by Linear and Goal Programming. J. Operat. Res. Soc. 34: 9-16.

Briend A, Darmon NF, urguson E, Erhardt JE (2003). Linear Programming: A mathematical Tool for Analyzing and Optimizing Children's Diets During the Contemporary Feeding Period. J. Ped. Gastroel. Nutr. 36:12-22.

CIA statistics. Central Intelligence Agency WORLD FACTBOOK.Available at www.cia.gov/library/publications/the-world-factbook/rankorder/2224rank.html.Accessed on December, 2012.

Colavita C, D'Orsi R (1990). Linear programming and paediatric dietetics. Br. J. Nutr. 64:307-17.

Conforti P, D'Amicis A (2000). What is the cost of a healthy diet in terms of achieving RDAs? Public Health Nutrition: 3(3), 367-373.

Darmon N, Ferguson E, Briend A (2002). A Cost Constraint Alone Has Adverse Effects on Food Selection And Nutrient Density: An Analysis of Human Diets by Linear Programming. J. Nutr. 64:3771.

FAOSTAT. Food and Agriculture Organization Statistical database. Available at www.faostat.fao.org. Accessed on November 20, 2009.

FAO/WHO Expert Group (2003). Requirement of Ascorbic Acid, Vitamin D, Vitamin. B12, Folate and Iron. Food and Agriculture Organization.

Ferguson EL, Darmon N, Andre B, Emachandral M (2003).Food-Based Dietary Guidelines can be Developed and Tested Using Linear Programming Analysis.Am. Soc. NutritionalSci. pp.951-957

Foytik J (1981). Very low-cost nutritious diet plans designed bylinear programming. J. Nutr. Educ., 13: 63-6.

Garille SG, Gass IS (2001). Stigler's Diet Problem Revisited. Operat. Res. 49:1-13

Ghana Districts. Available at www.ghanadistricts.com/districts.Accessed November 29, 2009.

Lancaster LM (1992).The Evolution of the Diet Model in Managing Food Systems. Interfaces 5:59-68

Muller O, Krawinkel M (2005). Malnutrition and Health in Developing Countries. Can. Med. Assoc. J. 173: 279-286.

Smith VF (1959). Linear programming models for the determination of palatable human diets. J. Farm Econ., 31:56-72.

Smith L, Haddad L. (2000). ExplainingChild Malnutrition in Developing Countries:A Cross Country Analysis. International Food Policy Research Institute, Washington DC, Research Report. p.111.

Stigler GJ (1945). The Cost of Subsistence. J. Farm Econ. 27:303-314

World Bank (2008).World Development Indicators (WDI).Available at www.worldbank.org.Accessed on November 29, 2009.

Ziraba AK, Fotso JC, Ochako R (2009). Overweight and obesity in urban Africa: A problem of the rich or the poor? BMC Public Health.

17

The livelihood effects of landless people through communal hillside conservation in Tigray Region, Ethiopia

Melaku Berhe[1] and Dana Hoag[2]

[1] Department of Resource Economics, Agricultural Extension and Development, College of Dryland Agriculture and Natural Resources, Mekelle University, P.O.Box 231, Mekelle, Ethiopia.
[2]Department of Agricultural and Resource Economics, Colorado State University, Fort Collins, CO 80523-1172, USA.

In the *Tigray* region of northern Ethiopia, landless people contributed to the existing land degradation by exploiting the economic possibilities of natural resources from communal hillside areas. This has been practically observed in the area that the landless people have been depending on the available natural resources to supplement their means of living through sales of timber, fire-wood and charcoal. To address this problem, the *Tigray* Regional State has distributed denuded hillside areas to the landless people. It was believed that renovating bared mountain hillsides through conservation practices could serve as a means to create livelihood sources for the landless poor. This study has been inspired to investigate whether the introduction of communal hillside distribution to the landless people has resulted in livelihood and environmental improvements in the *Tigray* Regional State. Six districts were randomly selected namely; *Kola-Tembien, Hintalo-Wejerat, Kilte-Awlalo, Degua-Tembien, Alaje and Ofla* which all represented by 450 sampled respondents (418 males and 32 females). The respondents were interviewed using semi-structured questionnaires including ideas from group discussants and key informants. Results revealed that landless in all the districts applied conservation methods mainly of stone bund, trench and tree plantation. Their main livelihood sources using the hillside areas were; production of honey, fruits, livestock products, timber, vegetables, fuel-wood and animal fodder. Estimated results further indicated that supporting services given by forest experts and local authorities, credit access, membership in the village development committee, respondents' perception to land degradation and their educational levels were the major inducing factors that affect landless people to participate in hillside conservation.

Key words: Conservation, degradation, hillsides, landless, livelihood, sustainable, people.

INTRODUCTION

At global level, natural resource degradation on mountains slopes is widely believed to be one of the causes of environmental damage that expedites the adverse effects of climate change. The increased effect of this degradation coupled with climate change (Sanchez and Leakey, 1997; Havstad et al., 2007) has led to the decline in agricultural productivity in the Global South. The consequent loss in land productivity results in global economic and social crisis that is the dominant threat to the security and well-being of populations of most countries of the Global South (Fitsum et al., 1999; Hengsdijk et al., 2005; Atakilte et al., 2006). As much of the sources of land damage have prevailed on steep communal hillside areas and the available resources are fully accessible to the advantage of everyone, the issues of environmental management have become priority concerns in these countries (Gebremedhin et al., 2003). Cognizance of resource mismanagement due to the presence of externalities arising from communal resource overutilization, two collective action theories proposed by Hardin (1968) and Ostrom (1990) have developed over time. The mantra of each theory is based on the existing common resource challenges and thus it seems reasonable to view the analysis of this study in light of both theories.

From the time that Hardin (1968) published his article "The Tragedy of the Commons", many scholars understood that communal resources were often severely exposed to maximum overexploitation by free riders within any given community. For Hardin, it was assumed that individuals proceed to satisfy their self-interest by over utilizing the common resources with no regard for others, which would eventually damage the communal resource bases (Hardin, 1968; Welch, 1983). Their underpinning proposition was based on the absence of individual incentives to utilize communal resources in efficient ways. According to this theory, common pool resources can be managed through central stewardship either by a government or through privatization. The reason cited for this is that individual incentives evolving from numerous benefits are associated with property rights over long term, and thus privately owned resources are more sustainable than publicly owned resources.

An increasing number of scholars oppose Hardin's theory and base their reasons for their opinion that the source of the problems highlighted in 'tragedy of the commons' did not originate from the failure of common property ownership, but rather from institutional and policy failures to manage communal resources, nor the individual's mismanagement to enforce internal decisions for collective actions at communal level (Ostrom, 1990; Beaumont and Walker, 1996; Poteete and Ostrom, 2000; Forsyth, 2006). These scholars conceived that decentralized collective management of common property resources by users could be an appropriate system for overrating the 'tragedy of the commons' (Agrawal, 2001; Ostrom, 2008). As viewed by Trivers (1971), the tragedy of commons can be solved via people's dedication to manage communal resources altogether. As people exist within social bonds, they interact closely, including experience sharing among each, common cooperation against free-riders, and creation of local bylaws and binding regulations.

Furthermore, Axelrod and Hamilton (1981) contended that relationships among people usually strengthened due to their learning behaviour for cooperation in mutuality sense. In this sense, global environmental damages can be addressed via international collective actions (Ostrom, 2008).

The above debates over how communal resources can compatibly be governed by local community and which approach can plausibly be applicable for renovating the denuded resources have been studied widely. However, there is scant information on distributed mountainous hillside areas and how landless people utilize communal hillside resources and how the resources can be brought back into their natural green scenery. Such similar cases are apparently observed in the Tigray region of northern Ethiopia where land degradation is the major reason for low level of land productivity and devoid of vegetation (Hengsdijk et al., 2005; Wolde et al., 2007). Many landless people depended on the remnants of communal hillside forest patches by selling fire-wood and charcoal, timbering, traditional mining, cutting tree branches and herbaceous woods to feed their animals. This has further spirally escalated the ill-effects of land degradation (Hurni et al., 2005; Carolyn and Asenso, 2011). Due to their dependency on communal forest patches, lot of erosion is taking place and the areas are getting degraded year by year. In the upper catchment hillside mountain areas where many people cut-down live trees for firewood and charcoal sales, the position of forestland is being further exposed to severe land degradation (Badege, 2009). In other land areas in which mining and logging of trees have been taking place, large trees are almost lost, only bushes and stony degraded areas can be seen. At present, big trees can be observed only near to churches and mosques.

The increasing problem of landlessness in the areas has put pressure on the local administrators to rethink about the sustainable use of non-arable hillside mountains. The strategy designed was distributing the bared hillside areas to the landless people; after first establishing structures for soil and water conservation through community mobilization (Yifter et al., 2005; Carolyn and Asenso, 2011). Such hillside distribution to the landless poor started in 1999 in the Tigray region. As the result, landless people have conserved the areas by planting fruit trees, growing fodder for their cattle and engaging in honey production. The distribution of non-arable hillside areas to the landless people has two advantages; the areas which were previously found denuded are recently getting renovated, and, the landless people whose means of living previously depending on fuel-wood sales and traditional mining have started shifting to the production of fodder trees, fruits, vegetables, honey and commercial timber trees like eucalyptus.

However, those landless people who do not have access to other means of living are still dependent on the hillside communal forests. Since they contributed to deforestation and land degradation, it is reasonable to address the problem by creating long-term linkages between livelihood sources and hillside conservation so as to enable the landless people to utilize the hillside areas in sustainable ways. In a community of having shortage of arable land like in the Tigray Region, linking hillside distribution to the landless people and conservation is crucial. Therefore, a clear understanding on the livelihood effects of landless people's participation on hillside conservation is helpful in bringing long-lasting hillside renovation. In doing this, the objectives of the study were:

(1) To examine perception of landless people on land degradation
(2) To verify the contributions of chosen hillside rehabilitation activities to the livelihood of landless people.
(3) To identify factors that affect the participation of landless people on hillside conservation

MATERIALS AND METHODS

Sampling design

Field survey was done in the Tigray Regional State of northern Ethiopia during 1 April to 30 May, 2013. In the study area, many landless people were reaping benefits by practising hillside rehabilitation. Following Atakilte et al. (2006), the six research sites selected for this study were categorized on the basis of their agro-climatic classifications. Areas below 1500 m above sea level are considered as lowland (Quola), areas situated at 1500 to 2300 m above sea level are medium altitude (Weina-Degua); and the areas over 2300 m above sea level are highland (Degua). Based on these classifications, Quola-Tembien is located at an altitude of 800 to 1500 m above sea level in a lowland district. While Degua-Tembien (midland) is situated at an altitude of 1200 to 2100 m above sea level, Kilte-Awlalo and Hintalo-Wejerat (midland) are found at an altitude of 1500 to 2540 m above sea level. The high land areas of Alaje and Korem are also positioned with a proxy altitude range of 2300 to 3140 m above sea level.

Across all climatic zones, the Tigray Regional government is the first region that has undertaken distribution of bared communal hillside areas to landless people to enable them access to various livelihood sources in a sustainable way. In Tigray region, there are 35 woredas (districts) that have distributed non-arable hillside areas to the landless people. During sampling, criteria were used so as to distinctively identify the districts that have fully practised the hillside rehabilitation from those did not. They were:

(1) Districts that distributed hillside areas to landless people
(2) The presence of landless people whose livelihood sources depend on the conserved and improved hillside areas
(3) Landless people who have got training regarding hillside rehabilitation.

Six districts that have fulfilled the criteria were selected and the survey was conducted across three stratified agro ecological zones (lowland, midland and highland). From each agro ecological zone, individual sample representatives were selected using simple random sampling technique. Following Chand et al. (2012), the

required sample size was estimated at 99% confidence level and below 1% error commitment as shown below:

$$ n = \frac{NZ^2 P(1-P)}{N.e + Z^2 P(1-P)} $$

Where: n = is the sample size, N = is the population size, Z= Confidence level at 99%, Z=2.57, P= Estimated population proportion (0.5), e = is the error level (0.003).

Based on the sampling estimation made out of the total 1808 population size, the required sample size was 450. Doing a proportionate stratification from the total 1679 males and 129 females, the representative sampled households were 418 males and 32 females drawn from each of the three agro ecological zone (117 from lowland, 166 from midland and 167 from highland) as shown in Table 1.

Data sources and collection

This study was based on data obtained from both secondary and primary sources. Due to the wide ranging implications of the involvement of landless people into hillside conservation to generate their income, primary data were broadly collected by mixing both qualitative and quantitative methods. The following methodological approaches were employed to address the objectives. In order to bring together logical information addressing the research objectives, eight key informants were taken. The selection of the key informants was based on their experiences, better technical knowledge on hillside conservation, representatives of both men and women landless people and village leaders drawn from each district. In the presence of the chosen discussants, the whole thing was open for discussion and the informants were participating to criticize, correct, or point out, and answer in any way based on the context of their villages. The eight key informants participated in the in-depth interviews were: animal expert, two experienced farmers (a male and a female), forest expert, representative elder, leader of Farmers' Association, Women's Association and leader of development committee.

Dependent and independent variables

Dependent variable

The dependent variables are two consecutive hillside conservation actions. The first one is participation of landless people on hillside conservation and the next one is the amount of conservation done by the landless people, measured in meters of stone/soil bund.

Independent variables

Some of the independent variables are: income sources from the hillside areas, (household demographics such as gender, age, education, marital status of household head), active family size, value of livelihood assets gained from the hillside rehabilitation, cattle size, land size in the hillside, non-cattle tropical livestock unit, access to water sources, cost of conservation, cost on water use, new hillside conservation strategies, conservation methods, source of information, experience in hillside conservation, contact with extension agents, member of farmers association, satisfaction with improved tree varieties, access to off-farm activities, location of the hillside area, hillside income, non-hillside income, etc).

Table 1. Sample frame of the respondents.

Agroecological zone	Sex	Population	Sample
Lowland	Females	29	8
	Males	403	109
	Total	432	117
Midland	Females	60	15
	Males	844	151
	Total	904	166
Highland	Females	40	9
	Males	431	158
	Total	464	167
Total	Females	128	32
	Males	1680	167
	Total	1808	450

Data analyses

Both qualitative and quantitative data analyses methods were applied. Qualitative data analysis was carried out to capture information that was collected from key informant interviews, household surveys, and direct observations. As the project uses a mix of methods to understand the richness and complexity of the hillside conservation and their effects on the benefits of landless people, data triangulation was used to analyse, validate and verify the results. During triangulation, the results from different methods of qualitative and quantitative information were compared to strengthen the outcome of the project. To analyse the linkages between livelihood sources obtained from the hillside areas and the role of the landless people on hillside conservation, descriptive statistics such as measures of central tendency and dispersion were employed. The strength and direction of relationships between different selected dependent and independent variables were examined using statistical tests like chi-square to look at the associations between discrete variables and t-tests to compare the mean differences between continuous variables.

Econometrics techniques

To identify various factors that influence the involvement of landless people on hillside conservation the two-stage Heckman model was applied. Conservation activities can be influenced by various explanatory factors. Some of these factors can be household-behaviours, income levels and sources, resource availabilities, land management and institutional variables. In this view, it is possible to analyse the different factors that instigate landless people to participate in environmental rehabilitation. Households in the study areas undertake two decisions (such as decision of participation in hillside conservation and construct some amount of stone-bunds, trench, tree plantations, and others in the hill-side areas) as the activity provides them a certain threshold level of utility in terms of yield gained from the improved hillsides after rehabilitation, (fodder, honey or fruit produce or commercial trees).

The choice that the landless people have to make is based on the unobserved utility obtained from participation in those activities. These kinds of choice models assume that an individual household's choice is the result of his/her preference (Wooldridge, 2002). In such a scenario, some of the factors that influence the behaviour of landless people in conservation participation activities

may also influence their performances on the level of conservation or produce using the hillside. Analytical estimation of the outcome equation (hillside conservation in meters) alone would be, therefore, biased in the presence of sample selection. Sample selection may occur as a result of self-selection by research units (observation units – landless people in this case). The resulting bias (sample selection bias) emanates from the correlation between the error term and independent variables (Heckman, 1979; Verbeek, 2004). All these problems basically may arise from endogenous relationships among variables, measurement error of variables and missing cases of variables. Hence, the selection equation in the first stage of the Heckman two-stage model is accountable to capture factors affecting participation decision made by the landless people. This equation is used to construct a selectivity term known as the 'Inverse Mills ratio' which is added to the second stage 'outcome' equation' so as to explain factors affecting hillside conservation measured in meters. The inverse Mill's ratio is a variable for controlling bias due to sample selection (Heckman, 1979). The second stage involves including the Mills ratio to the amount of hillside conservation to be measured in meters and estimating the equation using Ordinary Least Square (OLS). Moreover, with the inclusion of extra term (inverse Mill's ratio) into the second stage, the coefficient in the second stage 'selectivity corrected' equation becomes to be unbiased (Wooldrige, 2002; Verbeek, 2004). Specification of the Heckman two-step procedure, which is written in terms of the probability of landless people to participate in hillside conservation, is given as follows:

$$\left. \begin{array}{l} z_i^* = \theta_i x_i + \varepsilon_i \\ z_i = 1 \quad \text{if } z_i^* > 0 \\ z_i = 0 \quad \text{if } z_i^* \leq 0 \end{array} \right\} \quad Selection \quad Equation$$

(1)

$$\left. \begin{array}{l} y_i^* = \lambda_i v_i + u_i \\ y_i^* = y_i \quad \text{if } z_i = 1 \\ y_i \text{ is not observed if } z_i = 0 \end{array} \right\} \quad Outcome \quad Equation$$

Where $(i = 1, 2, 3 \ldots n)$ for both equations

(2)

Table 2. Socio-economic characteristics of the landless people.

Socio-economic characteristics of landless people	Mean	Minimum	Maximum
Age in years	40.9	20	68
Experience in years	7.4	3	12
Cattle Holding (cows and oxen)	2.5	1	7
Family size	2.8	1	6
Net gain in Birr annually	783.3	20	2330

In this case, the amount of hillside conservation measured in meters represented by Equation (2) becomes the outcome equation- the variable on which we are interested to see the effect of various factors and the income gained from the hillsides. Equation (1) represents the decision for the participation activities of the landless people become the selection (precondition) equation. The overall selection model indicates that the extent of hillside conservation (y_i^*) is observed when a given landless household (i) participates in conservation activities, that is, $z_i = 1$. In the given model, in Equations (1) and (2), sample selection occurs when the correlation between the error terms of the two models, $corr(\varepsilon_i, u_i) = \rho$, is different from zero, assuming that the error terms ε_i and u_i are jointly normally distributed, independently of x_i and v_i, with zero expectations. Both x_i and v_i are the vectors of independent variables that affect the participation of landless households on hillside conservation. In the presence of the selection bias, typical models such as probit models are inefficient and OLS estimation is biased (Verbeek, 2004). Thus, the implication is that the selection problem should be corrected and we need a superior estimator for this. Based upon the specification of the dependent variable of the outcome equation, the two-step Heckman selection model is appropriate.

RESULTS AND DICUSSION

Socio-economic characteristics of respondents

According to the criteria profoundedly categorized by Jacobsen (1999), the *age* between 15-64 years was active labour force population, whereas people whose age less are than 15 years and the older people whose ages exceeding 64 years were grouped as economically passive and dependent. Following this, the results shown in Table 2 indicate that the mean age of the sample respondents was found to be about 41 years; implying the involvement of the landless people mainly from the active labor force group. Out of the total 450 interviewed respondents, male landless people were 418 and females were 32. Based on additional ideas obtained from the key informants and group discussants, the females in these areas were dominantly burdened with indoor family management tasks and cultural stereotypes which hindered their participation in hillside conservation to support their own livelihood.

This finding has similarity with the studies made by Chala et al. (2012) and FAO (2012) in the sense that females in Ethiopia have cultural hindrances that obstructed their involvement in various developmental activities outside their home. It was found that women were engaged in family management of daily house tasks such as cooking, washing and taking care of their children. In most cases, the men acted as the head of the household; in making money and satisfying the family demand. The survey result further revealed that the landless people in the study areas had an average experience of 7.4 years in hillside conservation with a minimum of 3 and maximum 12 years. Each respondent consisted of an average family size 2.8, and owned a mean number of cattle about 2.5 with a minimum 1 and maximum 7. The average annual net gain reaped by the respondents out of their participation in the hillside rehabilitation was estimated in Ethiopian Birr 783 with a minimum of 20 and maximum 2330.

Perception of landless people and their particiaption on hillside conservation

The landless people were asked to elucidate their perception towards the damages they imposed on the environment due to their dependence for fire-wood and charcoal sales by exploiting the remnant forest areas. The key informants and group discussants reported that degradation on hillside areas was perceived as a problem hindering livelihood improvement and agricultural productivity in the study areas. The data gathered from the field survey further confirmed that the landless people sensitized the exisitng damages they were imposing on the environment such as charcoal and fire-wood extraction. About 95% of the interviewed landless people witnessed that the severity of land degradation in the area was getting worsened year to year. These respondents largely examined the incidence of land degradation predominantly occurred in denuded mountain hillside areas. Of the landless people who reported the problems of land degradation in the hillside areas, more than 86% admitted that their dependence on the mountain hillsides accelerated the damages. The remaining 5% of the respondents did not notice the prevalence of land degradation in the hillside areas (Table 3).

Table 4 depicts the summary of hillside conservation done by the landless people all over the six *woredas* (districts). The landless households participated in conservation activities of soil bund, stone bund, tree plantation and a mix of trench and bunds to rehabilitate

Table 3. Perception of landless people on land degradation.

Perceive land damages	Worsened	No Change	Improved
Frequency	427	15	8
Percentage	95	3.3	1.7

Table 4. Summary of hillside conservation done by the landless people.

Conservation methods	Observation	Mean	Std. Dev.	Min.	Max.
Soil/Stone bund in meters	421	111.5	51.3	32	196
Trench (in meter square)	450	22	9.5	0	53
Tree plantation (Number)	402	36	19.2	6	81
Mix of trench and stone bund	356	72	44.3	86	108

Table 5. Ways to achieve livelihoods of landless people using hillside conservation by woredas (Districts).

Income Sources	Woredas (Districts)						Frequency	Percent
	Alaje	Hintalo-wejerat	Degua-Tembien	Kilte-Awlalo	Quola-Tembien	Ofla		
Sale of timber/Fodder	3	14	4	11	7	7	46	10.2
Sales of vegetables	5	15	7	7	5	3	42	9.3
Sale of honey	12	22	14	37	10	22	117	26
Livestock	8	26	10	28	10	20	102	22.7
Sale of fuel-wood	7	16	5	16	7	11	62	13.8
Others	8	24	5	18	11	15	81	18
Percent	9.6	26	10	26.5	10.5	17.3	100	100

the hillside areas. On average, about 111.5 m of soil/stone bund was implemented by the landless people. The average number of trees planted by the landless people was about 36. Some efforts made by the landless households to implement conservation practices using trench was accounted for 22 m^2 during the year 2012/2013. Applying different types of conservation methods, the landless people in the area were to rehabilitate the denuded hillside areas from which they generated their income sustainably. The amount of soil and water conservation done in the hillside areas indicates that the landless people who have obtained land grants did not perform sufficient hillside conservation as compared to the hillside areas given to them. Group discussants mentioned that the large portion of the hillside areas distributed to the landless people has not yet been conserved. Another study done by Gebremedhin et al. (2003) similarly found that in the Tigray Regional State, various conservation measures have been carried out by community mobilization mainly of stone terraces and bunds, micro-dams, trenches, tree planting, area enclosures, regulations for grazing lands, control of burning and applications of natural fertilizers like manure and compost.

Contributions of chosen hillside conservation to the livelihoods of the landless poor

Table 5 illustrates the major ways through which the landless people pursued to improve their livelihoods by applying various hillside conservation methods. There is a potential for improving the livelihoods of the landless people by restoring the degraded hillside areas through their participation in various conservation methods. As the result of their participation, various income sources were created by the use of hillside areas. Out of the total income, the portion obtained from the sales of honey accounted for 26%. This was followed by 22.7% of the income share generated from the livestock products. Using the hillside areas, about 10.2% of the total income was reaped from sales of timber like commercial eucalyptus trees, and 9.3% was from sales of vegetables. While the landless people still continue to generate 13.8% of the total income sources from the sales of fire-wood and charcoal, the remaining 18% was from other income sources such as daily labour wage, pity business, poultry and sales of fodder. It is indicated in the figures that different livelihood sources have served the landless people as sources of additional income to supplement

Table 6. Total Annual Gain in Birr from the Hillsides by Agro-ecology.

| Agro-ecology | Summary of Income from the Hillside Rehabilitation in Ethiopian Birr | | | | |
	Frequency	Mean	Standard Deviation	Minimum	Maximum
Lowland	117	778.2	436.1	32	2107
Midland	166	772.8	478.5	26	2327
Highland	167	797.2	433.6	20	2318

Table 7. Participation of landless people on hillside conservation using heckman regression.

| Explanatory variables | Soil/Stone-bund | | Explanatory variables | Soil/Stone-bund | |
	Coefficient	P-Value		Coefficient	P-Value
Tree satisfy	1.056088	0.646	Religious leader	-4.461652	0.700
Benefit hillside	-2.763954	0.311	Social committee	.2037514	0.956
Experience	3.842082	0.323	Village justice	-17.01396	0.291
Absence of demarcation	-8.161377	0.122	Perceived degradation	24.37742	0.000***
Dummy advice	6.878082	0.021**	Farm size	4.426882	0.215
dummy extension	4.436521	0.007***	education	-7.33115	0.001***
age	-.5748796	0.057	credit	8.600637	0.000***
seedlings	.2487836	0.054	_cons	133.3123	0.000
Development committee	5.945144	0.047**			

Note that *** and ** are significant at 1 and 5% respectively.

their means of living. Sales of honey and livestock played a considerable role in supplementing the landless people with additional incomes.

The proportion of fuel-wood comprising both charcoal and fire-wood (13.8%) serving as an income source for the landless people has important implication that about 62 landless people were found to be dependent on the natural resource forests. This may show how their dependency on the natural forests has imposed them to pursue on their short-term perspective, whereby they stick to deal with the immediate livelihood needs without considering the long-term effects of their actions on the natural resource base. This requires compatible intervention that can reshape the direction of the landless people towards honey production, commercial tree plantation, livestock rearing and vegetables which are eco-friendly livelihood alternative sources. In light of these findings, similar conclusions made by Habibah (2010) indicating that farmers can only be active participant in conserving natural resources if they find that it gives them any kind of perceived benefits. Hence, all the benefits from the hillside areas should be clearly categorized as environment friendly and non-friendly so that the landless people could be directed towards the sustainable pathways. The landless people whose income sources generated from the hillside areas in each woreda (district) is presented in Table 5. The income share of the landless people by districts as ways of their livelihood sources were: Quola Tembien (37.1%),

Hintalo-wejerat (26%), Ofla (17.3%), Degua-Tembien (10%) and Alaje (9.6).

The recorded annual income obtained by the landless people from the highland, midland and lowland agro-ecologies were on average 797.2, 772.8 and 778.2 birr, respectively (Table 6). The one way anova test revealed that there was no statistical and significance income differences among the three climatic zones. However, the big variations between the minimum and maximum income earnings in each agro-ecology indicates the need to intervene to narrow the disparities among the landless people.

Factors that affect participation of landless people on hillside conservation

Table 7 demonstrates the regression outputs of the two-stage Heckman Model to distinctively identify the major factors that induce the landless people to involve in hillside conservation. The coefficient of the inverse Mils ratio (lambda) using the Heckman first stage regression was statistically significant at 1% probability level (Prob > chi2 = 0.0000), indicating the presence of sample selection bias. After the correction of selection biases by including the inverse Mills ratio into the second stage of the Heckman model (OLS regression), the results were obtained as shown on Table 7.

Conservation was used as dependent variable which dominantly practiced by the farmers in the study area such as soil and stone-bund measured in meters. Regressing the dependent variable soil/stone bund on the explanatory variables, the variables like extension services given by the agriculture experts (dummy extension), membership of the development committee in the village (*Development committee*), perceived land degradation (*Perceived degradation*), access to credit services (Credit) and educational level (*education*) are statistically significant at 1% probability level. The variable, dummy advice refers to support given to the landless people mainly from the local authorities which is statistically significant at 5% as shown in Table 7. It indicates that the more the landless people received advices from the forest experts and local leaders; they would be inspired to apply more meters of soil and stone bund in the mountainous hillsides. For instance, the landless people that received advice regarding hillside conservation implemented about 6.9 more meters than those did not receive the services, where the other intervening variables held constant.

The implication may be that the advisory service provided to the landless people is helpful in facilitating hillside conservation. This conforms to the view of Kashwan (2013) in the sense that development agents offer technical advices in conservation and bringing workable collaboration between the entire community and the forest users. Similarly, practical lessons and experiences disseminated by the forest experts and local leaders in the study areas hasten the action of the landless people through which the hillside areas have become livelihood sources for the landless people. This further encourages them to share responsibilities in hillside conservation along with the community which eventually leads to reduce social costs. Accordingly, the landless people and the community at large will have intimate knowledge in renovating the hillside areas sustainably and are able to monitor and protect the area from any threats.

The participation of the landless people at various development activities within the village is another important factor affecting the level of hillside conservation by the landless people. Hence, the landless people having exposure to participate as a village development committee (*Development committee*) tended to apply 5.9 more meters of soil and stone bund than the ordinary people. Their participation as a member of development committee in the village may broaden their awareness about the severity of land degradation.

Hence, the landless people that perceived the existing land degradation (*Perceived degradation*) implemented more stone bunds of about 24.4 m than those did not perceive. With respect to educational level, the landless people having high level of years of schooling may tend to conserve their environment. But, the Two-Stage Heckman model regression output shows that additional one year schooling on average decreases the participation of the landless people by about 7 m of

stone and soil bunds in the hillsides (Table 7). The negative result indicates the decrease level for the landless people to participate in stone bund conservation. This was also supported by the key informants and group discussants in the sense that most of the landless people having higher level of education are assigned in various administration activities in the villages. In addition, some migrate to other places to search better income sources temporarily.

CONCLUSIONS AND SUGGESTIONS

Landless people have contributed to the existing land degradation by exploiting the economic possibilities of natural resources from the communal hillside areas. This has been practically observed in the hillside mountain areas from which landless people supplemented their livelihoods through sales of *fire-wood* and charcoal, timbering and fodder production, livestock rearing, growing fruit trees, and growing vegetable items. The study has been inspired to investigate whether the introduction of communal hillside distribution to the landless people has resulted in livelihood and environmental improvements in the Tigray Regional State. It was based to verify the idea that environmental rehabilitation in Tigray can be achieved via conservation of mountainous hillside areas with concurrent emphasis of supporting the livelihoods of the landless poor. The study revealed that landless people have tackled land degradation by applying various hillside conservation methods such as soil/stone bunds, trenches, tree plantation and zero grazing. However, lots of landless people have still relied on sales of fire-wood and charcoal to supplement their means of living. Besides, the people' participation on hillside conservation was found to be insufficient due to several restraint factors. The Heckman two-step regression output indicated that collaborative advisory services from forest experts and local authorities (dummy advice), membership in the village development committee (Development committee), landless people' perception on land degradation (Perceive degradation), extension services, credit accesses and educational levels are the major determinant factors that induce the landless people' participation on hillside conservation to improve their livelihoods. Therefore, the following actions can possibly be sound to use hillside areas sustainably:

(i) Provision of continuous support from local leaders and development agents by instigating the landless people to involve on hillside land grants and undertaking extensive conservation to improve their livelihood bases.
(ii) Build the capacity of the landless people through training, workshops, demonstrations, information dissemination, and experience sharing to increase their ability to utilize the hillside areas sustainably.

(iii) Enable the landless people to be fully detached from the sales of fire-wood and charcoal by providing substitutive income sources via communal hillside distributions.

(iv) Identify trees compatible to each agro-ecology and cultivate the hillside areas with trees that can bear fruits, and serve for animal feed.

(v) Plant bee forage, thereby increase honey yield with positive attitude for forest care and protection, which leads to sustainable job creation for landless people.

Conflict of Interests

The author(s) have not declared any conflict of interests.

ACKNOWLEDGEMENTS

Authors would like to thank the Livestock Climate Change Collaborative Research Support Program (LCC-CRSP) for funding this study and providing useful technical guidance all through this research. Further, we are grateful to the landless people, all discussants and agricultural experts for their cooperation in providing valuable information.

REFERENCES

Agrawal A (2001). Common property institutions and sustainable governance of resources. World Development: www/http. Annu. Rev. Anthropol. Annual reviews.org (Accessed on 31-01-2014).
Atakilte B, Gibbon D, Mitiku H (2006). Heterogeneity in land resources and diversity in farming practices in Tigray, Ethiopia. Agric. Syst. 88:61–74.
Axelrod R, Hamilton WD (1981). The evolution of cooperation. Science 21:1390-1396.
Badege B (2009). Deforestation and land degradation in the Ethiopian Highlands: A Strategy for Physical Recovery. Ethiopian e-Journal for Res. Dev. Foresight. 1:5-18.
Beaumont PM, Walker RT (1996). Land degradation and property regimes. J. Ecol. Econ.18:55-66.

Chala K, Taye T, Kebede D, Tadele T (2012). Opportunities and Challenges of Honey Production in Gomma District of Jimma Zone, South-West Ethiopia. J. Agric. Ext. Rural Dev. 4:85-91.
Chand BM, Bidur P, Upadhyay, RM (2012). Biogas option for mitigating and adaptation of climate change. Rentech Symposium Compendium. 1:1-5.
Carolyn T, Asenso O (2011). Responding to land degradation in the Highlands of Tigray, Northern Ethiopia: Discussion Paper 01142. Eastern and Southern Africa Regional Office. Addis Ababa, ILRI.
FAO (2012). Environment and natural resource management: adaptation to climate change in semi-arid environments experience and lessons from Mozambique. FAO, Rome, Italy. P. 71.
Forsyth T (2006). Sustainable livelihood approaches and soil erosion risks : who is to judge? [online]. London: LSEResearch Online. Available at: http://eprints.lse.ac.uk/archive/00000909 (Accessed on 27-01-2014)
Fitsum H, Pender J, Nega G (1999). Land degradation in the highlands of Tigray, and strategies for sustainable land management. Socio-economic and policy research working paper no. 25. International Livestock Research Institute.

http://www.ilri.cgiar.org/InfoServ/Webpub/Fulldocs/Work   P25/toc.htm (Accessed on 31-01-2014).
Gebremedhin B, Swinton BSM (2003). Investment in soil conservation in northern Ethiopia: the role of land tenure security and public programs. Agric. Econ 29:69-84.

Habibah A (2010). Sustainable Livelihood of the Community in Tasik Chini Biosphere Reserve: the Local Practices. J. Sustain. Dev. 3:184-196.
Havstad KM, Debra PCP, Rhonda S, Brownc J, Bestelmeyera B, Fredricksona E, Herricka J, Wrightd J (2007). Ecological services to and from rangelands of the United States. Ecological Economics. 64: 261-268.
Hardin G (1968). The tragedy of the commons. Science. 162: 1243-1248.
Heckman J (1979). Sample selection bias as specification error, econometrica. 45:155-161.
Hengsdijk H, Meijerink GW, Mosugu ME (2005). Modelling the effect of three soil and water conservation practices in Tigray, Ethiopia. Agric, Ecosyst.   Environ.   105:29-40.

Hurni H, Tato K, Zeleke G (2005). The implications of changes in population, land use, and land management for surface runoff in the upper Nile Basin area of Ethiopia. Mountain Res. Dev. 25:147-154.

Jacobsen JP (1999). Labor force participation. The Quart. Rev. Econ. Fin. 39:597-610.
Kashwan P (2013). The politics of rights-based approaches in conservation.   Land   Use   Policy.   31:613-626.

Ostrom E, Hess C (2008). "Private and Common Property Rights." Encyclopedia of Law and Economics. Northampton, MA: Edward Elgar. http://papers.ssrn.com/sol3/papers.cfm?abstract_id=1304699
Ostrom E (2000). Reformulating the commons. Suiss Pol. Sci. Rev. 6:29-52.
Ostrom E (1990). Governing the commons: The evolution of institutions for collective action. Cambridge: Cambridge University Press.

Poteete AR, Ostrom E (2004). In pursuit of comparable concepts and data about collective action. Agric. Syst. 82:215-232.
Sanchez RRB, Leakey RR (1997). Land use transformation in Africa: Three determinants for balancing food security with natural resource utilization. Eur. J. Agron. 7:15-23.

Trivers RL (1971). The evolution of reciprocal altruism. The Quart. Rev. Biol. 46:35-57.
Verbeek M (2004). A Guide to modern econometrics, 2nd ed. John Wiley and Sons Ltd., Chichester, England.
Welch WP (1983). The political feasibility of full ownership property rights: The cases of pollution and fisheries. Pol. Sci. 16:165-80.

Wolde M, Veldkamp E, Mitiku H, Nyssen J, Muys B, Kindeya G (2007). Effectiveness of exclosures to restore degraded soils as a result of overgrazing in Tigray, Ethiopia. J. Arid Environ. 69:270-284.

Wooldridge JM (2002). Econometric analysis of cross section and Panel data. The MIT Press, London, England.
Yifter F, Mitiku H (2005). Using local capacity for improved land resources planning: A Case from Ethiopia. From Pharaohs to Geoinformatics, FIG Working Week 2005 and GSDI-8. Cairo, Egypt April 16-21.

18

Assessment of farmers' perceptions and the economic impact of climate change in Namibia: Case study on small-scale irrigation farmers (SSIFs) of Ndonga Linena irrigation project

Montle, B. P. and Teweldemedhin, M. Y.

Polytechnic of Namibia, Windhoek, Namibia.

This paper examines perceptions of small-scale irrigation farmers (SSIFs) with regard to climate change and their adaptation strategies in terms of its effects. The Multinomial Logit (MNL) and the Trade-Off Analysis models were applied. Farm-level data was collected from the entire population of 30 SSIFs at the Ndonga Linena Irrigation Project in February 2014. Results from the MNL reveal that the gender, age and farming experience and extension services, yield and mean rainfall shift, are significant and positively related to the level of the farmers' diversification strategies. Trade-off analysis for multi-dimensional impact assessment (TOA-MD) model results project that climate change will have a negative economic effect on farmers, with 17.5, 25.95, 41.15 and 3.76% of farmers set to gain from climate change across 20, 30, 40 and 50% physical yield reduction scenarios respectively. Farm net return and per capita income are also expected to decline across all scenarios in future, while the poverty level is expected to rise. This study will have certain policy implications in terms of safeguarding the farmers' limited productive assets. Policy should target diversification.

Key words: Climate change, perceptions, small-scale irrigation farmers, multinomial model, trade-off analysis for multi-dimensional (TOA-MD), policy implications.

INTRODUCTION

Empirical evidence of climate change impact studies (Schulze et al., 1993; Du Toit al., 2002; Kiker, 2002; Kiker et al., 2002; Poonyth et al., 2002; Deressa et al., 2005; Gbetibouo and Hassan, 2005; Benhin, 2008) on the agricultural sector in Southern Africa show that climate change will adversely affect agricultural production, induce (or require) major shifts in farming practices and patterns, and will have significant effects on crop yields.

Available evidence indicates that Southern Africa is already experiencing climate change, with increases in surface temperature evident over both South and Southern part of the region (Kruger and Shongwe, 2004; New et al., 2006). In addition, the projected increases in temperatures and changes in precipitation timing, amount and frequency have critical implications of the agricultural sector.

The recent completed project on 'Impact of Climate Change in Southern Africa regional study, which involved five countries, that projected that Southern Africa will exceed 2°C of mean annual temperature and projected rainfall in the mid and late 21 century is variable and uncertain in terms of timing. Rainfall decreases are also projected during austral spring months, implying a delay in the onset of seasonal rains over a large part of the summer rainfall.

Future rainfall projections show changes in the scale of the rainfall probability distribution, indicating that extremes of both signs may become more frequent in the future. The changing climate is exacerbating existing vulnerabilities of the poorest people who depend on semi-subsistence agriculture for their survival; in particular is predicted to experience considerable negative impacts of climate change (SAAMIP, 2014). The latest report from the Intergovernmental Panel on Climate Change (IPCC, 2014) indicates that the effects of global warming are already occurring on all continents, however, few sectors are prepared for the risks that this change brings.

Namibia is among the countries that are most vulnerable to climate change in Sub-Saharan Africa. The climate is characterised by semi-arid to hyper-arid conditions and highly variable rainfall, although small stretches of the country (about 8% in total) are classified as semi-humid or sub-tropical (MAWF, 2010). Rainfall distribution across the country varies from an average of <25 mm per year in parts of the Namibian Desert to 700 mm in some parts of the Caprivi Strip, to the northeast.

Although the agricultural sector in Namibia contributes only about 4.1% to the GDP, it is regarded as an important part of the economy, as it employs 37% of the workforce and sustains 70% of Namibia's population as being fully or largely dependent on agriculture for their livelihoods (CBS, 2012). In comparison, for the year 2010, the fishing and fish-processing industry contributed 3.6% towards the GDP, while the mining and quarrying industry remained the highest contributor at 12.4% (CBS, 2012). Identifying new methods that can improve food security in Namibia with view towards developing an adoptive management strategy to mitigate the impact of climate induced risks that threaten to agriculturesector constitute among the most important government policy priority; due to the fact that as majority of the populations are sustenance farmers depend on the limited farming sources, further being climatic condition is characterized by semi-arid to hyper-arid conditions and highly variable rainfall. This nature of study may promote economic growth and poverty reduction, furthermore, can provide a policy formulation base that may benefit the agricultural sector.

This study form part of the broader Southern Africa Agricultural Model Inter-comparison and Improvement Project (SAAMIIP), focusing on the impact of climate change on maize farmers in Southern Africa (constituting Namibia, South Africa, Lesotho, Swaziland, Zimbabwe, Mozambique, Botswana and Malawi). Therefore, this study focuses mainly on the Kavango region of Namibia, which is the location of some significant crop irrigation incentive projects. In this area, small-scale irrigation farming is promoted through high-level government support in the form of "Green Scheme", as part of government's efforts to promote crop production for export in support of the economy (FAO, 2005). This irrigation project extract water from the perennial river, Kavango river hence the pressure on renewable water resources. This pressure is largely influenced due the demand for food and attempts to increase agricultural production (Valipour, 2014).

However, efforts are being explored for future water usage in benefit of this projects as agricultural water management is one of the most important parameters to achieve the sustainable development worldwide (Valipour, 2012). Pearl millet, maize, sorghum and cassava are among the dominant crops in the region, with approximately 95% of cultivated land being planted with millet and only small patches of mostly clay soils being used for maize and sorghum production (Mendelsohn, 2006).

The Okavango region is characterised by semi-arid conditions with an average rainfall of 550 mm per annum (October to April). The natural vegetation consists of fairly tall woodlands and tree savannahs. The dominant soil types are Kalahari sands, which are nutrient-poor aerosols with low water retention (NNF, 2010). The region is one of the most densely populated in Namibia, with the population of approximately 202,694 (Mendelsohn, 2006).

DATA ANALYSIS

Study area

The main study area, namely the Ndonga Linena Green Scheme Project, is located 80 km along the Rundu Katima Mulilo highway, at coordinates 17°57'20.41 S and 20°31'41.56 E, and at an elevation of 3,543 ft. All 30 small-scale irrigation farmers (SSIFs) involved in the project were included in the study (Figure A4). The soil type is mainly sandy soils with excellent drainage, while the average temperature is 22.4°C and the average rainfall is 577 mm annually. Most rainfall occurs during the month of February, with an average of 147 mm (Mendelsohn, 2006).

Data collection

Farm-level data was collected during February 2014 from the entire population of 30 SSIFs participating in the Ndonga Linena Irrigation Project in the Kavango region of Namibia. As a continuation of the broader research project, the study was based on interviews with the SSIFs through the use of a semi-structured and self-administered survey questionnaire, consisting of both closed- and open-ended questions.

Methodology

For purposes of this study, the Multinomial Logit (MNL) model and the Trade-Off Analysis for Multi-Dimensional Impact Assessment (TOA-MD) model were applied. To date, limited research has been conducted from a combined econometric, mathematical and simple calculation perspective, using quantitative analysis, to produce results able to assist policymakers, not only with regard to information on the impact of climate change, but also as a means to measure the perceptions of farmers in view of developing mitigation policy that takes into account the willingness of farmers to change their approaches and adopt new technology.

In analysing the economic impact of climate change and the relevant adaptation strategies, this study employed the TOA-MD model under different scenario considerations, as previously applied through SAAMIIP to intensively analyse the adoption of technology (Antle, 2011; Antle and Stoorvogel, 2006, 2008; Antle and Valdivia, 2006; Immerzeel et al., 2008, Claessens et al., 2012). With the TOA-MD model, farmers are assumed to be economically rational, meaning that they make decisions aimed at maximising expected value while being presented with a simple binary choice: They can continue to operate with production system 1, or they can switch to an alternative production system 2 (Antle and Valdivia, 2006). The logic of this analysis can be summarised as follows: Farmers are initially operating a base technology with a base climate – a combination defined as system 1. System 2 is defined as the case where farmers continue using the base technology under a perturbed climate. If some farmers are worse off economically under the perturbed climate, they are said to be vulnerable to climate change. Overall, vulnerability can be measured by the proportion of farmers that have been rendered worse off, and can also be defined relative to some threshold, such as the poverty line, in which case there is an indication of the number of households put into poverty by climate change (Antle and Valdivia, 2011).

Using the TOA-MD model, impacts that can be simulated include changes in farm income and poverty rates, as well as other environmental and social outcomes (Antle and Valdivia, 2011).

$$\omega = \text{system 1 value} - \text{system 2 value} \tag{1}$$

$$\omega = (P_1 Y_1 \alpha_1 - C_1) - (P_1 Y_1 \alpha_1 - C_1)$$

Where: P = price in system 1 and system 2 respectively; Y = production (yield) in system 1 and system 2 respectively; a = land use; C = production cost in system 1 and system 2 respectively.

$$\omega = V_1 - V_2 \text{ losses from CC} \tag{2}$$
V_1 = Value of CClim + XTech
V_2 = Value of FClim + XTech

To examine the econometric relationship between farmers' perceptions of climate variation and household characteristics, the study employed the MNL model to estimate the effects of explanatory variables on a dependent variable involving multiple choices with unordered response categories (Legesse et al., 2012). The MNL model works by denoting "y" a random variable taking on the values {1,2....j} for choices j, a positive integer, and denoting "x" a set of conditioning variables. Legesse et al. (2012) equated the model as follows:

$$P\left(y = \frac{j}{x}\right) = \frac{\exp(x\beta_j)}{1 + \sum_{x=1}^{j} \exp(x\beta_k)} \quad j = 1, \dots, j$$

Where β_j is K × 1, j = 1......., J.

The parameter estimates of the MNL model provide only the direction of the effect of the independent variables to the dependent variable, and the weakness of the model lies in its failure to quantify the actual magnitude of change or the probabilities of occurrences (Greene, 2000). However, the model does serve to interpret the effects of explanatory variables on the probabilities; hence the marginal effects need to be computed in some other way. In a study conducted in South Africa, Gbetibouo (2009) applied MNL specifications in order to model the climate change adaptation behaviour of farmers, involving discrete dependent variables with multiple choices.

The models used in this study were selected on the basis of their suitability in reaching conclusions about the use of resources at farm level and the adoption of suitable technology, in view of finding solutions to the issue of farmers' uncertainty regarding resource allocation into the future and their production capacity in the long run.

RESULTS AND DISCUSSION

Econometrical relationship between factors affecting climate change and farm household characteristics

Tables A1 to A3 depict a number of crop diversifications included in the model, in terms of model fitness and multiple logic model output respectively. Table A1 shows the level of diversification applied in the model, with farmers farming with one crop representing about 30%, farmers diversifying to two or three crops representing about 43%, and farmers farming with more than three crops representing about 27%, fitted to multiple logic regression analysis. Table A2 shows the model fitness, with likelihood ratio tests being significant and thus implying linear regression and a well-fitting model. Table A3 presents the model output.

The results of the analysis examining the factors influencing farmers' perceptions of climate change, as depicted in Table A3, reveal that the gender, age and farming experience of the household head, as well as extension services, yield and mean rainfall shift, have a positive and significant relationship with farmers' perceptions of climate change.

Farming experience

This variable was found to be statistically significant at the 5% level of significance and to be positively related, as shown by a p-value of 0.000. The estimated coefficient being positive implies that farming experience has a strong influence on farmers' level of diversification. Experienced farmers have an increased likelihood of diversifying their enterprises – as the level of experience increases by 1%, the level of diversification increases by 20% (Table A3). These results confirm the findings of Gbetibouo (2009) in a similar study of farmers' perceptions in South Africa – that is, experienced farmers have diverse skills in farming techniques and management, and are able to spread risk when faced with climate

Table A1. Level of diversification.

Variable		N	Marginal percentage
How many crops	Farm with one crop	9	30.0
	Diversify to 2 and 3	13	43.3
	more than three crops	8	26.7
Valid		30	100.0
Missing		0	
Total		30	
Subpopulation		30[a]	

The dependent variable has only one value observed in 30 (100.0%) subpopulations.

Table A2. Model fitting information.

Model	Model Fitting Criteria	Likelihood ratio tests		
	-2 Log likelihood of reduced model	Chi-Square	df	Sig.
Intercept only final	64.562			
	0.000	64.562	20	.000

Table A3. Relationship between independent variables and farmers perception to climate change.

Effect	Model fitting criteria	Likelihood ratio tests		
	-2 Log likelihood of reduced model	Chi-square	df	Sig.
Intercept	0.000[a]	0	0	.
Gender	5.574[b]	5.574	2	0.062
Householdsize	0.000[c]	.	2	.
Ageofhh	7.716[b]	7.716	2	0.021
EdulevelHh	0.000[c]	0	2	1
Farmingexperience	20.64	20.64	2	0
Anyextensionadvice	8.638[b]	8.638	2	0.013
Yieldha	22.653[b]	22.653	2	0
Farmsize	0.001[c]	0	2	1
AnylongtermshiftsinTemp	0.000[c]	0	2	1
Anylongtermshiftsinrainfall	431.780[d]	431.78	2	0

variability. Highly experienced farmers tend to have more knowledge of changes in climatic conditions and the relevant response measures to be applied.

Gender of household head

The decision to adapt to multiple crops through crop diversification was found to be statistically significant at the 10 % level of significance, with a p-value of 0.062, implying that in light of the time and labour required to diversify to multiple crops, it is likely to be more difficult for female farmers to diversify, and they are likely to require more support in this regard. In addition, it is

implied that cultural experience in terms of various management practices, and the ability to carry out labour-intensive agricultural innovations, might be challenges faced by female farmers.

Moreover, female-headed households might be slow to respond to changing climate conditions through the adaptation of diversification strategies due to the challenge posed by their customary household duties (e.g. childcare) and the fact that they are by nature less physically able to perform labour-intensive agricultural work. In addition, a variety of constraints play a role in the decisions made by farmers in this regard, including constraints with respect to available production technologies, biophysical or geophysical constraints, labour and input market constraints, financial and credit constraints, social norms, inter-temporal trade-offs, policy constraints, and constraints in terms of knowledge and skills (Teweldemedhin and Van Schalkwyk, 2010).

Age of household head

This variable was found to be significant at the 5% level of significance, with a p-value of 0.021 and a positive coefficient, implying that the age of the household head has a strong influence on the level of diversification. The older the farmer, the more experienced he/she is in farming and the more exposure he/she has had to past and present climatic conditions over longer periods of time. Mature farmers are better able to access the characteristics of modern technology than younger farmers, who might be more concerned about profit than the long-term sustainability of their operations. Similarly, Deressa et al. (2009) found that the age of the household head represents experience in farming, and that age is an indication of specialisation, because as the farmer matures he/she is more likely to grow more commercialised. The negative estimate coefficient for age implies the decision on diversification. It appears, therefore, that older and more experienced farmers are less willing to diversify their enterprise. Farmers with such characteristics might have acquired enough knowledge over time to deal with income and risk without diversification. However, the findings of Jarvie and Nieuwoudt (1988) and Vandeveer (2001, cited in Teweldemedhin and Kafidi, 2009) indicate that younger farmers, or those with less experience, are less likely to diversify their enterprise.

Extension advice

This variable was found to be statistically significant at the 5% level of significance, as shown by a p-value of 0.013, with a positive sign. This implies that extension advice has a strong influence on the ability of farmers to diversify their crops. Access to extension services increases the likelihood of perceiving changes in climate, as well as the likelihood of adapting to such changes through the creation of opportunities for the farmer to adopt suitable strategies that better suit the changed climatic conditions. This suggests that extension services assist farmers to take climate changes and weather patterns into consideration, through advice on how to deal with climatic variability and change. These results are in line with the findings Nhemachena and Hassan (2007), namely that access to information on climate change forecasting, adaptation options and other agricultural activities is an important factor in determining the farmers' use of various adaptation strategies.

Yield per hectare

This parameter was found to be statistically significant at the 1% level of significance and positively linked (p-value of 0.000). The magnitude and weight of this parameter of the estimated coefficient were found to be greater than the other parameters, implying that yield/ha has a strong influence on the level of crop diversification in effect. Where diversified crops are proven to have a greater yield per hectare than a single crop, with an associated advantage in terms of market opportunities, farmers are likely to have the ability to provide a unique product giving them a competitive advantage, which would be a good incentive for farmers to continue diversifying into even more crops, thus spreading the risk of vulnerability to the changing climate.

Mean rainfall shifts

This variable was found to be statistically significant at the 5% level of significance (p-value of 0.000). With the level of significance at 1% and a positive coefficient, it implies that rainfall has a strong influence on the level of crop diversification within the study area. An increase in the mean annual precipitation is associated with an increased probability of farmers changing their management practices, in particular by diversifying to crop varieties best suited to the prevailing and forecasted precipitation. Equally, a decrease in the mean precipitation would cultivate the farmers' technical knowledge in view of responding with sustainable measures in order to withstand the changing climate. Through this study, the farmers' priority solution areas were found to be moisture conservation and crop diversification.

Climate change impact

The farmers involved in this study were all found to be aware of the negative effect of climate variability on their production levels (Table A4). With regard to the farmers' long-term observations/perceptions of changing rainfall patterns, 78% of respondents perceived an increase in air temperature and 80% a reduction in rainfall (Table A5). With regard to direction and tendency, 78% claimed to have noticed an increase, while 13% had noticed a decrease and only 9% responded as not understanding the question about shift in temperature. Similarly, with regard to shift in rainfall patterns, 83% responded that they had noticed a decrease and 17% responded that they had noticed an increase in rainfall (Table A5). The farmers reported that they had been experiencing high temperatures, with negative effects on their crops (wilting, stunted growth and subsequent poor yields).

Table A4. Perceived impact of climate change.

Variable	YES	NO
Do you notice long-term impact of climate change?	30 (100%)	0
Do you notice shift in temp?	23 (77%)	7 (23%)
Do you notice rainfall shift over time?	24 (80%)	4 (20%)

Table A5. Direction and magnitude of perceived temperature and rainfall shifts.

	Consistent (%)	Decrease (%)	Increase (%)	Do not understand (%)
Perception of mean temperature shifts	0	13	78	9
Perception of mean rainfall shifts	0	83	17	0

Table A6. Perceived adaptation strategies to climatic variation ranking (1 – top priority and 7 – bottom priority options).

Variable	1 (%)	2 (%)	3 (%)	4 (%)	5 (%)	6 (%)	7 (%)	Total
Early planting	7	0	3	17	7	7	60	100
Use of hybrid seeds	0	10	3	3	27	43	13	100
Mixed farming	43	10	7	7	3	20	10	100
Conservational tillage/moisture conservation	3	23	30	17	20	0	7	100
Switching farming system (crop to livestock)	63	20	7	0	7	3	0	100
Information on meteorological service	33	7	17	27	10	3	3	100
Crop diversification	3	0	7	30	30	23	7	100

Furthermore, the farmers mentioned that the average annual rainfall had dropped dramatically in recent times, posing a threat to their operations due to their reliance on the Okavango River as a source of irrigation water.

Adaptation strategies to climatic variations

Table A6 presents the perceived adaptation strategy options identified by the farmers in the study area. Switching the farm system (for example to livestock) and adopting a mixed farming system was identified by 63 and 43% of farmers respectively as their future vision for coping with climate change variability, while the remaining options were selected by less than 7% of respondents. Conservation was identified as second on the list of priorities by 23% of respondents, while 60% of respondents selected early planting as the last option. These results imply that the level of understanding and awareness amongst farmers is lacking.

In a study by Lorenzoni and Langford (2005) using group discussions, respondents were asked to express their level of concern about climate change and their belief in human influence on climate. The findings of that study revealed that most of the participants possessed detailed knowledge of the issue, which they invariably related to their personal perceptions and interpretation. Through much discussion of the influence of human activities on the climate and the consequent need for behavioural and lifestyle changes, the aforementioned participants differentiated among various institutions, organisations and governmental levels with regard to the responsibility of reducing the impact of climate change, as well as those who should be entrusted with this responsibility.

As a solution, changing the crop planting date would be cost effective, but would require good technical knowledge and up-to-date information on the best time to plant. Furthermore, the use of improved crop species and crop diversification in response to climate change would require some measure of scientific input, technical knowledge and access to information by the farmers. The implication of this finding is that for climate change adaptation strategies to be effectively adopted by small-scale farmers, they should not have to face any heavy financial burden. Awareness and capacity building in terms of climate change adaptation options, as well as the provision of the necessary farm inputs, should also be incorporated into the adaptation options for small-scale farmers.

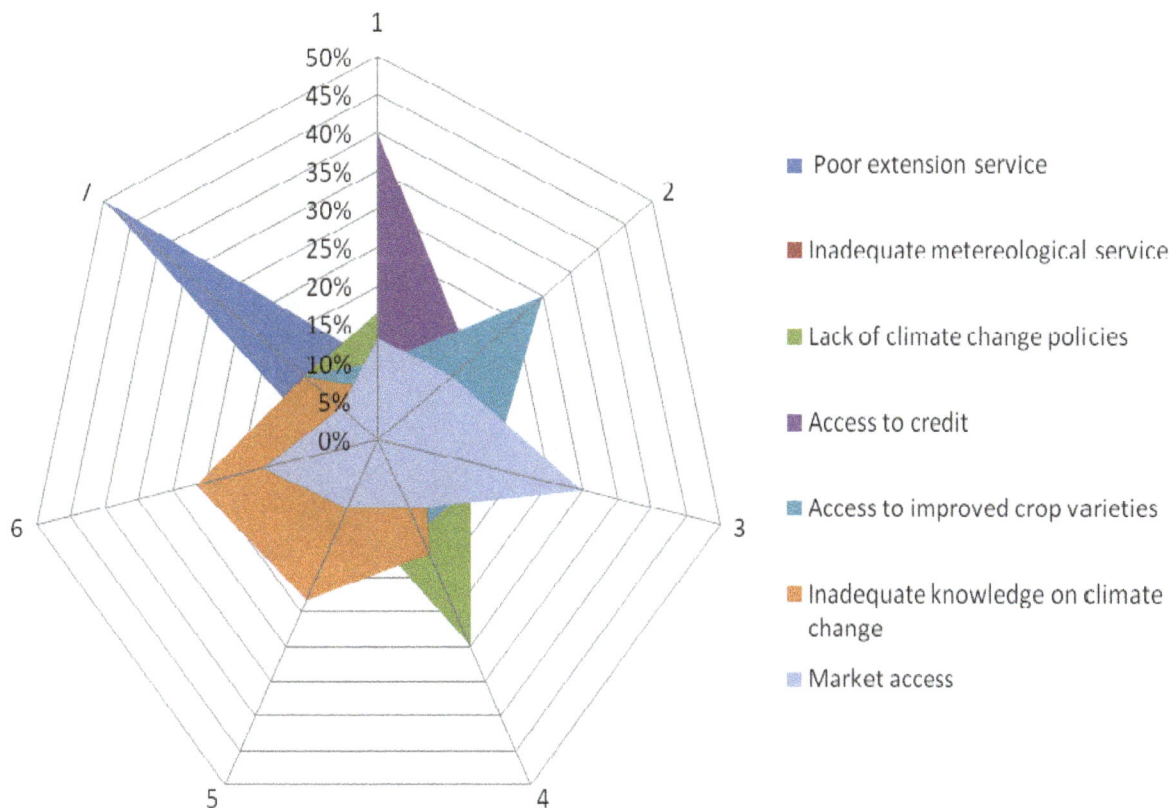

Figure A1. Major constraints in adapting to climate change and variability (7 – most constraining and 1 – least constraining).

Constraints to climate adaptation

Figure A1 depicts the constraints involved in making the necessary adjustments to climatic variations between seasons. The most significant constraint identified by respondents was poor extension services (50%), followed by lack of access to credit (40%), and inadequate meteorological services and lack of climate change knowledge (26%). Market access was also identified as a major constraint (Figure A1).

Economic analysis of the impact of climate change

Assumptions adopted by the study

Table A7 depicts the crop and climate model simulation for SAAMIIP (including South Africa, Botswana and Namibia) in respect of the percentage of the mean net return impact for the maize production summary, calibrated to the climate and crop model. On average, the five climate scenarios presented are predicted to experience a future rise in temperature (+2.0 to +3.5°C), accompanied by greater variability in rainfall. Future rainfall/precipitation projections are less consistent, with different climate models revealing different projections in the Southern African region.

In achieving these results, different crop management practices (planting date, soil depth, fertiliser application and harvesting date) were identified for use as inputs into crop modelling. Sequential climate modelling, followed by crop modelling, yielded a projection of a negative economic impact across five different climate scenarios, with an average net impact of 12.73, 34.07 and 48.16% for South Africa, Botswana and Namibia respectively. However, it is important to note that the case study in South Africa was focused on commercial farmers, while the studies in Botswana and Namibia were focused on small-scale farmers and thus yielded results that are more applicable to the small-scale irrigation farmers involved in the study at hand.

The key findings mentioned above were used to develop four different scenarios for the farmers of Ndonga Linena, in terms of modelling the economic impact within the study area. In summary, the following assumptions were considered in the application of the TOA-MD model:

1. In the absence of a climate and crop model simulation based on the above key findings of SAAMIIP resulting

Table A7. Percentage of mean net return impact on maize farmers per country.

Country	CCMS4		GFDL		HadGEM_2ES		MIROC-5		MPI-ESM		Average
	APSIM	DSSAT	APSIM	DSSAT	APSIM	DSSAT	APSIM	DSSAT	APSIM	DSSAT	
South Africa	-11.15	-8.25	-14.72	-8.89	-16.56	-11.18	-10.72	-9.25	-22.25	-14.30	-12.73
Botswana		-60.80		-30.60		-22.51		-24.40		-32.05	-34.07
Namibia	-44.36	-46.53	-54.00	-54.89	-36.93	-38.73	-60.33	-44.79	-52.25	-48.78	-48.16

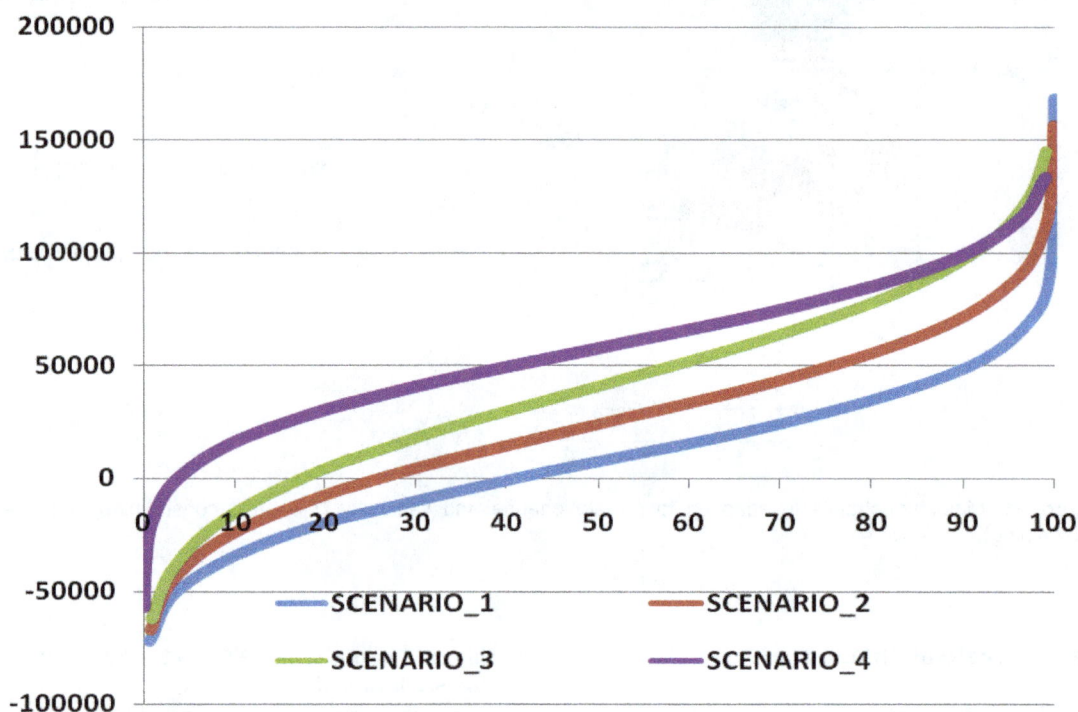

Figure A2. Economic impact of climate change.

from climate variability/shift, the four main scenarios of 20, 30, 40 and 50% reduction in yield were considered.
2. The price of maize was based on the current local market of N$ 4,000.00 per ton.
3. The costs were based on the current average production costs of the Ndonga Linena farmers (N$ 4,500.00 per ha) and annual fixed costs (N$ 3,000.00 on average per ha), assumed to remain constant in future.
4. The poverty line was assumed to be US$ 2.00 per day, at the current exchange rate (US$ 1.00 equivalent to N$ 10.00), at N$ 7,200.00 per year.
5. The current average maize farm size was set at 5.85 ha per person, assumed to remain constant in future.
6. The total farm area of all farmers participating in the Ndonga Linena project was set at 164 ha, assumed to remain constant in future.
7. Average household size was set at 2.75 members per household, assumed to remain constant in future.

Empirical results on the economic impact of climate change

These results, which are based on the four scenarios mentioned above and presented in Figure A2, show the effects of different climate scenarios on the adoption rates for new technologies, as well as potential income gains and losses. Figure A2 also shows that climate change is projected to have a negative economic effect on the net return of approximately 82.5% of farmers under scenario 1 (20% reduction in physical maize yield), while only 17.5% of farmers ware projected to gain under climate change conditions. Furthermore, under scenario 2 (30% reduction), only 25.95% of farmers would gain on their net return under climate change conditions, while 74.05% would lose. In scenario 3, the impact is projected to be 41.15% of farmers gaining and 58.8% of farmers losing under climate change conditions. With scenario 4,

Table A8. Poverty level, farm net return and per capita income.

Scenario	Poverty line percentage		Farm net return		Per capita income	
	System-1	System-2	System-1	System-2	System-1	System-2
Scenario_1	5.179457	11.26902872	166184.493	86252.34	96116.209	49885.809
Scenario_2	5.32445	19.35320648	157788.302	49008.305	91260.099	28344.958
Scenario_3	5.674843	37.4199866	150389.909	21657.499	86981.087	12526.058
Scenario_4	8.318014	65.23855689	129,260.72	(4,569.63)	74,760.59	(2,642.94)

US$ 1.00 equivalent to N$ 10.00.

the projected impact sees 3.76% of farmers gaining and 96.24% losing on their net return under climate change conditions.

Impact of climate change on poverty level, farm net return and per capita income

Table A8 presents the poverty rates resulting from farm households switching from system 1 to system 2 under climate change conditions, as well as the change in future farm net return due to changing climate conditions. As expected, the different climate scenarios produce different poverty impacts on the farm. The results show that overall poverty rates under system 1 (base system) are lower than those under system 2, meaning that the poverty level would rise in future due to the impact of climate change across all four scenarios. The farm net return and per capita income is shown to be sensitive to climate change (Table A8), with farmers projected to lose more in future as a result of the impact of climate change across all four scenarios. For example, under system 1, the net return would be N$ 166,184.50, compared to N$ 86,252.34 under system 2 (scenario 1). Similarly, scenarios 2, 3 and 4 under system 1 show a net return of N$ 157,788.30, N$ 150,389.90 and N$ 129,260.72 respectively, compared to N$ 49,008.31, N$ 21,657.00 and N$ 4,569.63 respectively under system 2. Per capita income was shown to decrease across all four scenarios in future (Table A8).

Required assistance in coping with climate change

Table A9 presents the perceived assistance required by farmers to cope with climate change and variability. The majority (30%) of respondents identified an early warning service as the most important requirement in coping with climate variability, followed by training (27%) and access to information (23%). Figure A3 depicts a spider diagram, with the provision of the necessary information on climate variability being the dominant requirement, followed by credit access and availability, and lastly training and

early warning.

Conclusion

Factors affecting farmers' perceptions of climate change

The main findings of this study revealed that farmers are aware of climate change and have perceived major shifts in temperature and rainfall on their farms. Household characteristics such as gender, age and farming experience of household head, yield/ha, rainfall shifts and extension advice were all found to have a positive and significant influence on the farmers' perceptions of climate change in the region of the Ndonga Linena project. Education level, household level, farm size and temperature shifts were found to be statistically insignificant in terms of influencing farmers' perceptions of changing climate conditions. On the other hand, all farmers in the study area claimed to have perceived changes in climate conditions and major shifts in mean rainfall and temperature. Furthermore, the farmers identified major constraints in adapting to climate change, namely poor extension services, lack of access to credit, and lack of information on climate change. The priorities identified by the farmers in terms of adaptation strategies for the future include mixed farming systems, early planting, and moisture conservation.

Government and development partners should therefore plan effective intervention programmes to build the farmers' resilience to climate change and also reduce their vulnerability to the impact thereof. This could be done through frequent training on adaptation strategies suited to their operations and the provision of subsidised input requirements. In addition, the following key points are important to consider:

Technology adoption

This is the key to realising a dramatic improvement in agricultural productivity, as proven through the Green

Table A9. Perceived required assistance to cope with climate change (1 – top priority and 5 – bottom priority options).

Variable	1 (%)	2 (%)	3 (%)	4 (%)	5 (%)
Training of farmers	27	23	10	37	3
Early warning system	30	17	30	20	3
Credit to farmers	13	40	20	27	0
Information availability	23	20	43	13	0

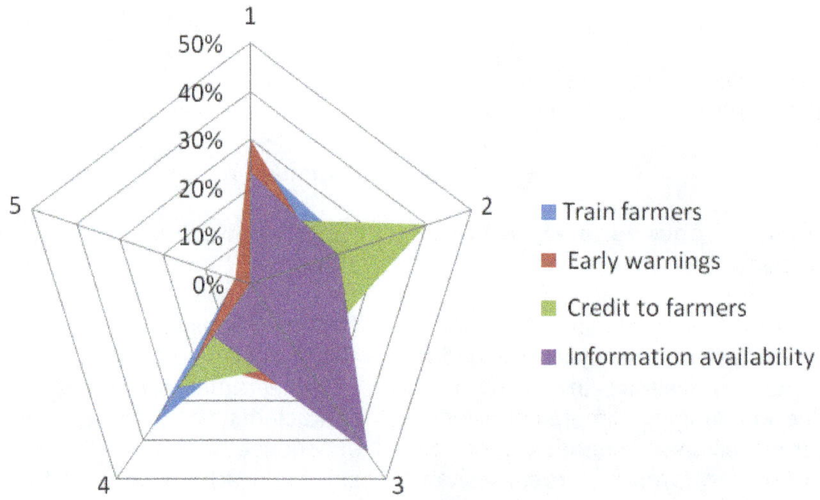

Figure A3. Required assistance to cope with climate change.

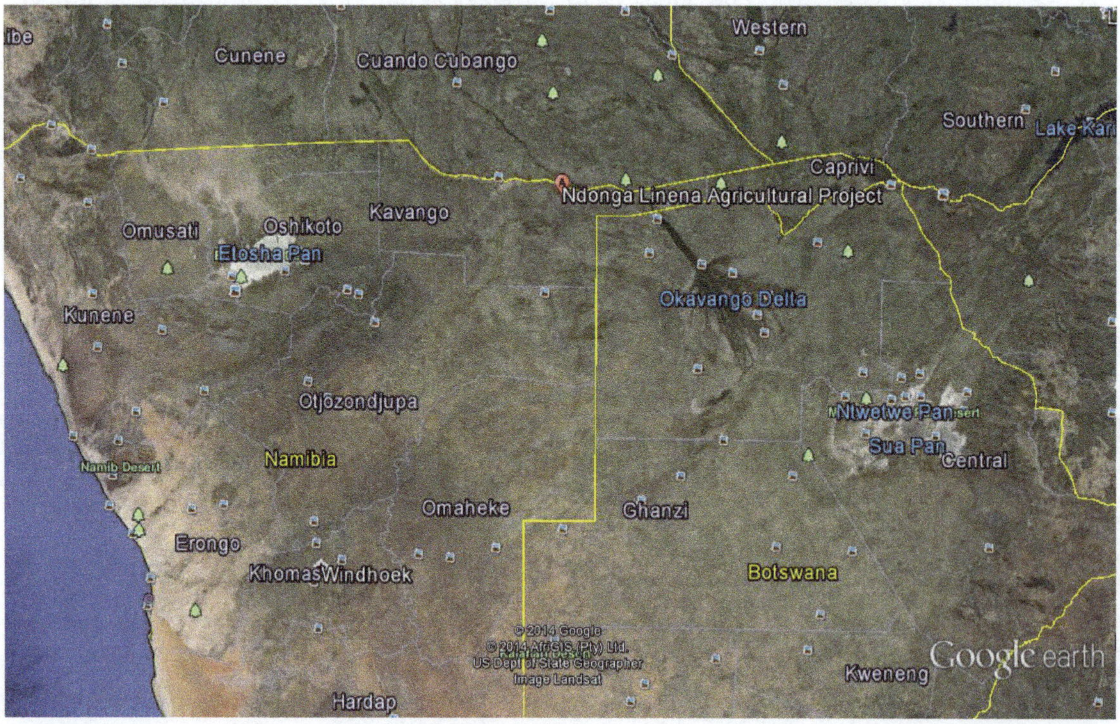

Figure A4. Location of study area. Source: Google Earth (2014).

Revolution with the development and dissemination of new technologies (or new seed varieties) invented through scientific research. Given the current low application rate of new technologies in the study area, there seems to be ample room among these small-scale farmers to improve and enhance their productivity through the adoption and adaptation of technologies (including the application of suitable fertilisers of the right quality and in the right quantity, and the use of improved/hybrid seeds). Furthermore, technological innovation is not a unilateral activity and must be amplified across the entire agricultural supply chain in Namibia.

Experimental site

This site is where the farmer is not currently testing the application of the correct fertilisers (in the correct quantity and of the correct quality) or seeds yielding better productivity. This allows for the cost of production to be determined/estimated. Therefore, it is highly recommended that farmers continuously test their input application prior to use on the farm as a whole.

Social learning

Social learning is a key determinant of the rate of diffusion of new technologies and hence productivity growth. The application of social learning could serve as a platform for easy and rapid learning with regard to available technology and any other risks faced by the farmers.

Economic impact of climate change

By assessing the impact of climate change at farm household level, the study revealed that net farm income and poverty rate is sensitive to climate change. The TOA-MD model applied for the economic analysis of the future impact of climate change revealed that climate change would have a negative economic impact on farmers' livelihoods, as very few farmers would gain from climate change. The poverty level would rise and net farm return would drop, translating into losses for farmers. Moreover, per capita income would also decrease in future. The study found a need amongst farmers for the necessary assistance to cope with climate change in the study area. Among the priorities identified were the need for government intervention to assist in terms of coping with climate variability, information availability, credit accessibility, training and early warning systems.

An important recommendation derived from the study results is that extension support/personnel knowledgeable on risks related to climate change should work closely with farmers to capacitate and prepare them

to cope with climate change. Farmers should be aware of the specific interventions to be put in place and at what magnitude in order to prepare for future climate change conditions. It is recommended that farmers practice the sustainable utilisation of resources such as water, through moisture conservation, to minimise the risks posed by the depletion of resources, and they should adjust their farming practices accordingly.

Conflict of Interest

The authors have not declared any conflict of interest.

ACKNOWLEDGEMENTS

This research was carried out as part of the Southern Agricultural Africa Inter-Comparison and Improvement Project (SAAMIIP), through the Polytechnic of Namibia. The financial support of SAAMIIP is gratefully acknowledged. The opinions expressed and conclusions arrived at in this article are those of the authors.

REFERENCES

Acquah-de Graft H (2011). Farmers' perceptions and adaptation to climate change: A willingness to pay analysis. J. Sustain. Develop. Afr. 13(5):150-161.

Antle JM, Stoorvogel JJ (2006). Predicting the supply of ecosystem services from agriculture. Am. J. Agric. Econ. 88(5):1174-1180.

Antle JM, Stoorvogel JJ (2008). Agricultural carbon sequestration, poverty and sustainability. Environ. Dev. Econ. 13:327-352.

Antle JM (2011). Representative Agricultural Pathways for Model Inter-Comparison and Impact Assessment. Available online at: <www.agmip.org/RAPs>.Accessedon21/5/2014.

Antle JM, Valdivia RO (2006). Modelling the supply of ecosystem services from agriculture: A minimum-data approach. Australian J. Agric. Res. Econ. 50:1-15.

Benhin JKA (2008). South African crop farming and climate change: An economic assessment of impacts. Global Environ. Change 18:666-678.

CBS (2011). National Accounts 2008 [Internet]. Namibian Central Bureau of Statistics (CBS). Available. from: http://www.npc.gov.na [Accessed September 2013]

Claessens L, Antle JM, Stoorvogel JJ, Valdivia RO, Thornton PK, Herrero M (2012). A method for evaluating climate change adaptation strategies for small-scale farmers using survey, experimental and modeled data. Agric. Syst. J. 111(2012):85-95.

Deressa T, Hassan R, Poonyth D (2005). Measuring the economic impact of climate change on South Africa's sugarcane growing regions. Agrekon 44(4):524–542.

Deressa TT, Hassan RM, Ringle C (2010). Perception and adaptation to climate change: The case of farmers in the Nile Basin of Ethiopia. J. Agric. Sci. 149: 23-31.

Du Toit AS, Prinsloo MA, Durand W, Kiker GA (2002).Vulnerability of maize production to climate change and adaptation in South Africa.

Combined Congress: South African Society of Crop Protection and South African Society of Horticultural Science, Pietermaritzburg, South Africa.

FAO (Food and Agricultural Organisation) (2005). National Medium Term Investment Programme. Support To Nepad–Caadp Implementation. Windhoek: FAO.

Gbetibouo G (2009). Understanding farmers' perceptions and adaptations to climate change and variability: The case of the Limpopo Basin, South Africa. Washington, DC: IFPRI Discussion P. 00849.

Gbetibouo GA, Hassan RM (2005). Measuring the economic impact of climate change on major South African field crops: A Ricardian approach. Global. Planet. Change 47:143–152.

Greene WH (2000). Econometric analysis. 4th Ed. New Jersey: Prentice-Hall.

Immerzeel W, Stoorvogel JJ, Antle JM (2008). Can payments for ecosystem services secure the water tower of Tibet. Agric. Syst. 96(1-3):52-63.

IPCC (Intergovernmental Panel on Climate Change) (2014) Climate change 2014: Impacts, adaptation, and vulnerability. IPCC WGII AR5 Summary for Policymakers.

Jarvie EM, Nieuwoudt WL (1988). Factors influencing crop insurance participation in maize farming. Agrekon. 28(2):11-16.

Kiker GA, Bamber IN, Hoogenboom G, Mcgelinchey M (2002). Further Progress in the validation of the CANEGRO-DSSAT model.Proceedings of International CANGRO Workshop, Mount Edgecombe, South Africa. PMid:12184583

Kiker GA (2002). CANEGRO-DSSAT linkages with geographic information systems: applications in climate change research for South Africa. Proceedings of International CANGRO Workshop, Mount Edgecombe, South Africa.PMid:12184583

Kruger AC, Shongwe S (2004). Temperature trends in South Africa: 1960-2003. Int. J. Climatol. 24:1929-1945.

Legesse B, Ayele Y, Bewket W (2012). Smallholder Farmers' Perceptions and Adaptation to Climate Variability and Climate Change in Doba District, West Hararghe, Ethiopia. Asian J. Empir. Res. 3(3):251-265.

Lorenzoni I, Langford I (2005). Climate Change Now and in the Future: A Mixed Methodological Study of Public Perceptions in Norwich (UK). CSERGE Working Paper ECM 01-05. University of East Anglia, Norwich.

MAWF (Ministry of Agriculture, Water and Forestry) (2010). Extension Staff Handbook: Crop Production. Windhoek: Directorate of Crop Production.

Mendelsohn J (2006). Farming systems in Namibia. Windhoek: Research and Information Services of Namibia.

New M, Hewitson B, Stephenson DB, Tsiga A, Kruger A, Manhique A, Gomez B, Coelho CAS, Masisi DN, Kululanga E, Mbambalala E, Adesina F, Saleh H, Kanyanga J, Adosi J, Bulane L, Fortunata L, Mdoka ML, Lajoie R (2006). Evidence of trends in daily climate extremes over southern and west. Afr. J. Geophys. Res. P. 111. D14102,

Newsham AJ, Thomas DSG (2011). Knowing, farming and climate change adaptation in North-Central Namibia. Global Environ. Change. 21(2):761-770.

Nhemachena C, Hassan R (2007). Micro-level Analysis of Farmers' Adaptations to Climate Change in Southern Africa. IFPRI, Environment and Production Technology Division. Washington, DC: International Food Policy Research Institute.

NNF (Namibia Nature Foundation) (2010). Land Use Planning Framework for the Kavango Region of Namibia within the Okavango River Basin. Windhoek: NNF.

Poonyth D, Hassan RM, Gbetibouo GA, Ramaila JM, Letsoalo MA (2002). Measuring the impact of climate change on South African agriculture: A Ricardian approach. A paper presented at the 40th Annual Agricultural Economics Association of South Africa Conference, Bloemfontein, 18–20 September.

Southern Africa Agricultural Model Intercomparison and Improvement Project (SAAMIIP) (2014). Impacts of Projected Climate Change Scenarios on the Production of nutritionally important crops in Southern Africa: South Africa, Botswana, Namibia, Swaziland and Lesotho. A Final Report Submitted to ICRISAT/Agricultural Model Intercomparison and Improvement Project (AgMIP), Project P. 120.

Schulze RE, Kiker GA, Kunz RP (1993). Global climate change and agricultural productivity in Southern Africa. Global Environ. Change 3(4):330–349.

Teweldemedhin MY, Van Schalkwyk HD (2010). The impact of trade liberalisation on South African agricultural productivity. Afr. J. Agric. Res, 5(12):1380-1387.

Valipour M (2014). Land use policy and agricultural water management of the previous half of century in Africa. Applied Water Science: Springer Berlin Heidelberg. 4:1-29.

Valipour M (2014). Pressure on renewable water resources by irrigation to 2060. Acta Advan. Agric. Sci. 2(8):32-42.

Valipour M, Montazar AA (2012). An evaluation of SWDC and WinSRFR Models to Optimize of infiltration Parameters in Furrow Irrigation. Am. J. Scient. Res. 69:128-142.

19

Estimating the impact of a food security program by propensity-score matching

Tagel Gebrehiwot[1] and Anne van der Veen[2]

[1]Laboratory for Social Interactions and Economic Behaviour (LSEB), University of Twente, P. O. Box 217, 7514 AE Enschede, Netherlands.
[2]Faculty of Geo-Information Science and Earth Observation, University of Twente, P. O. Box 217, 7514 AE Enschede, Netherlands.

Reducing poverty and improving household food security remains an important policy objective for rural development in the semi-arid areas of many countries in Africa. Many development programs have been introduced in efforts to bring the cycle of poverty and food insecurity to an end. This paper investigates the impact of a food security package (FSP) program in improving rural household's food consumption in Tigray region, Northern Ethiopia. An empirical analysis based on a propensity score matching (PSM) method, which is a popular approach to estimate causal treatment effects, is employed. Using kernel-matching estimation technique, program beneficiaries were matched with non-beneficiaries. The results show that the program has had a significant effect on improving household food calorie intake. The findings indicated that the program raised the food calorie intake of beneficiary households by 41.8% above that of individuals not involved in the program. Sensitivity analysis also indicated that the observed estimate of impact is not vulnerable to hidden bias or selection on unobservables.

Key words: Propensity score, matching, selectivity bias, average treatment effect, impact, evaluation.

INTRODUCTION

It is increasingly being recognised that improving food security is a basis for reducing poverty and hunger, but also for economic development. Despite notable progress in economic growth and welfare improvement in developing countries over the recent decades, food security has not been attained in most developing countries. In particular, food insecurity continues to form a deep seated problem in several sub-Saharan African (SSA) countries. A recent Food and Agriculture Organization of the United Nations (FAO) report indicates

that the number of undernourished people in Africa still remain high at 226.7 million (FAO, 2014). Even now, countries in the Horn of African are overwhelmed by heightened food security crises, making the problem of food security an issue of great concern to governments and the international community.

Like other SSA countries, Ethiopia is one of the least developed countries in the world according to all measures of poverty. Despite the country has made progress in economic growth over recent decade, food insecurity is still evident. The 2012-2014 FAO assessment report estimated 32.9 million of the Ethiopian people are undernourished, indicating food shortage as an on-going problem in the country (FAO, 2014). The country's food production is highly vulnerable to the influence of adverse weather conditions as the agricultural sector is totally dependent on rainfall. Previous studies reported that a 10% decline in the amount of rainfall below the long-term average leads to a 4.4% reduction in the country's national food production (Webb et al., 1992). Furthermore, drought has increasingly occurred over the recent decades, as has the proportion of the population adversely affected by it. Consequently, the country has been dependent on food aid to bridge its huge food gap. Devereux (2006) reported that, even in a year where rainfall is favourable, around 4 to 5 million Ethiopians depend on food aid, reflecting how deep-rooted food insecurity is in the country.

The causes of food insecurity problems in Ethiopia are complex and interrelated. Lack of governance and misdirected economic policies during the military regime (1974-1991), unfavourable weather fluctuations, high dependency on rainfed agriculture, and failure to bring about economic transformation have all contributed negatively to the country's agricultural performance in past decades (Gebremedhin, 2006). Declining soil fertility, land degradation, and shrinking landholding due to population pressure had contributed to the deterioration food production. These and other factors are responsible for the country's struggle to ensure food security.

Hence, ensuring food security is one of the top national priorities and forms the cornerstone of the sustainable economic growth and poverty reduction strategy in Ethiopia. To this effect, the current government has embarked in November, 2002 an aggressive economic reform program. Policies that tackle food insecurity at household level are seen as the most effective way to reduce poverty. The integrated household food security package (FSP) program is among the programs introduced for this purpose. The program aims to secure food at household level by diversifying the income base of the poor through provision of credit for a range of activities. Large amounts of money and effort have been spent by the government and multi-lateral development bodies to reduce the problems of widespread rural food insecurity and thus improve people's access to food. However, program implementation is not an end in itself. The question of how the FSP program affects the targeted beneficiaries should be evaluated after a certain period of time to investigate whether the program actually contributed to household's food security.

Despite the FSP program has been implemented in Tigray, Northern Ethiopia, over the recent decade, to our knowledge no attempts has ever been made to systematically evaluate its impacts on household food consumption. Abebaw et al. (2010) studied the impact of food security program on household food consumption in two villages of the Amhara region in the North-western part of Ethiopia using propensity-score matching. However, Abebaw et al. (2010) only provided the average impact of the food security program but did not attempt to analyse the sensitivity of their estimated impact to selection bias. In practice, there may be unobserved variables that simultaneously affect the outcome, and the assignment into program beneficiary. In such circumstances, a 'hidden bias' may influence the robustness of the matching estimators (Rosenbaum, 2002). As Ichino et al. (2006) have suggested, the presentation of matching estimates should therefore be accompanied by sensitivity analysis since propensity-score matching cannot fully account for selection bias. This apparent limitation of Abebaw et al. (2010) provides us with the starting point of this article.

The main objective of this paper is to evaluate the impact of the FSP program upon improving rural household food consumption in Tigray using a propensity score matching (PSM) method. We build up our research on the works of Abebaw et al. (2010). In this paper, we adopt the definition of food security by Siamwalla and Valdes (1980) that is, the ability of households to meet target levels of consumption on a yearly basis.

The household food secirty package program (FSP)

Tigray is one of the most drought-prone areas of Ethiopia, and faces recurrent droughts and food shortages. Most smallholder farmers face sizeable food deficits every year and are vulnerable to recurrent drought shocks. Poverty reduction and ensuring food security is Tigray's most significant development challenges.

The household oriented extension package program known as the integrated household FSP was launched in 2002 (Desta et al., 2006). This program was developed within the framework of the federal government's overall development policy and food security strategy, but addresses the specific and complex problems and causes of food insecurity that plague the region. To this end, a twin-track strategy was employed with target beneficiaries to redress short-term food deficits, while building up sufficient self-help capacity to allow the rural population to attain self-reliant food security in the long term (TFSPC, 2005).

Accordingly, the FSP program has been widely introduced in Tigray. The intention of the program is to secure food at household level by diversifying the income base of the poor through provision of credit for a range of activities in a package. It also provides income transfers through public works. To this end, identifying the basic abilities of the poor and providing the required financial resource, technical assistance and training to engage in their choice of activities is the prime concern of the program.

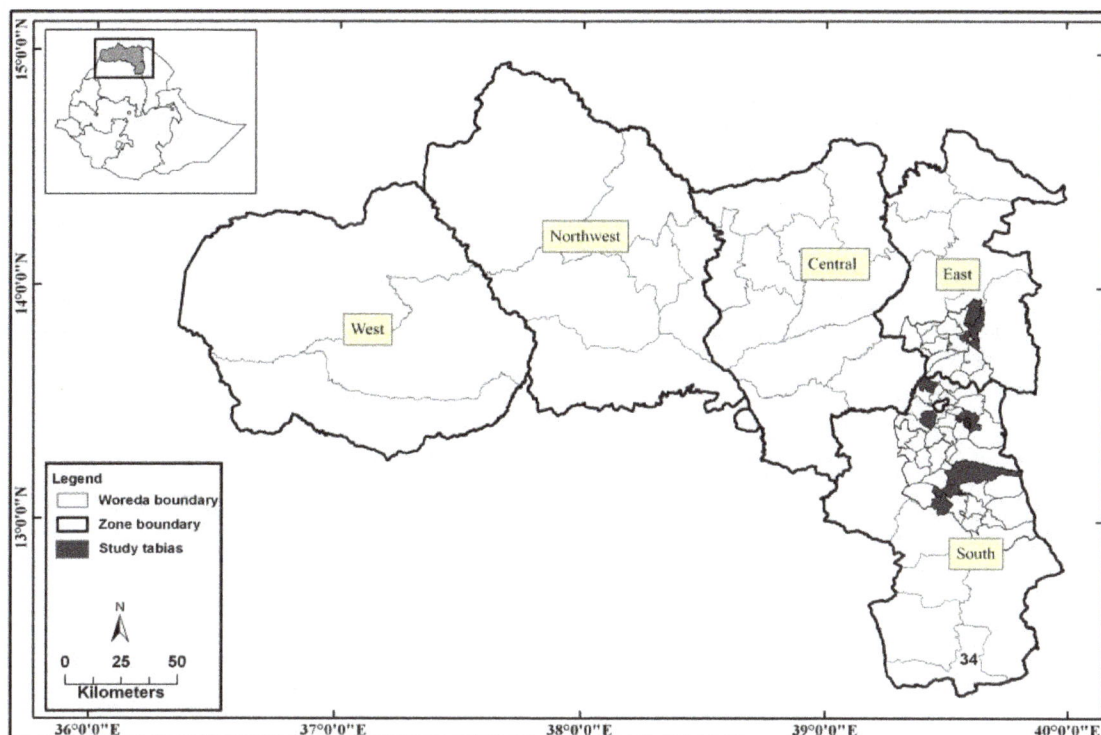

Figure 1. Administrative of map of Tigray region and location of the study villages.

The selection process of a household into the program is clearly defined in the Productive safety net program implementation manual. In each village (locally called *tabia*), beneficiary households were first selected by the local administration (food security task force) based on pre-defined criteria (TFSPC, 2005). Local communities also have discretion to identify food-insecure households based on local knowledge (Coll-Black et al., 2011). Poverty status as expressed by the household's livestock (households without cows and oxen were given priority), land holding size and quality, and severity of food insecurity are the main criteria for selecting households into the program (TFSPC, 2005). After a household is selected for the program, financial support as a loan for a range of activities is provided as a package. Households thus participate in one or more program activities, including vegetable and fruit production, livestock production (oxen and cows), small animals (sheep and goats), poultry, and beehives (Nega, 2008).

The FSP program was thus expected to address the rural household's risks of not having access to sufficient food through increasing food production and promoting employment. Provision of credit to the poor is expected to stabilize consumption and promote self-employment in off-farm activities. The program was also expected to increase household's livestock ownership and provide access to draft power that has been the long-time constraint of the agrarian society in Tigray region (TFSPC, 2005; Nega, 2008).

MATERIALS AND METHODS

The study area

Tigray is one of the regional States of Ethiopia and is located in the northern part of the country, covering a total area of 53,000 km². Geographically, it lies between latitudes 12°15' N and 14°57' N, and longitudes 36°27' E and 39°59' E (Figure 1). In the year 2007 the region had a population of 4.4 million with a population growth rate of 2.5% per annum (CSA, 2008). The climate of the region is characterized by large spatial variations in rainfall. The mean annual monsoon rainfall of the region is estimated to be 473 mm, representing 84% of the annual rainfall in the region (Gebrehiwot et al., 2011).

Tigray mainly relies on rainfed agriculture. The tremendous importance of this sector to the regional economy can be gauged by the fact that it directly supports about 82% of the population in terms of employment and livelihood.

Data sources and variable definitions

The data for the study was derived from a household survey conducted in three rural districts from January to February, 2011, and included 400 farm households randomly drawn from 9 villages. A three-stage sampling techniques was employed to draw the samples. Three districts were first chosen: two districts (Enderta and Kilte Awelaelo) from the FSP program areas and one (Hintalo Wajirat) from the non-FSP districts. Second, 4 villages from the program area were purposively chosen. Five comparable non-program villages from Hinatlo Wajirat districts were chosen based on their similarity in social, economic and agro-climatic characteristics with the program villages. Finally, random sampling was employed to draw a total sample size of 189 and 211 farm households from the program and non-program villages, respectively.

To generate the data, a structured household questionnaire was administered, with a household defined as a group of people in a housing unit living together as a family and sharing the same kitchen. The survey captured information related to demographic characteristics, asset endowment, food consumption, economic activities, wealth and income, expenditure on food and non-food items, and access to basic infrastructures and agricultural services. The sample households were asked to report food items consumed in kind and amount, purchased or otherwise, by their families during the week preceding the survey visit. The physical quantities consumed by a household were then converted into food calories adjusted for household age and sex composition using the national food composition table compiled by the Ethiopian Health and Nutrition Research Institute (EHNRI, 2000).

Enumerators with knowledge of the local language and experience with socio-economic surveying were recruited locally, and trained based on the content of the questionnaire. Prior to the actual fieldwork, the questionnaire was pre-tested. During the survey field work, close and regular supervision was made.

The food security outcome indicator

Determining the food security status of households can help public officials and policy makers to evaluate the effectiveness of existing programs. However, as with other social programs, identifying and quantifying the causal effect of a program on household food security is not straightforward (Abebaw et al., 2010). Identifying an appropriate food security indicator is thus a difficult issue as not all characteristics of food security can be captured by any single outcome indicator (Maxwell et al., 1999; Hendriks, 2005).

Maxwell and Frankenberger (1992) reported 25 broadly defined indicators. In the work by Maxwell and Frankenberger, a distinction is made between process indicators describing food supply and outcome indicators describing adequate food consumption and food access. Chung et al. (1997) found that there is little correlation between a very large set of process indicators and measures of food security outcomes. von Braun et al. (1990) described outcome indicators as proxies for adequate food consumption measured directly as food expenditure and caloric consumption.

Similarly, different organisations and government agencies use different food security indicators depending on their primary objectives. Per capita food intake per day in kilocalories is used as the indicator for food security for regional and global assessments. For example, according to FAO (2003), at national level a per capita food intake of less than 2,200 kcal/day is taken as indicative of a very poor level of food security. The most common methods of poverty measurement have also used the nutritional norm and defined a poverty line in terms of minimum calorie requirements (Greer and Thorbecke, 1986; Ahmed et al., 1991; Ravallion and Bidani, 1994). Swindale and Ohri-Vachaspat (2005) also reported that the percentage of minimum daily food calorie requirements consumed provides a good indication of overall household food security.

For this study, food calorie intake which is one of the most direct indicators related to food security and nutritional security (Hoddinott and Skoufias, 2004; Gilligan and Hoddinott, 2007) was considered as an outcome indicator to measure the impact of FSP program. In Ethiopia, food poverty is defined in terms of food calorie intake (MoFED, 2006). This implies that this indicator has direct relevance to local conditions and the food security context, which is identified as one of the criteria by Davies et al. (1991). As is also reported by Baker (2000), establishing measurable indicators that correspond directly to planned interventions is a key step in social program impact evaluation.

Empirical approach

A valid measure of the impact of a household FSP would be to compare the outcomes in households receiving FSP benefits with the presumed outcomes that had the same households and not received any benefits. Assessing the impact of any intervention thus requires making an inference about the outcome that would have been observed had the program participants not participated. Following Heckman et al. (1997) and Smith and Todd (2001), let Y_1 be the mean of the outcome conditional on participation, that is, membership of the treatment group, and let Y_0 be the outcome conditional on non-participation, that is, membership of the control group. The impact of participation in the program is the change in the mean outcome caused by participating in the program, which is given by:

$$\Delta Y = Y_1 - Y_0, \tag{1}$$

Where Δ is the notation for the impact for a given household.

The fundamental problem of evaluating this individual treatment effect arises because for each household, only one of the potential outcomes either Y_1 or Y_0 can be observed, but Y_1 and Y_0 can never be observed for the same household simultaneously. This leads to a missing-data problem, which is the heart of the evaluation problem (Smith and Todd, 2005). The unobservable component in Equation 1, be it Y_1 or Y_0, is called the counterfactual outcome. Measuring impact as the difference in mean outcome between all households involved in the FSP and those not involved may thus give a biased estimate of program impact. Since there will never be an opportunity to estimate individual treatment effects in Equation 1 directly, one has to concentrate on sample averages for the impacts of a treatment.

Average impact of the treatment on the treated (ATT), which focuses explicitly on the effect on those for whom the program is actually introduced, is the most commonly used evaluation parameter. In random program assignment, the expected value of ATT is defined as the difference between expected outcome values with and without treatment for those who actually participated in the program (Heckman et al., 1998), which is given by:

$$\Delta Y_{ATT} = ATT(\Delta Y \mid X; Z=1) = E(Y_1 - Y_0 \mid, Z=1) = E(Y_1 \mid, Z=1) - E(Y_0 \mid, Z=1), \tag{2}$$

Where Z is an indicator variable indicating whether a household i actually received treatment or not: Z_i being equal to 1 if the household is a beneficiary of FSP and 0 otherwise. X denotes a vector of control variables. Data on program beneficiaries identify the mean outcome in the treated state E (Y_1|X, Z=1). The mean outcome in the untreated E (Y_0|X, Z=1) is not observed, and a proper substitute for it has to be chosen in order to estimate ATT. As noted earlier, the FSP program followed a non-random process in targeting its beneficiaries. As Gilligan and Hoddinott (2007) have noted, this gives rise to a biased estimate of program impact and the procedure in Equation 2 should not be applied in our case. Applying PSM approach is therefore the most appealing approach to estimate the impact of the program for our study.

Propensity score matching (PSM)

The majority of the literature on evaluation methodology is centred on the use of matched-comparison evaluation techniques, which are among quasi-experimental design techniques generally considered a second-best alternative to experimental design (Baker, 2000). The propensity score is defined by Rosenbaum and Rubin (1983) as the conditional probability of receiving a treatment given pre-treatment observable characteristics. Let P = Pr (Z=1| X) denote the probability of participating in the FSP program, that is, the propensity score. PSM constructs a statistical comparison group by matching observations on the FSP participants to non-participants for similar values of propensity score. PSM estimators are based on two assumptions:

i) That non-participants provide the same mean outcomes as participants would have provided had they not received the program. This reflects a major strand of evaluation literature that focuses on the estimation of treatment effects under the assumption that the treatment satisfies some form of exogeneity (Imbens, 2004). Thus, testing is important to check if a household's characteristics within its group are similar.

$$E(Y_0|P, Z = 1) = E(Y_0|P, Z = 0) = E(Y_0|P) \quad (3)$$

ii) That households with the same Z values have a positive probability of P being both participants and non-participants [the common support assumption; Heckman et al. (1999)]:

$$0 < P < 1 \quad (4)$$

If assumptions (i) and (ii) are both satisfied, then, after conditioning on P, the Y_0 distribution observed for the matched non-participant group can be substituted for the missing Y_0 distribution for participants. Under these assumptions, the ATT of the program can be estimated as:

$$\text{ATT} = E(Y_1 - Y_0 \mid Z = 1)$$
$$= E(Y_1 \mid Z = 1) - E_{P|Z=1}\{E_Y(Y_0 \mid Z = 1, P)\}$$
$$= E(Y_1 \mid Z = 1) - E_{P|Z=1}\{E_Y(Y_0 \mid Z = 0, P)\} \quad (5)$$

Where the first term on the right-hand side of the last expression can be estimated from the treatment group and the second term from the mean outcomes of the matched (on P) comparison groups.

Based on Baker (2000), and Heckman et al. (1997, 1998) criterion, the PSM will provide reliable and low-bias estimates of FSP program impact because: (i) similar questionnaire was used to elicit data from beneficiaries and non-beneficiaries, (ii) the dataset came from farm households with similar socio-economic and demographic conditions as well as a similar economic environment, (iii) the propensity score was estimated by using the sample households' observable characteristics that were relevant for both participation in the program and for the outcome variable of interest, and (iv) the dataset has a larger sample of non-beneficiaries households.

In implementing the PSM, an empirical model has to be specified to derive the propensity score. For the FSP program, we estimated the propensity score for participation in the program with a logit model using observable variables that included both determinants of participation in the program and factors that affected the outcome. Once we estimated the propensity score that appeared to capture the similarities, we used these similarities to match each beneficiary with his/her closest non-beneficiary. We performed several tests to select a preferred estimator and chose the estimator that yielded statistically identical covariate means for both groups (Caliendo and Kopeinig, 2008). Moreno-Serra (2009) indicated that a good matching estimator is expected to retain relatively larger observations for evaluating the impact of a program. We implemented a kernel-matching estimator using the PSM algorithm with the software package STATA 12 to compute the average impact of the program among FSP households based on the above indicators. Morgan and Winship (2007) argued that kernel-matching, introduced by Heckman et al. (1998) appears to

be the most efficient and preferred algorithm.

Finally, the PSM approach cannot fully account for selection bias or unobservable characteristics. In practice there may be unobserved variables that simultaneously influence treatment allocation as well as potential outcomes (Becker and Caliendo, 2007). In such circumstance, a 'hidden bias' might arise that influence the robustness of the matching estimators (Rosenbaum, 2002). Thus, the bias due to selection on unobservables remains as its drawback. Hence, following Rosenbaum (2002) we performed sensitivity analysis to examine the vulnerability of the estimated impact to unobservables.

Conditioning variables for program participation

In PSM, it is desirable to condition the match on variables that are highly associated with the outcome variables (Heckman and Navarro-Lozano, 2004). Smith and Todd (2005) noted that there is little guidance available to researchers on how to select the set of conditioning variables used to construct the propensity score. Thus we focussed on finding a set of conditioning variables that were highly associated with program eligibility and the outcome variable. Fortunately, our data set contained a set of conditioning variables to control program participation decisions.

As described earlier, the FSP program is intended to serve the food insecure households. One way of judging the welfare level of rural households in the study region would be on the basis of assets owned. Hence, we included the two basic assets in the Ethiopian rural economy, land and livestock owned. Lack of these assets was associated with program eligibility. Pre-intervention demographic variables such as type of household headship, age of household head, family size, number of children under five and dependency ratio associated with program eligibility and the outcome variables were also included.

Furthermore, we included as a control variable the households' proximity to basic physical infrastructure. With this rich set of control variables (Table 1) and relatively large and comparable sample sizes (in both the treatment and the comparison group), we could capture many of the determinants of participation typically unobservable to researchers.

RESULTS AND DISCUSSION

Descriptive

Participation in the FSP program, the dependent variables in the impact assessment analysis, takes the value of 1 if a household participates in the program and 0 otherwise. Summary statistics of FSP participants and non-participants are presented in Table 2. About 26% of the participating individuals were women. As presented in Table 2, household FSP program beneficiaries and non-beneficiary had significant differences on certain pre-intervention characteristics, which are elicited using respondents recall. The main differences between the two groups of households were in particular observed with respect to family size, dependency ratio, size of land, livestock ownership, and distance to all-weather roads and to the nearest market. As compared to non-beneficiary households, FSP program beneficiary households' had smaller number of livestock and oxen ownership and smaller size of land.

Table 1. Variable description and measurement.

Variable	Type	Measurement
Dependent variable, treated	Dummy	1 if yes-participants of FSP, 0 otherwise
Explanatory variables		
Sex of household head	Dummy	1 if head is male, 0 otherwise
Age of household head	Continuous	Age of the household head in years
Education	Dummy	1 if he/she can read and write, 0 otherwise
Farm size	Continuous	Size of the household in numbers
Children under 5 years	Integer	Number of children under five
Dependency ratio	Continuous	Ratio of dependent members to the productive age group
Land holding size	Continuous	Hectare
Livestock ownership in TLU[a]	Continuous	Tropical Livestock Unit
Oxen ownership	Continuous	Tropical Livestock Unit
Value of agricultural equipment owned	Continuous	Ethiopian Birr
Distance to the market	Continuous	Walking distance in minutes
Distance to all-weather road	Continuous	Walking distance in minutes

Table 2. Summary statistics: characteristics of beneficiaries and non-beneficiaries.

Variable	Sample households N = 400		FSP beneficiary HHs N = 189		FSP non-beneficiary HHs N = 211		Difference		t-value
	Mean	STD	Mean	STD	Mean	STD	Mean	STD	
Sex	0.77	0.42	0.74	0.44	0.80	0.40	-0.06	0.04	-1.42
Age	39.04	12.22	39.61	13.57	38.52	10.87	1.09	2.70	0.88
Education	0.46	0.49	0.46	0.49	0.48	0.50	-0.02	-0.01	-0.40
Family size	5.30	1.77	4.98	1.76	5.57	1.72	-0.59	0.04	-3.38***
Dependency ratio	1.28	0.81	1.37	0.94	1.21	0.66	0.16	0.28	1.94*
Land size	0.96	0.47	0.72	0.39	1.19	0.42	-0.27	-0.15	-5.54**
Livestock ownership	2.35	1.61	1.15	0.95	3.44	1.20	-2.29	-0.25	-21.26***
Oxen	1.35	1.04	0.79	0.87	1.85	0.91	-1.06	-0.04	-11.90***
Value of agri. Equip.	230.19	187.13	141.05	73.56	307.35	221.42	-166.30	-147.86	-10.29***
Distance to all-weather road	35.12	11.76	28.76	7.86	40.83	11.75	-12.07	-3.89	-12.19***
Distance to the nearest market	43.36	19.57	35.00	11.82	50.85	21.98	-15.85	-10.16	-6.80***

* = Significant at 10%; ** = Significant at 5%; and *** = Significant at 1%.

Table 2 also clearly depicts that FSP and non-FSP households had a food calorie intake of 2512 and 1748 cal, respectively indicating that households' in the FSP program are better off. Abebaw et al. (2010) reported similar findings.

Nonetheless, descriptive result cannot explain whether the observed difference in calorie intake between the two groups of household is due to FSP program or other exogenous factors. Indeed identification of a casual effect cannot be made before accounting for the effects of confounding factors.

Propensity score estimate

Prior to non-parametrically estimating the impact of the scores required specification justifying that a household

had been included in the FSP. Thus, we had to respect the conditional independence assumption that the covariates are exogenous and unaffected by the program. Caliendo and Kopeinig (2008) noted that the basic idea of matching is to compare a beneficiary with one or more non-beneficiaries who are similar in terms of a set of observed characteristics. This requires predicting the propensity scores for each individual using a logit or a probit model. In this study, we used a logit model to predict the probability that a household participates in the food security program; in this model, different household characteristics are included as regressors.

[a] The total number of livestock ownership is measured in Tropical Livestock Units (TLU), an index that aggregates different types of livestock a household owned into a single number. It is calculated using the following weighing index factors from ILRI (1990): cow = 0.8, sheep and goat = 0.09, donkey = 0.36, horse and mule = 0.8, 0x = 1.1

Table 3. Logit estimates for participation in the FSP program (n = 400).

Logit specification	Model
Sex of household head	1.519* (2.08)
Age in years	-0.461* (1.85)
Education	-1.387** (3.15)
Farm size	-0.189 (-0.13)
Number of children under 5 years	1.185** (3.85)
Dependency ratio	0.387 (1.32)
Size of land holding	-3.198** (4.88)
Livestock ownership in TLU	-1.772** (5.18)
Oxen ownership in TLU	-1.026*** (3.79)
Value of agricultural equipment's	-0.158** (5.72)
Distance to the nearest market	-0.102** (5.21)
Distance to all weather road	-0.045** (3.48)
Constant	12.19** (8.17)
Log likelihood	-87.07
Pseudo R2	0.27
Chi2	400.07**
P	0.000

Dependent variable equals 1 if household participated in the FSP program and 0 otherwise. Absolute value of z-statistics are in parentheses. * and ** significant at probability levels of 10 and 1%, respectively.

Chaouani (2010) argued that the functional form of propensity score is chosen based on the results of the logit estimation of the probability of going public. We tried various alternative specifications and chose the logit model presented in Table 3 because it seemed to be the more significant and robust specification. The 'common support' restriction was imposed to improve the quality of the matches and the balancing property was set and passes the balancing tests at the 95% level of statistical significance. Hence, we ensured that the mean propensity score was not different for the treatment sample and the sample of comparison observations at various levels of propensity scores. Significant coefficients in the estimated equation implied that FSP and non-FSP households were different with respect to the corresponding variable.

As indicated in Table 3, size of landholding, livestock ownership, oxen ownership and proximity to an input and output markets significantly influenced household participation in the FSP program. As expected, participation in the program was negatively and significantly influenced by the value of agricultural equipment owned. Distances to all-weather roads and to a market were also directly correlated with a household's participation in the program.

The estimated mean propensity score using the main specification for the whole sample was 0.472 (with a standard deviation of 0.453) implying that the average probability of participating in the FSP program for all individual households was 47%.

Average impact of participation in the FSP

Using estimated propensity scores for the program from the model specification in Table 3, the impact of the integrated FSP program on household calorie intake is estimated with kernel-based matching. We also estimated the FSP impact using other matching estimators particularly the nearest neighbor (NN) matching estimator, to assess the robustness of the results. Matching with replacement was performed. The latter minimized the propensity-score distance between the matched comparison units and the treatment unit, each treatment unit being matched to the nearest comparison unit, even if a comparison unit was matched more than once. This is important in terms of bias reduction. By contrast, when matching without replacement, and with few comparison units similar to the treated units, one may be compelled to match treated units to comparison units that are quite different in terms of the estimated propensity score. This increases bias, but could improve the accuracy of the estimates (Mendola, 2007). Dehejia and Wahba (2002) have reported that the results of matching without replacement are potentially sensitive to the order in which the treatment units are matched.

Table 4 presents estimates of the average impact of participation in the FSP. Overall, matching estimates show that the FSP program has a positive and robust effect on household food calorie intake. The findings indicate that the program improved household's food

Table 4. FSP program impacts on households' food calorie intake, matching estimates (n = 400).

Outcome variable	Model specification
Household food calorie intake	772.19* (6.13)
Observations	
FSP households	97
Non-FSP households	211

Absolute values of t statistics on ATT are in parentheses. * Significant at probability levels of 1%.

calorie intake by 772.19 kcal/day per adult equivalent unit. This means that, if we selected someone to be in the FSP (that is, provided with access to a loan for a package of activities and training), his/her food calorie intake would on average increase to 41.8% above that of individuals not involved in the program. This suggests that the FSP program has a causal influence on total food consumption when individuals are matched according to relevant socio-demographics, assets and other covariates. In a population made up of poor households where the major income-earning asset is human labour, increased calorie intake may imply increase productivity, increased income and hence increased nutrition (Aromolaran, 2004). Nega (2008) similarly reported that the importance of the food-for-work and food security program for the chronically poor and transiently poor households in Tigray region. Abebaw et al. (2010) also found a positive impact of the FSP on household consumption in two villages of the Amhara region in the Northwest part of Ethiopia.

An explanation for this significant effect of the FSP program may be: first, the household-level FSP program is a coordinated one involving key players in the rural development of the region, in particular the Regional Bureau of Agriculture and Rural Development, the Food Security Coordination Office and the Dedebit Credit and Saving Institute - the leading locally operating microfinance institute in Ethiopia. Second, the nature of the program provided better opportunities for the beneficiaries to engage in their choice of activities and obtain the required resources, technical assistance and training. Third, the number of development agents assigned to each village centre also increased from one to three over recent decade.

Sensitivity analysis

As indicated, the PSM approach cannot fully be controlled for unobservable characteristics. As Ichino et al. (2006) have suggested, the presentation of matching estimates should be accompanied by sensitivity analysis. Accordingly, we checked the sensitivity of the estimated treatment effects to selection on unobservables using the

bounding approach developed by Rosenbaum (2002). We applied the 'mhbounds' procedure by Becker and Caliendo (2007) in STAT programs to aid in the construction of Rosenbaum bounds for the sensitivity testing. This procedure uses the matching estimates to determine the confidence intervals of the outcome variable for different values of Γ (gamma)[1]. Γ captures the degree of association of an unobserved characteristic with the treatment and outcome required for it (the unobserved characteristic) to explain the observed impact (Duvendack and Palmer-Jones, 2011). DiPrete and Gangl (2004) indicated that, if the lowest Γ, which encompasses 0, is relatively small (say < 2), then one may state that the probability of such an unobserved characteristic is relatively high and the estimated impact is therefore sensitive to the existence of unobservables.

Table 5 reports the Mantel-Haenszel (mh) bounds results, showing that under the assumption of no hidden bias, when $\Gamma = 1$, the Q_{mh} test statistic indicates a highly significant treatment effect for improved food security program intervention on household food calorie intake. The two bounds in the Mantel-Haenszel output table (Table 5) can be interpreted in the following way: The Q_{MH+} statistic adjusts the MH statistic downward for positive (unobserved) selection. In our case, positive selection bias occurs when those most likely to participate tend to have higher food calorie intake even without participation in the program, and given that they have the same χ vector of covariates as the individuals in the control group. This effect leads to an upward bias in the estimated treatment effect[2]. The effect is significant under $\Gamma = 1$ and becomes even more significant for increasing values of $\Gamma > 1$ if we have underestimated the true treatment effect. The Q_{MH+} reveals that the study is insensitive to hidden bias at the 5% significance level. The sensitivity analysis thus indicates that the observed results on the impact of food security program on households' food calorie intakes are insensitive to

---

[1] Γ is the ratio of the odds that the treated have this unobserved characteristic to the odds that the controls have it.

[2] The Q_{MH-} statistic adjusts the MH statistic downward for negative (unobserved) selection.

Table 5. Mantel-Haenszel bounds for outcome = food calorie intake.

Γ	Q_{MH+}	Q_{MH-}	P_{MH+}	P_{MH-}
1	3.057	3.057	0.0012	0.0012
1.1	1.931	4.738	0.0586	0.0003
1.2	1.468	4.877	0.0336	0.0073
1.3	1.027	5.229	0.0132	0.0111
1.4	0.759	5.444	0.0289	0.0106
1.5	0.368	6.088	0.0324	0.0146

Source: MH Bounds using STATA 12. Γ = 1 ≈ No 'hidden' bias; Q_{mh+} : Mantel-Haenszel statistic; Q_{mh-}: Mantel-Haenszel statistic; p_{mh+} : significance level; and p_{mh-}: significance level.

selection on unobservable or hidden bias.

Conclusions

Reducing poverty and improving household food security is an important policy objective for rural development in the semi-arid areas of many countries in Africa. While much has been achieved in reducing rural poverty in recent years, the problem of food insecurity is still evident. It is thus pertinent to understand whether food security program contribute to household's food security. Systematic evaluation of the FSP program is therefore necessary in order to grasp how successful implemented household food security program has been. We used a survey data of 400 rural households in the Tigray region in Northern Ethiopia to analyse the impact of the most widely implemented household FSP program. To examine the impact of the program, observed outcomes were compared with the outcomes that would have resulted had the targeted group not participated in the program. We estimated the impact of the FSP program on calorie intake using PSM as a method of estimating the counterfactual outcome for program beneficiaries. Use of PSM ensured that the program beneficiaries and the comparison group shared almost exactly the same characteristics so that selection bias could be mitigated in the sample.

The findings indicate that the FSP program had a significant effect on improving household food calorie intake of poor farm households in the region. After matching participants in the FSP program with non-participants on the basis of some socio-demographic characteristics, asset and other variables, we found that the level of food calorie intake of the FSP program participants was 41.8% higher than the intake of households not involved in the program. Sensitivity testing of the results carried out using Rosenbaum bounds indicated that the observed estimate of impact is not sensitive to hidden bias or selection on unobservables. Thus, this study appears to have the successfully captured and used variables associated with provision of the program.

We concluded that the impact of pro-poor focussed programs, and the FSP program in particular, indeed show the insight that appropriate development policies and programs have a role to play in improving food security outcomes and reducing poverty in rural areas where most of the poor live. However, like all studies, ours is not without limitations. First, our analysis is limited to cross-sectional data. This limits the observation of short and long-term fluctuations in household food consumption level, and food calorie intake in particular. Accordingly, the seasonal dimension to household food security, and particularly food calorie intake, is not considered. Second, the PSM approach cannot fully eliminate bias caused by unobserved confounders and the bias due to selection on unobservables remains as its drawback. These limitations should be kept in mind when evaluating the conclusions of our study.

Conflict of Interest

The author(s) have not declared any conflict of interest

ACKNOWLEDGEMENTS

The authors would like to thank the editor and three anonymous referees for their valuable comments. We appreciate the survey households for their willingness to participate in this study. However, the authors alone are responsible for the ideas expressed in the paper.

REFERENCES

Abebaw D, Yibeltal F, Kassa B (2010). The impact of a food security program on household food consumption in Northwestern Ethiopia: A matching estimator approach. Food Policy 35:286-293.

Ahmed AU, Haider AK, Sampath RK (1991). Poverty in Bangladesh: Measurement, decomposition and intertemporal comparison. J. Dev. Stud. 27(4):48-63.

Aromolaran AB (2004). Household income, women's income share and food calorie intake in South Western Nigeria. Food Policy 29(5):507–530.

Baker JL (2000). Evaluating the impact of development projects on poverty. A handbook for practitioners. The World Bank, Washington, DC.

Becker SO, Caliendo M (2007). Sensitivity analysis for average treatment effects. Stata J. 7(1):71-83.

Caliendo M, Kopeinig S (2008). Some practical guidance for the implementation of propensity score matching. J. Econ. Surv. 22(1):31-72.

Chaouani S (2010). Using propensity score matching and estimating treatment effects: An application to the post-issue operating performance of French IPOs. Int. Res. J. Financ. Econ. 48:73-93.

Chung K, Haddad L, Ramakrishna J, Riely F (1997). Identifying the food insecure: The application of mixed-method approaches in India. International Food Policy Research Institute (IFPRI), Washington, DC.

Coll-Black S, Gilligan DO, Hoddinott J, Kumar N, Taffesse AL, Wiseman W (2011). Targeting Food Security Interventions When "Everyone is Poor": The Case of Ethiopia's Productive Safety Net Programme. ESSP II Working Paper 24, International Food Policy Research Institute.

CSA (2008). Summary and Statistical Report of the 2007 Population and Housing Census. Federal Democratic Republic of Ethiopia, Population Census Commission, Addis Ababa, Ethiopia.

Davies S, Buchanan-Smith M, Lambert R (1991). Early Warning in the Sahel and Horn of Africa: The State of the Art; a review of the literature. Institute of Development Studies, Brighton, UK.

Dehejia RH, Wahba S (2002). Propensity score matching methods for nonexperimental causal studies. Rev. Econ. Stat. 84(1):151-161.

Desta M, Haddis G, Ataklt S (2006). Female-headed households and livelihood Intervention in four selected Woredas in Tigray, Ethiopia. DCG Report No. 44.

Devereux S (2006). Distinguishing between chronic and transitory food insecurity in emergency needs assessments. World Food Program, Emmergency Needs Assessment Branch.

DiPrete TA, Gangl M (2004). Assessing Bias in the estimation of causal effects: Rosenbaum bounds on matching estimators and instrumental variables estimation with imperfect instruments. Sociol. Methodol. 34(1):271-310.

Duvendack M, Palmer-Jones R (2011). High Noon for Microfinance Impact Evaluations: Re-investigating the Evidence from Bangladesh. MPRA Paper No. 27902, University of East Anglia, Norwich, UK.

EHNRI (2000). Food Consumption Table for Use in Ethiopia. Ethiopia Health and Nutrition Research Institute (EHNRI), Part III, Addis Ababa, Ethiopia.

FAO (2003). Household Food Security and Community Nutrition. Agriculture and Consumer Protection Department. Available at http://www.fao.org/ag/agn/nutrition/household_en.stm.

FAO (2014). The State of Food Insecurity in the World. Strengthening the enabling environment for food security and nutrition. Rome.

Gebrehiwot T, van der Veen A, Maathuis B (2011). Spatial and temporal assessment of drought in the Northern highlands of Ethiopia. Int. J. Appl. Earth Obs. 13(3):309-321.

Gebremedhin TS (2006). The analysis of Urban poverty in Ethiopia. University of Sydney, Australia.

Gilligan DO, Hoddinott J (2007). Is there persistence in the impact of emergency food aid? Evidence on consumption, food security, and assets in rural Ethiopia. Am. J. Agric. Econ. 89(2):225-242.

Greer J, Thorbecke E (1986). Food poverty profile in applied to Kenyan Smallholders. Econ. Dev. Cult. Change 35(1):115-141.

Heckman JJ, Hidehiko Ichimura, Todd PE (1997). Matching as an econometric evaluation estimator: Evidence from evaluating a job training programme. Rev. Econ. Stud. 64(4):605-654.

Heckman JJ, Hidehiko Ichimura, Todd PE (1998). Matching as an Econometric Evaluation Estimator. Rev. Econ. Stud. 65(2):261-294.

Heckman JJ, LaLonde RJ, Smith JA (1999). The economics and econometrics of active labor market programs. Handbook of labor economics, 3:1865-2097.

Heckman JJ, Navarro-Lozano S (2004). Using matching, instrumental variables, and control functions to estimate economic choice models. Rev. Econ. Stat. 86(1):30-57.

Hendriks SL (2005). The challenges facing empirical estimation of household food (in) security in South Africa. Dev. S. Afri. 22(1):103-123.

Hoddinott J, Skoufias E (2004). The impact of PROGRESA on food consumption. Econ. Dev. Cult. Change 53(1):37-61.

Ichino A, Mealli F, Nannicini T (2006). From Temporary Help Jobs to Permanent Employment: What Can We Learn from Matching Estimators and their Sensitivity? Forschungsinstitut zur Zukunft der Arbeit (IZA) Discussion Paper No. 2149.

Imbens G (2004). Nonparametric estimation of average treatment effects under exogeneity: A review. Rev. Econ. Stat. 86(1):4-29.

Maxwell D, Ahiadekeb M, Levinc C, Armar-Klemesud M, Zakariahd S, Grace MaryLampteyd GM (1999). Alternative food-security indicators: revisiting the frequency and severity of 'coping strategies'. Food Policy 24:411-429.

Maxwell S, Frankenberger T (1992). Household food security: Concepts, indicators, measurements: A Technical Review. International Fund for Agricultural Development (IFAD), Rome.

Mendola M (2007). Agricultural technology adoption and poverty reduction: A propensity-score matching analysis for rural Bangladesh. Food Policy 32:372-393.

MoFED (2006). Plan for Accelerated and Sustained Development to End Poverty, 2005/06-2009/10, vol. I. Ministry of Finance and Economic Development (MoFED), Addis Ababa, Ethiopia.

Moreno-Serra R (2009). Health program evaluation by propensity score matching: accounting for treatment intensity and health externalities with an application to Brazil. Health Econometrics and Data Group, The University of York.

Morgan SL, Winship C (2007). Counterfactuals and Causal Inference. Methods and Principles for Social Research. Cambridge University Press, Cambridge.

Nega F (2008). Poverty, asset accumulation, household livelihood and interaction with local institutions in northern Ethiopia. PhD Dissertation, Katholieke Universiteit Leuven, Faculteit Bio-ingenieurswetenschappen, P. 191.

Ravallion M, Bidani B (1994). How Robust is a Povert Profile? World Bank Econ. Rev. 8(1):75-102.

Rosenbaum PR (2002). Observational Studies. 2nd ed. Springer, New York.

Rosenbaum PR, Rubin DB (1983). The central role of the propensity score in observational studies for causal effects. Biometrika 70(1):41-55.

Siamwalla A, Valdes A (1980). Food insecurity in developing countries. Food Policy 5(4):258-272.

Smith J, Todd P (2001). Reconciling conflicting evidence on the performance of propensity-score matching methods. Am. Econ. Rev. 91(2):112-118.

Smith J, Todd P (2005). Does matching overcome LaLonde's critique of nonexperimental estimators? J. Economet. 125:305-353.

Swindale A, Ohri-Vachaspat P (2005). Measuring Household Food Consumption: A Technical Guide. Washington, D.C.: Food and Nutrition Technical Assistance (FANTA) Project, Academy for Educational Development (AED).

TFSPC (2005). Productive Safety Net Program Implementation Manual. Tigray Food Security Project Co-ordination (TFSPC) office.

von Braun J, Kinteh K, Puetz D (1990). Structural adjustment, agriculture and nutrition: policy options in the Gambia. Working Papers on Commercialization of Agriculture and Nutrition No. 4. International Food Policy Research Institute, Washington, DC.

Webb P, von Braun J, Yohannes Y (1992). Famine in Ethiopia: Policy implications of coping failure at national and household levels. International Food Policy Research Institute, Washington, DC.

20

Determinants of postharvest losses in tomato production in the Offinso North district of Ghana

Robert Aidoo , Rita A. Danfoku and James Osei Mensah

Department of Agricultural Economics, Agribusiness and Extension, Kwame Nkrumah University of Science and Technology (KNUST), Kumasi, Ghana.

The aim of this study was to examine the determinants of postharvest losses in tomato production in the Offinso North district of Ghana. A standardized structured questionnaire was used to collect data from 150 farmers who were selected through a combination of purposive and simple random sampling techniques. We used descriptive statistics to summarize the characteristics of the respondents. Multiple regression analysis was conducted to examine the determinants of postharvest losses in tomatoes. A typical tomato farmer in the district was found to be a male of 44 years, married, with a household size of five and had attained basic level of education. On average, farmers cultivated tomatoes on a farm size of about 5 acres and had about 20 years of farming experience. The study showed that farmers obtained 1,159.21 kg of tomatoes in the major season and 962.78 kg in the minor season on an acre of land, out of which 40 and 14% were lost, respectively. From the perspective of the farmers, the primary sources of losses were rot and bruises caused by poor handling, diseases and pest attack. From the regression analysis, gender of the farmer, household size, farm size, days of storage, membership of Farmer Based Organization (FBO) and type of tomato variety cultivated were found to significantly influence the level of postharvest losses incurred. Female gender, farm size and days of storage were found to be positively associated with losses in tomato production. However, household size, membership of FBO and cultivation of improved varieties were found to reduce postharvest losses, *ceteris paribus*. Lack of storage facilities, high cost of production and limited access to credit were found to be the critical constraints faced by tomato farmers. The study recommended the formation and joining of FBOs, periodic training and education of farmers on the cultivation of improved varieties of tomatoes as well as training on proper handling of tomato fruits to reduce postharvest losses.

Key words: Tomato, postharvest losses, regression analysis, Ghana.

INTRODUCTION

Tomato is an important cash crop in the forest, transitional and savannah zones of Ghana (Norman, 1992). It forms a very important component of food consumed at the household level as evident in the fact that many Ghanaian dishes have tomatoes as a component ingredient (Tambo and Gbemu, 2010). Tomato production is a source of livelihood and income for a greater number of people in the Offinso North district in Ghana as well as agents involved in its distribution and marketing throughout the country.

Vegetables like tomato are usually harvested when they are fresh and high in moisture and are thus distinguished from field crops, which are harvested at the mature stage for grains, pulses, oil seeds or fibre. This high moisture content of such vegetables makes their handling, transportation and marketing a special problem particularly in the tropics (Sablani et al., 2006).

The quality and nutritional value of fresh produce like tomato are affected by postharvest handling and storage condition (Sablani et al., 2006). Tomato losses can be caused by a wide variety of factors, ranging from growing conditions to handling at retail level. Many postharvest losses are direct result of factors such as high field temperatures on crops before harvesting, pests and diseases attack, among others.

In Ghana, there has been serious attempt at improving the production capacities of farmers to increase tomato production (Yeboah, 2011). However, the sector is plagued with huge levels of post-harvest losses. Robinson and Kolavalli (2010) indicated in their research report that postharvest losses are highest for tomatoes and lettuce which record up to 20% after 5 days of harvesting. Out of the 510,000 metric tons of fresh tomato fruits produced annually in Ghana, the country losses about 153,000 metric tons (30%). In 2011, the Offinso-North district produced about 19,550 metric tons of tomatoes but lost about 31% due to postharvest losses (MoFA, 2011).

The tomato production sector in Ghana has failed to reach its maximum potential in terms of yields as compared to other countries as well as improving the livelihoods of those households involved in the production of the crop. Average yields remain low, typically under 10 tons/ha, due partly to postharvest losses (Robinson and Kolavalli, 2010). Not only are these losses clearly a waste of food, but they also represent a waste of human effort, farm inputs, and scarce resources such as water (World Resource Institute, 1998).

Many factors have been hypothesized in the professional literature to be very important determinants of postharvest losses in tomato. Inappropriate storage facilities and rough handling during harvesting result in bruising and increased possibilities of contact of the produce with the soil which leads to contamination with organisms. Long distances from farms to markets as well as insufficient storage conditions can lead to losses to the tomato produce (Chandy, 1989). Adarkwa (2011) reported that improper harvest and postharvest practices result in losses due to spoilage of the product before reaching the market, and loss of quality attributes such as appearance, firmness, taste and nutritional value. A study by Babalola et al. (2010) showed that the longer the distance from farm to the market, the greater the losses experienced due to congestion of the tomato fruits and the resultant build-up of heat. Mujib et al. (2007) also

noted that type and quantity of labour used in harvesting played a vital role in postharvest losses. Skilled labourers pick and handle the produce with care and hence do little damage to the fruit. They, therefore, recommended the use of trained labourers if postharvest losses are to be minimized. Tomato fruits should be harvested at mature green state for long distance marketing and full ripen stage for fresh consumption in order to reduce postharvest losses (Moneruzzaman et al., 2009). The variety of tomato cultivated affects the level of postharvest losses experienced by farmers as different varieties have different characteristics such as firmness, disease resistance, among others, which impact on postharvest losses. Orzolek et al. (2006) recommended that tomato producers should harvest mature fruits in the morning when the temperature is cool to reduce losses.

In Ghana, attempts at explaining the underlying causes of postharvest losses in tomato production have largely remained in the realm of speculation and conjecture. However, empirical information on the main causes of these losses are required if solutions are to be found for this critical problem in tomato production. Therefore, this study was designed to examine empirically, the factors that influence the level of postharvest losses of fresh tomatoes at the farm level. Specifically, the study sought to determine the level of postharvest losses experienced by tomato producers and the key factors that account for these losses.

METHODOLOGY

Study area

The study was conducted in the Offinso North district of the Ashanti Region of Ghana. Offinso North is located in the extreme North-Western part of the region and lies within longitude 1°45'N and 1°65'W. The district has a population of about 56,881 (GSS, 2010), with a total land area of 1,008.3 km². The current farming population is around 30,000 comprising 15,030 males and 14,970 females. The district lies within the wet semi-equatorial zone of Ghana with a bi-modal rainfall regime and a mean monthly temperature of 27°C. Offinso North district is the leading tomato producing district in the Ashanti region. Tomato is grown all over the district with heavy concentration at Akomadan, Afrancho, Nkenkaasu, Asuoso, Nsenua and Mantukwa communities. The average annual production is over 19,000 metric tons of tomato fruits. Each year over 30% of tomato fruits goes waste with some farmers refusing to harvest due to very low market price for the commodity. Total land area under tomatoes cultivation is estimated at about 20,049 ha. Tomato is produced throughout the year in the district in valley bottoms and with small scale local irrigation schemes (MoFA, 2011).

Method of data collection and analytical procedure

Primary data was obtained from tomato farmers through personal interviews with the use of a standardized structured questionnaire. In consultation with Agricultural Extension Agents (AEAs) at the district, a list of communities noted for tomato production was prepared and a simple random sampling technique was used to

Table 1. Summary statistics of the socioeconomic characteristics of respondents.

Variable	Minimum	Maximum	Mean	Std. deviation
Age (years)	19.00	62.0	44.00	9.44
Household size	1.00	13.00	5.00	1.82
Years of education	0.00	19.00	5.00	4.16
Annual income (GHȻ)	478.00	8000.00	3303.40	1880.45
Farm size (Ha)	0.20	8.00	2.12	1.84
Land owned by household (ha)	0.00	18.00	7.75	4.68

Source: Survey Data (2013).

select six communities including: Akomadan, Afrancho, Nkenkaasu, Asuoso, Nsenua and Mantukwa. A list of tomato producers at the community level was obtained and a systematic random sampling technique was used to select 25 farmers from each community. The questionnaire used for the interview sought information on general characteristics of respondents, production information, postharvest losses and constraints faced by tomato producers. Interviews were done in the local language in order not to create any language barrier. Key informant interviews (with Agricultural Extension officers and Researchers at Crops Research Institute) were also conducted to gather technical information on tomato production in order to verify and validate the accuracy of some information supplied by farmers.

Descriptive statistics such as arithmetic mean, standard deviation as well as frequency distribution tables and charts were employed to summarize the characteristics of the respondents. Economic value of fresh tomato fruits lost was obtained by multiplying the physical quantity of fruits lost by the average prevailing market price. Multiple regression analysis was employed to determine the main factors that influence postharvest losses. The model used was specified in the double logarithmic form as:

$$Ln\ PHL = b_0 + b_1 LnX_1 + b_2 LnX_2 + b_3 LnX_3 +$$
$$b_4 LnX_4 + b_5 LnX_5 + b_6 LnX_6 + b_7 LnX_7 + b_8 LnX_8 + \mu$$

Where Ln denotes natural logarithm; PHL = postharvest losses (kg); X_1 = time of harvest after maturity (days); X_2 = type of labour used for harvesting (1 = family labour; 0 if otherwise); X_3 = time between harvesting and selling of produce (days); X_4 = variety of tomato grown (1 = if improved variety; 0 if otherwise); X_5 = farm size (acres); X_6 = distance from farm to market (km); X_7 = member of Farmer Based Organization (FBO) (1 = Yes; 0 = No); X_8 = Quantity of fruits harvested (kg); μ = error term.

The double logarithmic functional form is usually preferred in empirical analysis since coefficients are easy to interpret; it also has the added advantage of reducing the incidence of multicolinearity. The model was estimated using the ordinary least squares method. A five-point likert scale was used to assess the constraints faced by tomato producers in the district.

RESULTS AND DISCUSSION

Characteristics of farmers

Tomato production in the Offinso district was found to be dominated by males; only 23% of the respondents were females. However, most of these males work together with their spouses on their tomato farms. A typical tomato farmer was found to be about 44 years, with basic level of

education and a household size of five people (Table1). Out of about 7.6 ha of farm land owned by a typical farmer, about 2.1 ha were put under tomato cultivation, implying that farmers are largely small to medium scale producers. Annual income at the household level was estimated to be GHC3303.40 (US$1573.05) which translates to about GHC660.68 (US$314.61) per capita per annum. It can be inferred from the figure that on average tomato farmers are quite poor since they live under US$2.00 per day per capita.

Causes of postharvest losses

Farmers were provided with several options to select the main cause of postharvest losses in tomato production. From their ranking, postharvest losses resulted largely from rot and bruises (mechanical damage) which were mainly caused by on-farm activities (Figure 1). Farmers reported that rot resulted from over-use of spraying chemicals (herbicides and insecticides), excess watering and contact of fruits with the soil. Bruises, however, resulted from poor staking and poor handling during harvesting and sorting. From the perspective of the farmers, the three most critical secondary factors that impacted heavily on postharvest losses in tomato production were lack of ready market for produce, unreliable means to transport produce to market and longer distances from producing centres to market centres (Table 2). It can be inferred from the table that farmers consider marketing issues as the main cause of postharvest losses in tomato production. Things within their control such as time of harvest, type of variety grown and harvesting technique adopted were rather considered to have low or minimal impact on postharvest losses.

Analysis of tomato output, revenue and postharvest losses

Table 3 summarizes information on production, losses and revenues obtained from tomato production during the 2012 cropping season (Detailed results are in the Appendix). The results indicate that the average land

kok

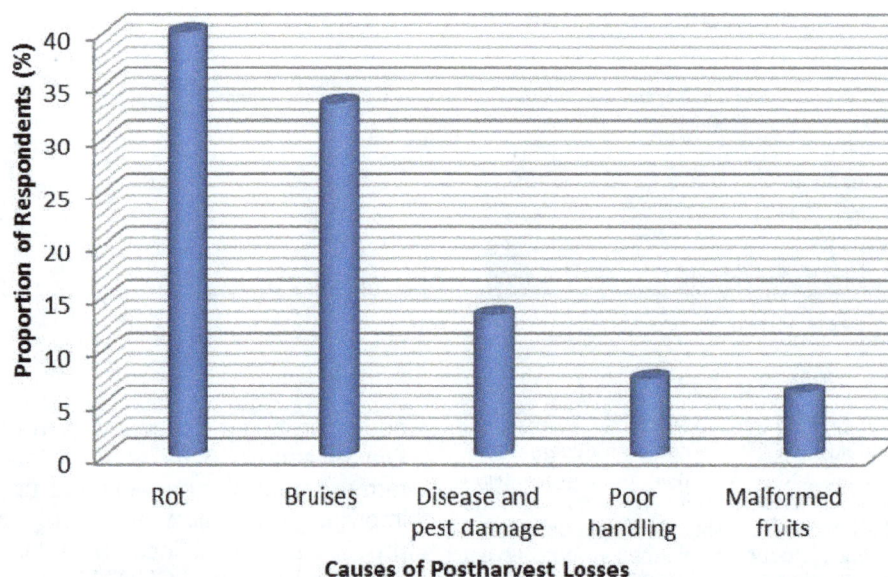

Figure 1. Primary causes of postharvest losses in tomato production. **Source:** Generated from field data (2013).

Table 2. Contribution of secondary factors to postharvest losses.

Factor	Very high (5)	High (4)	Moderate (3)	Low (2)	Very low (1)	Mean score	Rank
Lack of market avenue	91	44	14	1	-	4.50	1st
Unreliable means of transport	21	65	48	14	2	3.59	2nd
Longer distance to market	6	67	48	22	7	3.30	3rd
Untimely harvest	3	45	75	24	3	3.14	4th
Type of variety used	14	18	92	26	-	3.13	5th
Poor harvesting technique	15	33	66	27	9	3.12	6th

Source: Generated form field data (2013).

Table 3. Analysis of tomato output and postharvest losses for the 2012 cropping season.

Variable	Major season	Minor season
Land area (ha)	2.12	2.02
Output (kg)	6,143.80	4,871.68
Quantity of output lost (kg)	2,437.46 (39.7%)	690.83 (14.2%)
Quantity sold(kg)	3,706.34	4,180.77
Unit price (GH¢ /100 kg)	56.51	97.33
Revenue obtained (GH¢)	2,094.45	4,069.22
Value of losses (GH¢)	1,377.41	672.39
Potential revenue (GH¢)	3,471.86	4,741.61

Source: Generated from field data (2013).

area put under tomato cultivation was about 2 ha during both major and minor seasons. On average, the total output of fresh tomato obtained in the major season was 6,143.80 kg compared to 4,871.68 kg in the minor season. Average yield was estimated at 2,898 kg/ha for major season and 2,412 kg/ha for the minor season. Quantity of output lost during the major season was 2,437.4 kg and its value in monetary terms was GH¢

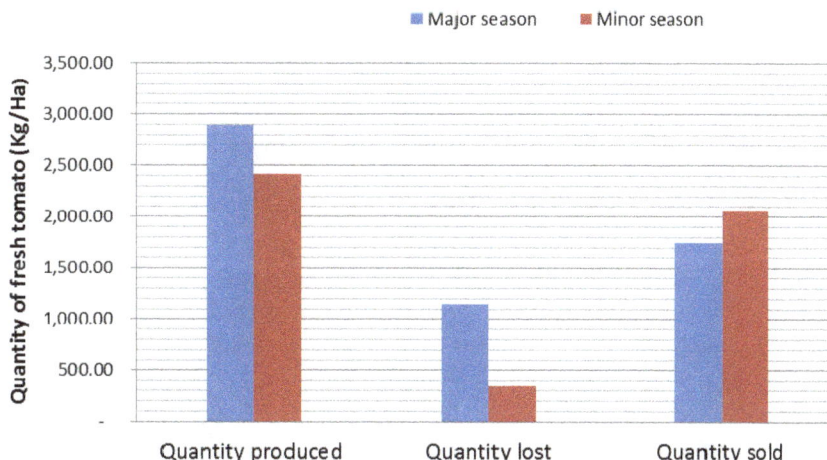

Figure 2. Analysis of tomato yield and postharvest losses per hectare. **Source:** Generated from field data (2013).

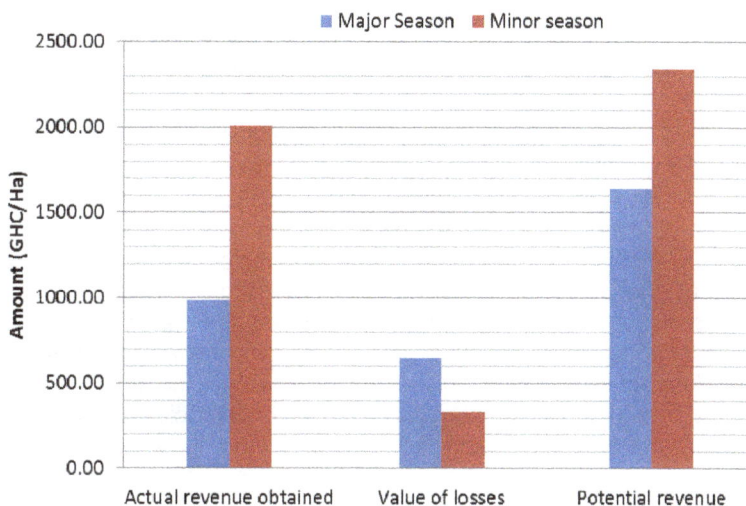

Figure 3. Analysis of actual and potential revenue from a hectare of tomatoes. **Source:** Generated from field data (2013).

1,377.41. This represents a loss of 40% of the harvested produce. Quantity of output lost during the minor season was 690.83 kg, which was valued at GH¢ 672.39, representing about 14% of the harvested produce. On average, quantity sold during the major season was found to be 3,706.34 kg valued at GH¢ 2,094 and that for the minor season was 4,180.77kg at a value of GH¢ 4,069.22.

Potential revenue that could have been generated in the absence of postharvest losses was estimated at GH¢ 3,471.86 for the major season and GH¢ 4,741.61 for the minor season. This means that farmers lost about 40% of the potential revenue from tomato production during the major season and 14% during the minor season.

Figures 2 and 3 indicate that on per hectare basis, quantity of tomato fruits lost during the major season was

about 1,150 kg, valued at about GHC649.72. In the minor season, only about 341 kg of tomato fruits (valued at GHC332.21) was lost per hectare cultivated. This implies that due to postharvest losses, tomato farmers received only 60% of the potential revenue during the major season and 86% during the minor season per hectare (Figure 3).

Determinants of postharvest losses

Table 4 gives a summary of the results obtained from the multiple regression analysis. The adjusted coefficient of determination (R^2) was 0.42 indicating that 42% of the variation in the quantity of tomato fruits lost during and after harvesting was explained by the specified variables

Table 4. Regression estimates of the determinants of tomato losses.

Variable	Coefficient	Std. error	t	p>t
Constant	6.0879***	0.2463	24.72	0.000
Gender (1 = male; 0 = female)	-0.2675***	0.0951	-2.81	0.006
Ln Household Size	-0.0638**	0.0242	-2.63	0.010
Ln Education (years.)	0.0158	0.0117	1.35	0.181
Ln Farm size	0.0312**	0.0147	2.12	0.036
Ln Days to storage	0.0551**	0.0243	2.27	0.025
Ln Extension contact (per month)	-0.0145	0.0132	-1.10	0.276
Membership of FBO (1 = yes 0 = no)	-0.6081***	0.0988	-6.15	0.000
Ready market (1 = yes; 0 = no)	-0.0978	0.1097	-0.89	0.374
Ln Distance to market	0.0049	0.0151	0.32	0.744
Improved variety (1 = Yes; 0 = No)	-0.1505*	0.0884	-1.70	0.091

R^2 = 0.424; F= 9.78; (Significant at 1%); SER = 0.127 (*, ** and *** denote 10, 5 and 1% significant levels, respectively)
Dependent variable: $Ln\_$ quantity of tomato fruits lost.

in the model. The F-statistic was found to be significant at 1%, which implies that all the explanatory variables had a significant joint impact on the level of tomatoes lost after harvest.

Gender and household size were the demographic variables that had a significant effect on postharvest losses in tomato production. Female farmers were found to be more prone to high levels of losses than their male counterparts. This contradicts the findings of Babalola et al. (2010) who concluded that there was little or no gender inequality in tomato farming and hence no effect of gender on postharvest losses. Tomato harvesting is very labour intensive. Generally, male-headed households tend to have many man-hours available and more time for tomato harvesting and other farm activities compared to their female counterparts who are naturally not too strong but also have household/family responsibilities to attend to. All things being equal, women tend to use longer period for fruit harvesting which then causes high levels of postharvest losses.

Household size was found to have a significant negative relationship with the level of postharvest losses incurred. Farmers who had larger household sizes tended to have lower levels of postharvest losses because they have relatively high amount of family labour that help with tomato harvesting for the process to be faster and efficient, ceteris paribus. Farm size had a significant positive effect on the level of postharvest losses recorded by farmers. Larger farms usually have higher output levels which require high amount of labour for harvesting and carting. When the household has labour constraint and there is a little delay from traders, huge volumes of tomato fruits are usually lost by farmers. This finding is consistent with findings of Babalola et al. (2010) who reported that the larger the area put under cultivation the higher the quantity harvested and chances of losses due to poor handling and lack of proper storage. Increase in the quantity of fruits to be harvested as

a result of larger farm size results in increase in postharvest losses because of poor storage facilities and the high labour requirement to carry out the harvesting on time.

The number of days harvested tomato fruits are stored till time of sale was also found to have a significant positive effect on losses experienced. This is consistent with a priori expectation because tomato is highly perishable due to its shorter shelf life. Membership of FBO had a negative correlation with the level of postharvest losses incurred. This means that farmers who join or are members of FBO's have lower probability of experiencing postharvest losses as they link up with trader associations who buy their produce after harvesting. Babalola et al. (2010) also noted that farmers who join agricultural cooperatives would obtain some form of assistance in selling their produce and invariably have lower postharvest losses.

Cultivation of improved varieties (that is, improved zuarungu and pectomech) was associated with lower levels of losses as these varieties have certain advantageous qualities that the local varieties do not have. Such qualities as firmness, disease resistance, longer shelf life and thick skin help the fruits to withstand pressure during harvesting and maintain quality during storage. This finding is in consonance with the finding by Moneruzzaman et al. (2009) who noted that the variety of tomato cultivated goes a long way to indicate the level of postharvest losses experienced by a farmer.

Constraints faced by tomato producers

Table 5 shows that tomato producers in the study area face a number of challenges. On a five-point Likert scale, lack of storage facility was ranked as the most important and critical constraint facing tomato producers in the Offinso North district. Overall cost of tomato production

Table 5. Constraints in tomato production.

Constraint	Very high (5)	High (4)	Moderate (3)	Low (2)	Very low (1)	Mean	Rank
Lack of storage facilities	115	30	4	5	-	4.8	1st
High cost of production	78	51	15	5	1	4.3	2nd
Limited access to finance	47	77	21	5	-	4.1	3rd
Lack of market	25	79	38	8	-	3.8	4th
Unreliable transport	18	66	47	18	1	3.5	5th
Lack of technology	12	30	80	28	-	3.2	6th

Source: Field survey (2013).

was considered to be very high and therefore, ranked as the second most important constraint faced by farmers. Farmers considered limited access to finance/credit as the next important production constraint. A survey by MoFA (2011) also indicated that lack of storage facilities, high cost of production, limited access to finance, unreliable transport and lack of technology were serious constraints that tomato farmers in Ghana are faced with.

Conclusion

The study has shown that postharvest losses are very significant in tomato production in the Offinso North district. The male gender, household size, membership of FBOs and cultivation of improved varieties (*pectomech* and improved *zuarungu*) were associated with lower levels of postharvest losses. However, farm size and number of days the produce is stored before sale were found to be associated with higher levels of postharvest losses in tomato production. Largely, a number of the underlying causes of the huge losses are within the control of the tomato farmer. When these factors are managed well, there will be reduction in postharvest losses, and food availability would be increased without necessarily cultivating an additional hectare of land. Through formation of FBOs, farmers can establish small processing centres that would process tomato into purees and other alternative products when there is no ready market for the fresh fruits. The extension unit of the Ministry of Food and Agriculture should sensitise and create awareness about the improved tomato varieties available (that is, *pectomech* and improved *zuarungu*) to increase their adoption rate in order to minimise postharvest losses. Farmers should be encouraged to stager production/plan production in stages to allow for harvesting in stages which comes with reduced labour requirements and reduced postharvest losses. Periodic training in harvesting and proper handling of harvested tomato fruits should be organized for farmers. Private entrepreneurs should also be encouraged to invest in the tomato industry by building appropriate cold storage facilities at the district level to help farmers store their harvested produce before they are taken to the market. This will help reduce losses that occur at the farm level.

Conflict of Interests

The author(s) have not declared any conflict of interests.

REFERENCES

Adarkwa I (2011). Assessment of the postharvest handling of six major vegetables in two selected Districts in Ashanti Region of Ghana; An MSc Dissertation submitted to the School of Graduate Studies, Kwame Nkrumah University of Science and Technology, Kumasi, Ghana.

Babalola DA, Makinde YO, Omonona BT, Oyekanmi MO (2010). Determinants of postharvest losses in tomato production: A case study of Imeko – Afon local government area of Ogun state. J. Life. Phy. Sci. 3(2):14-18.

Chandy KT (1989). Post-Harvest loss of fruits and vegetables. J. Agric. Environ. Educ. 4(3):1-14.

GSS (2010). Report on the 2010 Ghana Population and Housing census, Ghana Statistical Service (GSS), Accra, Ghana.

MoFA (2011). Annual Report for Offinso North District, Ministry of Food Agriculture, Ghana.

Moneruzzaman KM, Hossain ABMS, Sani W, Saifuddin M, Alenazi M (2009). Effect of harvesting and storage conditions on the postharvest quality of tomato (*Lycopersicon esculentum* Mill) cv. Roma VF. Austr. J. Crop Sci. 3(2):113-121.

Mujib ur R, Naushad K, Inayatullah J (2007). Postharvest Losses in Tomato Crop (A Case of Peshawar Valley). Sarhad. J. Agric. 23:4.

Norman JC (1992). Tropical vegetable crops. Elms Court: Arthur H. Stockwell Ltd.

Orzolek MD, Bogash MS, Harsh MR, Lynn F, Kime LF, Jayson K, Harper JK (2006). Tomato Production. Agricultural Alternatives Pub. Code # UA291. pp. 2-3.

Robinson JZE, Kolavalli LS (2010). The case of tomato in Ghana: Productivity. Working Paper # 19. Development and Strategy Governance Division, IFPRI, Accra, Ghana: Ghana Strategy Support Program (GSSP) GSSP. pp. 1-9.

Sablani SS, Opara LU, Al–Balushi K (2006). Influence of bruising and storage Temperature on vitamin C content of tomato. J. Food Agric. Environ. 4(1):54-56.

Tambo JA, Gbemu T (2010). Resource-use efficiency in tomato production in the Dangme West District, Ghana. Conference on International Research on Food Security. Tomato from related wild nightshades.

World Resources Institute (1998). Disappearing Food: How Big is Postharvest Losses? Published by Stanford University Press. P. 167.

Yeboah AK (2011). A Survey on postharvest handling, preservation and processing methods of tomato (*Solanum lycopersicum*). [http://www.dspace.knust.edu.gh:8080/jspui/bitstream.html. Accessed: 10th January, 2013].

21

Climate change and household food insecurity among fishing communities in the eastern coast of Zanzibar

Makame O. Makame[1] , Richard Y. M. Kangalawe[2] and Layla A. Salum[1]

[1]School of Natural and Social Sciences, State University of Zanzibar, P. O. Box 146, Zanzibar-Tanzania.
[2]Institute of Resource Assessment, University of Dar es Salaam, P. O. Box 35097, Dar es Salaam; Tanzania.

This paper examines the local vulnerability of households in two study communities in the east coast of Zanzibar focusing on food security, which is negatively impacted by climate variability and change. Findings have indicated that overall the local people in eastern coast of Zanzibar are insecure with respect to most major sources of food. Households solely dependent on natural resources through farming, fishing, livestock and poultry farming, have been found to be more vulnerable to food insecurity as these activities are facing considerable uncertainties associated with climate change and variability as well as other stress factors. Agricultural failure resulting from various factors, including local climate variability, coupled with uncertainty of fishing has many pushed households towards increasing dependence on market for their staple food supplies. Therefore, this enhances the household's vulnerability to food insecurity especially among households with low purchasing power. With increasing demand of fisheries resources in urban areas associated with the expanding tourism industry in the study area the price for fisheries resources has increased, causing the poor, including the fishers, to consume less fish and other seafood, and thereby limiting their dietary protein intakes.

Key words: Agriculture, climate change, coastal communities, fisheries based livelihoods, food insecurity, food accessibility, vulnerability, Zanzibar.

INTRODUCTION

The Fourth Assessment Report of the IPCC confidently contends that the observed climate variability and predicted changes in climate will potentially impact food and water security in Africa (Boko et al., 2007). Evidence in support of this argument include the considerable incidents of famine, food insecurity and water stress across Africa, which are partly associated with the variability of climate and the domination of El Niño Southern Oscillation (ENSO) events on the regional climatic patterns (Dai, 2011; Droogers, 2004). Similarly, more than 40% of people in Africa go to bed without enough nourishing food (Cordell et al., 2009). The east coasts of both islands of Zanzibar are frequently affected by localised food shortages and, are sensitive to even moderate abnormalities of rainfall. For instance, in 2010-2011 more than 7,000 people in Micheweni district, in north-east Pemba, where Kiuyu Mbuyuni (one of the study site) is located, did not have enough food (Said, 2011). This was caused by high fluctuations in rainfall which started around 2006 and which affected crop production. Indeed, even without climate variability, access to food for the majority of the households along the drier east coasts of both islands is problematic and is one of the major food security problems of Zanzibar (Boetekees and Immink, 2008). Rose (1994), cited in Walsh (2009) argues that even during the best years, malnutrition along the east coasts is widespread. This study therefore provides the data needed to inform future interventions to reduce poverty and vulnerability and to help to accomplish future sustainable development goals set to take off after 2015 when the current millennium development goals (MDGs) expire.

The definition of food security provided by FAO during the World Food Summit in 1996, and applauded by many, recognises food security as "when all people, at all times, have physical and economic access to sufficient, safe and nutritious food to meet their dietary needs and food preferences for an active and healthy life" (Ericksen, 2008). Unlike previous definitions, this one highlights the role of food availability in connection to the accessibility of food for understanding food security at all levels (Ericksen, 2008). Since the 1970s, food security as a concept has evolved and has been defined extensively across disciplines because of its multi-disciplinary nature and complexity. However, it is now widely recognised that food security comprises four components: food availability, accessibility, stability and utilisation (Ziervogel and Ericksen, 2010; Balaghi et al., 2010; Ericksen, 2008).

Food availability is determined by the ability of households to produce, distribute and exchange food, while access to food is determined by affordability (purchasing power), allocation and preferences (social and cultural determinants influencing consumers). Utilisation is influenced by the nutritional value of the food, its social value and by food safety (Ziervogel and Ericksen, 2010; Ericksen, 2008). Indeed, all components of food security are tightly connected to various global and local determinants and thus they are sensitive to a number of stressors that may include environment, politics, ethics, employment, choices, land alienation and/or land grabbing, land degradation and climate variability and change (Chakrabortya and Newton, 2011; Ziervogel and Ericksen, 2010; Merino et al., 2012; Barnett, 2011; Wang, 2010). This highlights the fact that food insecurity is unevenly distributed both between and within social systems, as interactions between these determinants vary both between and within social systems or decision units, such as the household. For example at the household level, food insecurity may also be triggered by household choices and preferences influenced by livelihood security. A household may choose to go hungry to preserve assets and future livelihoods (Ericksen, 2008; Maxwell, 1996).

Climate variability and change is an additional pressure on food security and affects all four components of food security in many ways. Erratic rainfall, floods, increasingly warm conditions, increasing intensity and frequency of drought and storms and sea level rise (estimated at 1-2 mm/year) are likely to increase the problem of coastal and ocean problems in Zanzibar (Zanzibar Revolutionary Government, 2009) and affect livelihoods, purchasing power, distribution systems, health, freshwater availability for farming and domestic use, important agricultural areas and marine resources, and ultimately affect the stability of food resources (Hanjra and Qureshi, 2010; Ziervogel and Ericksen, 2010; Ericksen, 2008). Therefore, the poor, who have low coping strategies and those who are dependent on climate sensitive ecosystems, are highly vulnerable to food insecurity.

METHODOLOGY

Study areas

This study was conducted in Kiuyu Mbuyuni, in the north-eastern parts of Pemba Island, and Matemwe, in the north-eastern parts of Unguja Island (Figure 1). Pemba and Unguja islands together form the island nation of Zanzibar which is part of the United Republic of Tanzania and located offshore Tanzania mainland coast. Zanzibar experiences two rainy seasons, the long rainy season locally known as *masika* is usually received in March, April and May and short rainy season locally known as *Vuli* in October, November and December. In between these two seasons the islands experience summer season (dry period) locally known as *kiangazi* in January, February and March and winter seasons locally known as *pupwe* in Unguja and *mchoo* in Pemba in June, July and August. The annual average rainfall along the east of both islands where the study sites are located is around 1400 mm, while the central and western parts receive up to 2000 mm per annum. The rainfall of 1400 mm cannot be considered low, however, recent studies (Walsh, 2009; Mustelin et al., 2010) revealed that east coasts are experiencing variations in the distribution of rainfall, onset of the rainy seasons and general decline of rainfall received particularly during short rainy seasons. Figure 2 for example shows that a total of 11 out of 19 years experienced rainfall below average during short rainy seasons between 1992 and 2010. The average annual rainfall is 1678 and 1623 mm/year in Unguja and Pemba islands respectively. Both study sites fall in the coral rag agroecological zone, which is less fertile than other agroecological zones and get exhausted easily in terms of soil fertility under minimum pressure and erratic water supply (Walsh, 2009). Shifting cultivation has been the main methods of farming in these areas (Walsh, 2009). The combination of poor soils and variability in rainfall along the east coasts have long been considered as major factors for the frequent localised food shortage in these areas including the study sites (Walsh, 2009).

Zanzibar is endowed with coastal and marine resources such as beaches, coral reefs, crop and grazing land, mangroves and other forests, sea grass, seaweed farms, fishery resources, salt marshes and collectable seafood that form the foundation of livelihood activities and which are important for the coastal well-being and the nation at large (Zanzibar Revolutionary Government, 2009). Fishing and agriculture are traditional livelihood activities in these areas, but people's livelihood portfolios have changed over the last two decades. Livelihood activities such as seaweed farming and those related to tourism (for example, handcrafts) have started to play a considerable role in the rural economy (Lange and Jiddawi, 2009).

Data collection and sampling procedure

To understand the current situation of food security in the households, this study employed a household survey, where 200 households were randomly selected, 100 from each site. With

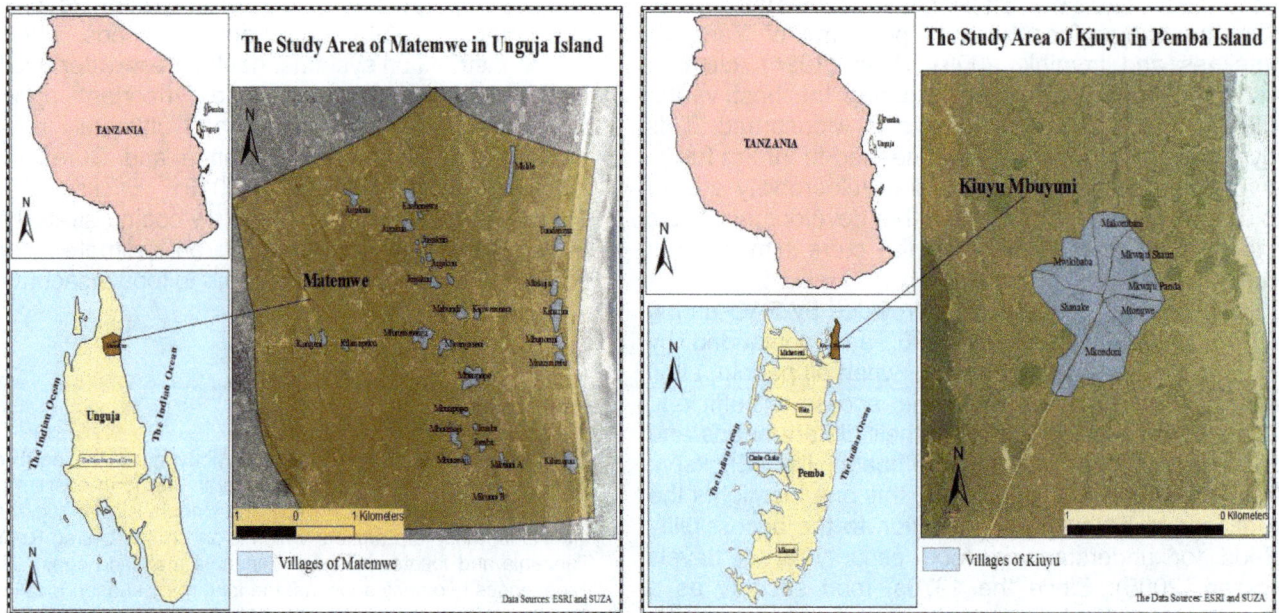

Figure 1. Map showing the locations of the study sites.

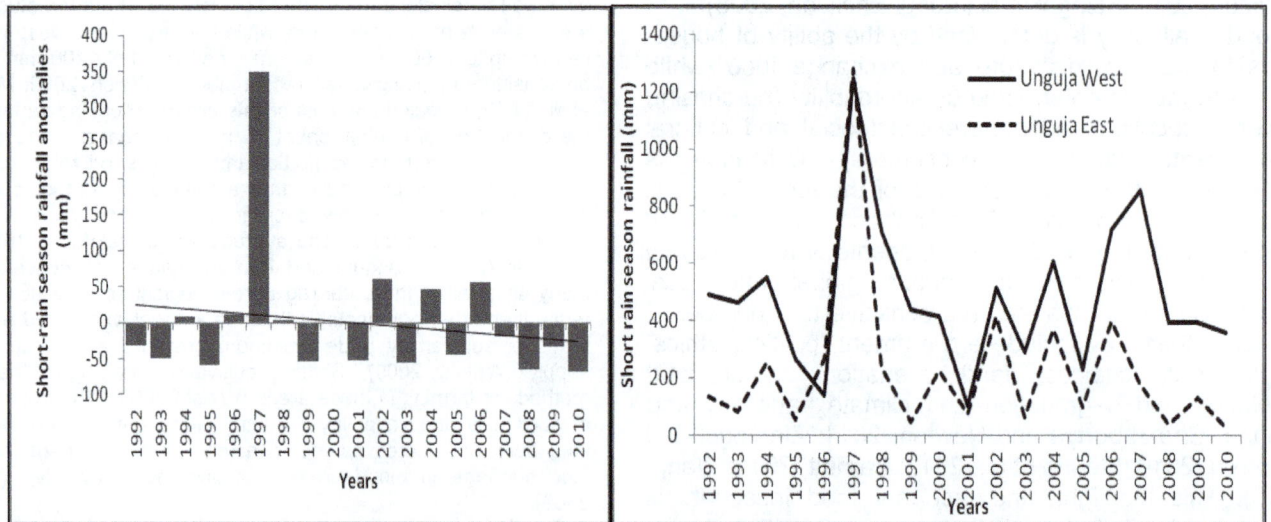

Figure 2. Left: Inter-annual variability of rainfall in the short-rain season/*Vuli* (October-December) on the east coast of Unguja from 1992-2010. Right: inter-annual variability of rainfall in short- rain season between Unguja West and East from 1992-2010 (Source: Makame 2013: pg 126-128). Note: The study did not compare rainfall received in Pemba West and East coasts as weather stations along the Pemba east coast were not operating reliably.

regard to availability and accessibility of main sources of food, the survey was designed to gather data on variables such as sources of major types of food, access to staple food, fish, and vegetables; the costs incurred by a household for staple food and fish per month; and accessibility of other types of food such as seafood, meat and chicken. The information collected provides insights into the current status of food security situation amongst households in the study sites. With regards to stability of staple foods, the survey was designed to capture data on consistency of food supply in the

households and seasonality. Lastly the survey aimed at understanding the various coping strategies employed by the households during the time of localised food shortages.

Data from the survey were analysed using the Statistical Package of the Social Science (SPSS) where descriptive statistics, including frequencies and percentages of respondents were determined across the four major themes of the study, namely availability, accessibility, stability and coping strategies during the time of crisis. The analytical results were disaggregated by study

Table 1. Percentage responses on availability of staple foods, fish and vegetables.

Adequate availability of food	Staple food		Fish		Vegetables	
	Kiuyu Mbuyuni (n=92)	Matemwe (n=97)	Kiuyu Mbuyuni (n=93)	Matemwe (n=92)	Kiuyu Mbuyuni (n=85)	Matemwe (N=96)
Yes	7 (8%)	17(18%)	26 (28%)	22 (24%)	9 (11%)	15 (16%)
No	85 (92%)	80 (82%)	67 (72%)	70 (76%)	76 (89%)	81 (81%)

Table 2. Pearson correlation results between inadequate availability of food and fish and livelihood diversification and family size.

Kiuyu Mbuyuni, Pemba			Matemwe, Unguja		
Types of food	Pearson Correlation	Livelihood diversification	Types of food	Pearson Correlation	Livelihood diversification
Food (N=94)	Correlation P value	-0.002 0.982*	Food (N=97)	Correlation P value	-0.022 0.834*
Fish (N=93)	Correlation P value	0.015 0.886*	Fish (N=92)	Correlation P value	0.121 0.249*

* Pearson correlation was not significant (p >0.05 level, 2 tailed).

sites to facilitate comparisons between the two sites.

RESULTS AND DISCUSSION

Availability and accessibility of food in the households

The major staple foods in the study area are cassava, sweet potatoes, rice, sorghum and maize meal. Respondents across the study sites were asked if they had enough staple foods, fish and vegetables throughout the year and the results showed that 85 out of 92 (92%) households in Kiuyu Mbuyuni and 80 out of 97 households (82%) in Matemwe experienced periods of inadequate availability (Table 1). With regards to fishery products the results also showed that majority of the respondents 72% (67 households) in Kiuyu Mbuyuni and 76% (70 households) in Matemwe experienced inadequate availability of fisheries products throughout the year. The proportion of households that experienced inadequate availability of fish is slightly higher in Matemwe than in Kiuyu Mbuyuni. This is probably influenced by the high demand triggered by tourism and the urban market in Zanzibar town. This is an issue for concern as both sites are considered as fishing villages and fisheries products are the major sources of cheap animal protein preferred and accessible by most people.

A large percentage of the households who perceived inconsistency in the accessibility and availability of staple food and vegetables (Table 1) may be influenced by the fact that the surveys were undertaken in the aftermath of the 2007-2010 periods which was characterised by prolonged dry conditions and declining rainfall (Figure 2) which impacted local farming and production. Vegetables, both wild and locally grown are sensitive to erratic rainfall, especially where the soil is poor. This is captured in the following quote from a respondent in Matemwe: "If rainfall becomes erratic we get a small amount of wild spinach in the bush, but these days even if we receive good rainfall and thus more wild spinach, we may not enjoy it because after a short while the plants are affected by pests. I remember in those early days we used to have massive coverage of wild spinach in the bush, to the extent of inviting people from the neighbouring villages to come and harvest". It is clear from this quotation that conditions have changed to the extent that local production of foodstuffs is being increasingly challenged by the changing climate.

A Pearson correlation was performed to understand whether there was a relationship between inadequate availability of food and fish, as observed and livelihood diversification and family size. The results revealed no relationship between these variables across the sites (p >0.05) (Table 2), suggesting that livelihood diversification and family size within the household do not necessarily reduce the risk of food insecurity.

The observed food insecurity mirrors the findings in the study by Walsh (2009) which showed that localised food shortages along the east coast of both major islands including the study sites, is attributed to poverty, unreliable rainfall, and poor soils. Unlike the 1971/72 famine, which

Table 3. Percentage responses on sources of major food types.

Source	Staple Food		Fisheries products		Vegetables	
	Kiuyu Mbuyuni (n=99)	Matemwe (n=96)	Kiuyu Mbuyuni (n= 100)	Matemwe (n= 91)	Kiuyu Mbuyuni (n=99)	Matemwe (n=99)
Buying	17	40	36	34	29	29
Own farm/fish	-	-	42	39	19	11
Buying+ own/ fish/ gardens	83	60	22	22	52	50
Relatives/neighbours	-	-	-	5	-	-
Wild	-	-	-	-	-	10

was influenced by both drought and the banning of food imports, recent food shortages may be linked to the low capacity of people to purchase or produce own food as even during good years, food insecurity and malnutrition are prevalent (Walsh, 2009). Furthermore, while local climate variability, affects locally grown crops, global climate change affects rice production in Asia, the major supplier of rice to Zanzibar (Peng et al., 2004). In terms of fish, households that were solely dependent on buying fish are more vulnerable compared to those who practice fishing as they cannot afford to consume fish on a daily basis because of competing prices offered by urban markets, particularly during the fishing off-seasons.

With regards to the relationship between households food insecurity and diversity of livelihood portfolio, the results from the present differ from a study conducted in northern Ghana which highlighted the positive and statistically significant impact of livelihood diversity particularly off-farm activities on household food security (Owusu et al., 2011). Although livelihood diversification in known as a coping strategy to food security (Barrett et al., 2001a,b), the observed low availability of food throughout the year is probably indicative of the failure of livelihood diversification to ensure food security in the study areas. This is mainly due to the fact that diversification of livelihoods was based on activities that are sensitive to normal seasonal variations in climate and to global market such as seaweed farming. Stress factors other than climate could have a role to play in influencing food insecurity in the area, especially since food shortage has traditionally been experienced even in years with good weather conditions. However, this was outside the scope of this study. Insights on other stress factors influencing food security in other parts of Tanzania are provided by Kangalawe et al. (2011) and Kangalawe (2012) from studies in the southern highlands, and Lyimo and Kangalawe (2010) in the semiarid zone of Tanzania.

Understanding sources of major food types in connection with availability and accessibility of food in the households

Respondents were asked 'where the household gets most of its food, fish and vegetables'. The results in Table

3 show that none of the households interviewed depended solely on the farm to meet their staple food demands throughout the year. The majority of the households (85%) in Kiuyu Mbuyuni, and more than half of the households in Matemwe were both buying and producing their staple food stuffs. The results indicated that more people still do some farming in Kiuyu Mbuyuni, probably due to the fact that the village is experiencing less competition on the land use compared with Matemwe. In Matemwe 40% of the respondents reported to solely depend on buying food stuffs from shops. This is probably associated with land scarcity due to increasing land value in the area caused by the expansion of the tourism industry. The reasons cited for the high dependence on food from shops included poor soils, seasonality of rainfall, land scarcity, lack of water for irrigation, pests and diseases, and the absence of suitable land for rice cultivation (Valipour, 2014a,b,c,d; Valipour et al., 2014). This dependence on purchased food is highlighted below in a comment from a respondent in Matemwe. "For five years now, a large part of my food comes from shops. Farming is like our religion - one must do it but truly speaking, we are getting nothing out of it. The soil is very poor and the short rainy seasons have disappeared lately" (Figure 2). While it is acknowledged in the above quote that there are other stress factors influencing agricultural production, such as soil fertility, the variations in the seasonality of rainfall has a considerable influence in altering the cropping calendar that farmer used to follow, hence affecting food availability and/or security during some seasons. Figure 3 that in both Unguja and Pemba islands the monthly rainfall is relatively low during some months of the year (particularly June, July and August), which may influence seasonal production of crops like vegetables, especially where no irrigation facility is available.

With regard to fish, a large proportion of the respondents across the study sites reported to catch their own fish for domestic consumption (Table 3). About 36 and 34% respondents in Kiuyu Mbuyuni and Matemwe, respectively, were buying most of their fish, whereas 22% in both sites were both buying and fishing for themselves. Reasons such as engagement in other works, old age and health, and seasonality of wind seasons were cited as barriers that prevented them from self-reliance in fishing.

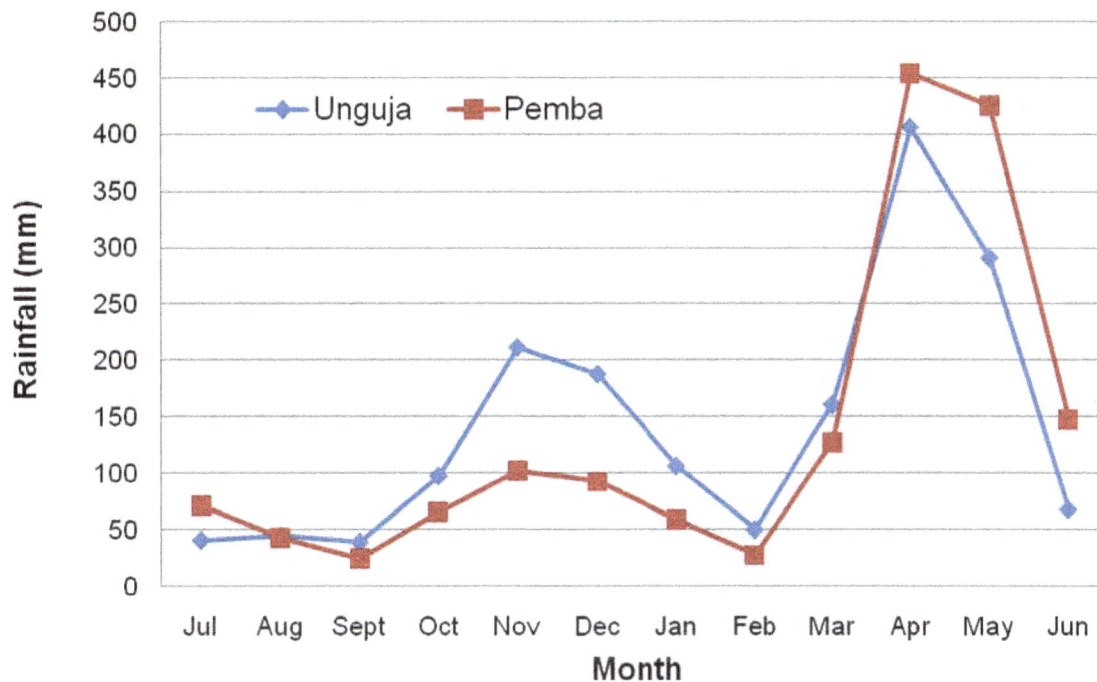

Figure 3. Monthly rainfall in Unguja and Pemba Islands.

fishing. The above observations further confirm the existence of multiple stress factors influencing food insecurity in the area. Interestingly however, 5% of the households in Matemwe were mostly dependent on remittances of fish and/or fish products from relatives and neighbours who practice fishing.

Perhaps the most striking results relate to the sources of vegetables. Unexpectedly, only 19 and 11% of the households in Kiuyu Mbuyuni and Matemwe, respectively, were largely dependent on their gardens for vegetables. About 29% in both sites bought most of their vegetables, while 53% in Kiuyu Mbuyuni and 50% in Matemwe bought and produced their vegetables (Table 3). The fact that a considerable proportion of respondents from both study sites were engaged in producing their own vegetables indicates that it has become important to diversify the sources of livelihoods in these predominantly fishing villages, especially given the various factors above that cause some of them not to engage in fishing. Nevertheless, most of these farmers/fishers have not been self-sufficient in vegetables because of the locally perceived poor soils, variations of rainfall, scarcity of land, diseases and pests and insufficient water for irrigation. Interestingly, however, 10% of the households in Matemwe draw many of their vegetables from the wild (Table 3), which indicates the need for continued protections and conservation of the source areas.

The results demonstrate that the majority of the households across the sites were using a combination of buying and producing their staple food, vegetables and fish. These results are inconsistent with other parts of

Africa, particularly with regard to staple food; for instance, in the rural district of Moma and Mabote in Mozambique more than 80% of households draw their food solely from their own farm plots (Hahn et al., 2009). Given the high levels of poverty within the households across the sites (Wash, 2009), concentrating on buying most food requirements, including vegetables, could be a major source of vulnerability to climate variability, food insecurity and social insecurity. Although over-dependence on small-scale farming for household food is always considered a source of vulnerability (Hahn et al., 2009; McDowell and Hess, 2012), the observed trend toward solely buying, diminishes the purchasing power, savings and access to assets for future adaptation to climate variability and change in the long run. For example, the reported localised food shortages in 2006-2007 (Walsh, 2009) and 2009-2011 (Said, 2011), especially in Kiuyu Mbuyuni were probably influenced by low purchasing power in the households as imported food was readily available in the food stores unlike during 1972 famine. This calls for a critical analysis of possible coping strategies and long-term adaptation options that these communities are using.

The findings from this study also suggest that a household's self-sufficiency with regard to the main types of food is challenged by a number of factors both climatic and non-climatic. Some of the explanations cited, such as scarcity of land for farming, poor soils and infrastructure for irrigation, are more powerful than the observed variability of rainfall for the last decade (Figure 2).

However, climate change has the potential to interact

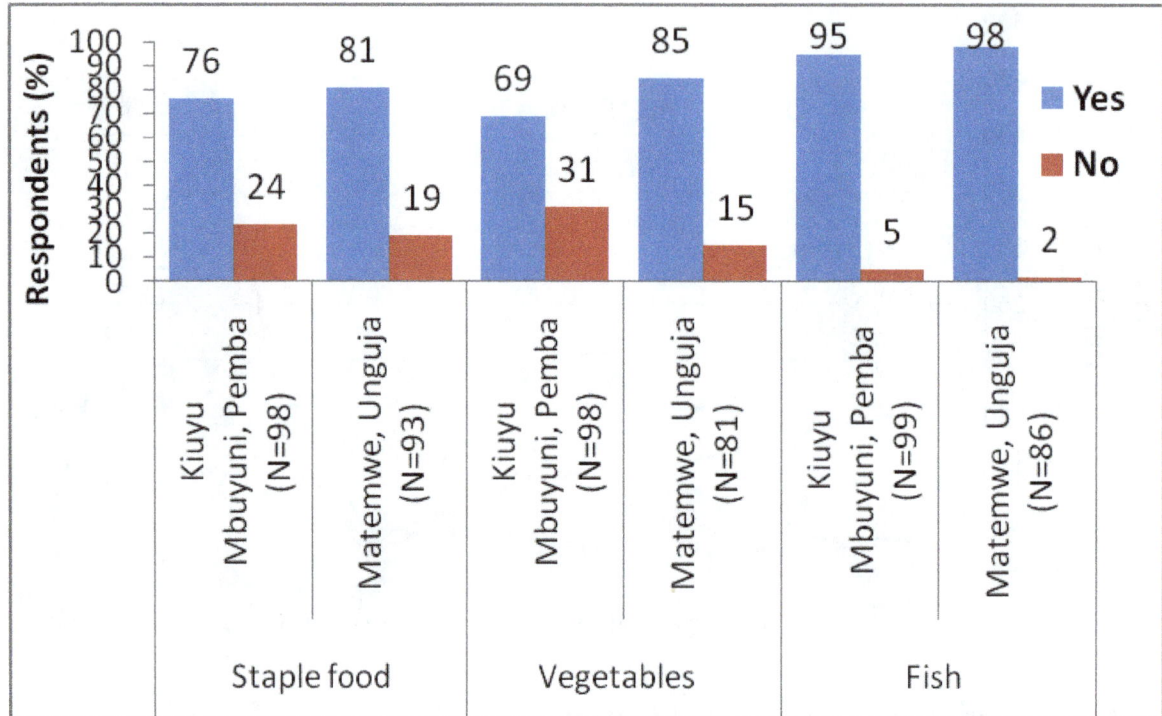

Figure 4. Responses on whether households ever experienced food/ fish/ vegetable instability over last five years.

adversely with these natural and developmental challenges, increasing vulnerability to food insecurity. Even in countries with massive land resources, these challenges threaten rural livelihoods and food security all over the developing world (Droogers, 2004; McDowell and Hess, 2012; Aggarwal et al., 2010; Ellis and Mdoe, 2003). Thus, reducing the severity of localised food shortages in small islands like Zanzibar requires a strict land use plan and increased access to irrigation facilities, better soil management through provision of farming inputs and enhanced access to assets related to fisheries.

Stability of food in the households

In order to understand the nature of food stability in the study areas, all respondents, including those producing fish for themselves, those buying, and those who combine buying and self-producing, were asked whether they had ever experienced an inconsistent supply of food/ fish/ vegetables in their household over the last five years. The results in Figure 4 show that the majority of the respondents across the study sites had experienced such inconsistencies and instability in the supply of fish, staple food and vegetables, which negatively influence their food security.

An inquiry to identify years or seasons when households were severely affected by supply inconsistency over the last five years showed similar responses to those on staple food and vegetables (Table 4). Most of the respondents mentioned the 2009-2010 period in which they faced both food and vegetable instability. During this particular period, both islands particularly along the eastern coast, experienced low rainfall (Figure 2), this resulted in localised food shortages in the study sites. Other periods identified in which households experienced difficulties in obtaining food and vegetables included the period in between 2007-2008 which experienced extended dry seasons, particularly along the east coast, that affected both staple crops and vegetables. Similar difficulties were reported to be experienced during the dry seasons of each year (Figure 3), during the south-easterly winds season locally known as kusi.

Since the majority of the households depend on fishing for their income to buy food, a considerable number of the households face food and vegetables instability during the fishing off-season (season of south-easterly winds). Therefore, the reported difficulties in obtaining food during this period can be translated as a lack of savings obtained during the fishing seasons. This again highlights the danger of over-dependence on purchased food stuff, as commented by one of the respondents in Kiuyu Mbuyuni: "Although we now obtain more money for selling just a small amount of fish catch, whatever we earn ends up in buying food, so we are facing difficulties of obtaining money for food during the off-season".

The majority of the households across the study sites believe that the insufficient availability of fish is most

Table 4. Years or seasons of difficulty in obtaining enough food.

Variable	Year/season	Kiuyu Mbuyuni	Matemwe	Total
	Staple food	**n=72**	**n=67**	**N=139**
Year or season shortage of food experienced	Dry season of each year	4	10	7.0
	South-easterly wind season each year	10	13	11.5
	2009-2010	83	75	79.0
	2007-2008	3	2	2.5
	Vegetables	**n=66**	**n=67**	**N=133**
Year or season shortage of vegetables experienced	Dry season of each year	17	36	26.5
	South-easterly wind season each year	-	2	2.0
	2009-2010	80	60	70.0
	2007-2008	3	3	3.0
	Fisheries products	**n= 93**	**n=86**	**N=179**
Season shortage of fish products experienced	South-easterly wind season of each year	97	94	95.5
	Rainy season of each year	3	4	3.5
	North-easterly wind season of each year	-	2	2

pronounced during the south-easterly wind season which normally lasts approximately four months in June, July, August and September. This windy season, is a period in which the winds blows from south-easterly direction to the north away from the Zanzibar coast and believed to drives fish away from the coast. More importantly, the wind hampers small vessels, the most common fishing vessels, from making fishing trips, mainly because these vessels cannot sail against the south-easterly wind on their way back from fishing trips. Thus many fishers remain at home during this season to minimise risk. During this period therefore very few boats operate (mostly motorised boats), and thus as a result of diminished supply, the demand increases and the price of fish products becomes too high for most people to afford. Even those households who were solely buying fish from the market face instability of fish consumption as they cannot compete with the prices paid in urban and tourism markets during this time of the year. Incidentally, this south-easterly wind season is also a period when the monthly rainfall is at the lowest (Figure 3), thus not being able to support significant crop production, especially of seasonal crops like vegetables. Consequently, the above two conditions together aggravate the food insecurity situation of the study sites during the respective months of June, July, August and September.

Even during normal fishing seasons (calm periods in April-May and October-November and some days of the north-easterly wind seasons in December-March), most of the fishers only operate during spring tides (approximately 17 or 20 days in each month); A similar example is given by Hill (2005) from his study in Vamizi island. Local experience indicated that a spring tide occur when the moon is either new or full and the difference

between high and low tide is the greatest. During the spring tides in the dry season the water is usually saline during with influx of brackish water. During this season the coastal areas are highly prone to cyclone-induced storm surges that may bring about the catastrophic damage (Chowdhury, 2010). In Zanzibar, the knowledge about the tidal cycles is crucial for fish vendors coming to buy fish. Tide may also affect inter tidal activities carried out during low water at spring tides such as collection of octopus, seashells, and sea cucumber (Zanzibar Revolutionary Government, 2009). Given the perceived decline in fish catch per fisher and the high demand for cash amongst fishers to buy food, many households experience food instability on monthly basis as far as the availability and accessibility of fish is concerned. This is demonstrated in the following remarks by one of the respondents in Matemwe: "the amount of fish supplied at home depends on the amount of fish caught; if we land more fish we will consume more fish, but if we land less we will consume less".

The foregoing analysis demonstrates that seasonality, coupled with variability of rainfall, lack of savings and of course, lack of off-farming and off-fishing activities affects the constant availability and accessibility of food. For example, although food instability and widespread malnutrition are common along the east coast, even in good years (Walsh, 2009) seasonal variation in rainfall intensifies the severity of food insecurity and nutritional status of the coastal communities (Makame, 2013). In Dinajpur, Bangladesh, food instability is far higher during the monsoon season than other seasons in the year (Hillbruner and Egan, 2008). Similarly, poor rural families in India are forced to cope with food insecurity mainly attributed to seasonal agricultural production caused by

Table 5. Percentage responses on the consumption of other sources of protein.

Variable	Seafood		Meat		Chickens	
	Kiuyu (n=100)	Matemwe (n=88)	Kiuyu (n=98)	Matemwe (n=88)	Kiuyu (n=99)	Matemwe (n=94)
Often	5	4	-	1	1	1
Sometime	42	38	14	32	86	65
Rarely	53	58	86	67	13	34

erratic rainfall (Agarwal, 1990). In assessing the risk of climate variability and change in two Mozambican communities, Hahn et al. (2009) also found that apart from other stress factors, climate variability and change, disasters such as floods and droughts have caused food instability for between three and eight months per year.

Availability and accessibility of other food types

In a situation where the consumption of fish in coastal villages is perceived to be declining because of seasonality, lack of technology, increasing demand and low access to storage facilities such as electricity and refrigeration, make the communities more vulnerable as they are not able to preserve the food stuffs for long. Respondents in this study were asked on how often their households consumed other foods, such as seafood, meat and chicken. Here, seafood comprised of crustaceans (crabs, prawns, shrimps, and lobsters), molluscs (various types of shellfish), cuttlefish and octopus. Meat comprised both beef and meat from goats and other small animals. Strikingly, the results in Table 5 show that more than half of the households in both the study sites rarely consumed seafood, while 42 and 38% of the households in Kiuyu Mbuyuni and Matemwe, respectively only consumed seafood sometimes. Seafood consumption, which was once regarded as an important source of additional protein in coastal villages has diminished considerably and become rare for the majority, because of its value to both tourists and urban dwellers. Consequently once caught these crustaceans and molluscs are sold to earn cash incomes. While the sales contribute to household ability to buy food staples, these seafood become inaccessible in regular diets of the household.

Although Pemba site has no tourism hotels, seafood is traded as far as Zanzibar town and Mombasa, Kenya. For instance, it was observed that in Pemba, octopuses are informally traded in Mombasa, Kenya. Despite the local belief that eating octopus increases male potency, fishers themselves cannot afford to eat them; they prefer to sell them in order to provide for household needs, including food and iron roofing materials for their homes. Local testimonies highlighted that "Currently one octopus can fetch up to USD 10, thus no one would dare to consume an octopus; after all, octopus is not a staple

food. Everyone would rather sell it in order to obtain money to meet other demands. Truly speaking, octopus has become a food for tourists and not for the poor". As similar testimony was given regarding other types of fish. It was narrated in Matemwe for instance, that "fish are available in Matemwe but people who are eating good fish are not natives. Most of them are tourists. People of this village cannot afford to buy fish. Villagers eat vegetable mostly. The only type of fish we afford to buy is dry anchovy (dagaa kavu). Octopus and squids are very expensive and none of the villagers can afford to buy them".

Table 5 shows also that most households in the study sites rarely consumed meat. Interestingly, the consumption of chicken has also inclined towards the rare category. As such meat or chicken have become part of the diet only during celebrations such as Eid celebrations (two Eid celebrations per year in the Islamic calendar) and during a wedding ceremonies. Although livestock and poultry keeping are common, especially in Kiuyu Mbuyuni, both cattle and chicken are used as a source of manure to improve the soil and as assets to sell when needed.

In this instance, it can be argued that Zanzibar coastal communities experience low accessibility, not only of primary sources of food (staple food, fish and vegetables) but also of other types of foods such as seafood, meat and chickens and are thus vulnerable to food insecurity. However, the observed low consumption of meat and chicken may be associated with household choices in order to increase assets (Erickson, 2008; Maxwell, 1996). For instance, a household may opt not to sell their cattle in order to solve an immediate but small problem (for example, a food shortage in the household) so that they can increase stock for future adaptations. With regard to other seafood, the observed low consumption is clearly linked to increasing demand both within and outside the country, especially in the tourism industry (Garcia and Rosenberg, 2010), and these food stuffs are no longer an important part of the diet for the majority of coastal communities. Globally, these commodities represent the most valuable fisheries exports (Bondad-Reantaso et al., 2012). In examining the role of crustaceans and aquaculture in global food security, Bondad-Reantaso et al. (2012) postulated that the high income obtained from selling crustaceans would enable producers to buy lower

Table 6. Percentage responses on coping strategies for food insecurity at the household level.

Coping Strategy	Kiuyu Mbuyuni (n=67)	Matemwe (n=60)	Total (N=127)
Coping with staple food insecurity			
Food loan	51	52	51.5
Food aid	3	-	3.0
Eating wild food	5	-	5.0
Sleeping without eating	20	8	14.0
Reducing volume per meal	19	38	28.5
Reducing number of meals	3	2	2.5
Coping with vegetables insecurity	Kiuyu Mbuyuni (n=73)	Matemwe (n=71)	Total (N=144)
Consuming fish	27	25	26.0
Buying from market or other village	22	17	19.5
Eating staple food without vegetables	48	42	45.0
Eating food with beans bought from shops	3	6	4.5
Eating dried wild spinach obtained during rain seasons	0	10	5.0
Coping with fisheries product insecurity	Kiuyu Mbuyuni (n=82)	Matemwe (n=84)	Total (N=166)
Eating fish stored in fridge	1	2	1.5
Eating beans, pigeon peas and vegetables	37	31	34.0
Buying from outside	1	-	1.0
Eating fresh sardine and mackerels	-	10	10.0
Eating dried small anchovies	30	38	34.0
Eating staple food only	32	19	25.5

value products and thus contribute to food security. However, changes in food patterns as observed in the study areas may have a negative impact on the nutritional status and health of coastal communities (Receveur et al., 1997; Kuhnlein et al., 2004). For example, Kuhnlein et al. (2004) found a significant correlation between obesity and changes in dietary patterns all over the world. One may argue that sacrificing consumption of various seafood, including octopus, to generate income, without replacing it with foods of equal nutritional value, may have negative consequences for the dietary patterns of the coastal communities.

Coping strategies for food instability at the household level

Periodic food shortages and famine are not new phenomena in the study areas. For example, in the 1971-1972 famine, local people used various strategies to cope (Walsh, 2009). The most frequently cited strategies for coping with staple food instability were loans from shops or neighbours, reducing the volume of the meal, sleeping without eating and reducing the number of meals (Table 6). Other coping strategies, only cited in Pemba, were eating wild food and accruing food aid. With regards to

vegetables, which in most cases are considered as optional, respondents also cited a wide range of coping strategies. These included consuming staple foods without vegetables, eating more beans bought from the market and consuming dried wild spinach. Eating dried wild spinach locally known as *mchunga* is more common in Unguja than in Pemba.

The findings mirror those in urban Uganda (Maxwell, 1996), in urban Accra, Ghana (Maxwell et al., 1999) and in an informal settlement in the Vaal Triangle, South Africa (Oldewage-Therona et al., 2006). In urban Uganda, for instance, people are reported to eat foods that were previously less preferred, limited portion size, borrowed food or money and skipped meals (Maxwell, 1996). Similar experiences are reported in some parts of North-western Tanzania where many people decline to eat other foods (such as maize and rice), except in periods of absolute food (banana) shortage (Mwisongo and Borg, 2002).

Unlike the 1971-1972 famine, during recent localised food shortages, the consumption of cultivated plants and wild food such as poisonous wild yam (*Dioscore sansibarensis Pax*), locally known as *chochoni* as a response to famine (Walsh, 2009) was marginal, probably due to the availability of imported foods in the shops, and because it was not difficult to a obtain a food loan from local shops because of the high social bonding capital. Although food insecurity is widespread along the eastern parts of Pemba and Unguja (Boetekees and Immink, 2008), strong social capital, and willingness to help each other and strong neighbourhoods, coupled with the availability of imported food in the shops, has probably helped reduce the severity of food insecurity, especially during droughts (Makame, 2013).

With regard to fisheries products, respondents also identified a range of strategies that helped them cope with insufficient fish products in their meals. The most cited coping mechanism was consuming dry anchovy and vegetables. About 10% of households in Matemwe replaced high-value fish species with low-value (based on the local perception of the consumers), cheaper species such as sardines and Indian mackerel. The prices for these species are generally affordable and thus they are a common food for the poor and needy all over the developing world (Albert and Marc-Metian, 2009). Interestingly, some households in both study sites were doing nothing to cope with insufficient fisheries products; they simply ate plain meals without either vegetables or dried anchovies. Given the observed low intake of other sources of protein, these households could become more sensitive to dietary problems. The "do nothing" segment of the population demonstrates the variance in vulnerability across social groupings.

For many, eating dried anchovies *(dagaa kavu),* particularly amongst the fishermen, is less preferred by the affluent population. However, the increasing price of fresh fish due to high demand in both urban and tourism markets, and the need for hard cash on the part of the fishers have forced households to rely heavily on dried anchovies as a replacement for fresh fish, even during fishing seasons. Kent (1998) concluded for example that "when fish decline and the price go up, poor people are forced to shift into inferior food, putting them at risk of missing important micronutrients". A dry anchovy probably contains as many important micronutrients for human health as fresh fish but competition between non-food uses and direct human consumption and global climate change (Albert and Marc-Metian, 2009) is threatening this small pelagic fish all over the globe. Indeed, increasing demand for dried anchovies in urban areas and in mainland Tanzania and neighbouring countries will, sooner rather than later, put dry anchovies out of the reach for the majority of the poor in Zanzibar. This will further intensify vulnerability to food insecurity for the majority because vegetables, peas and beans, both cultivated and wild, are sensitive to periodic drought.

Conclusions

Food security requires that all members in the household, at all times, have physical and economic access to sufficient, safe and nutritious food to meet their dietary

needs and food preferences for an active and healthy life. The overall picture emerging across the study sites is that local people are insecure with respect to major sources of food. Agricultural failure resulting from various factors, including local climate variability, coupled with uncertainty of fishing has pushed households towards buying most of their staple foods. This trend has affected food security tremendously due to low purchasing power, attributed to poverty. Increased demand in urban areas and the expansion of tourism industries within the study area and in neighbouring countries have increased the price for the limited fisheries resources, causing the poor, including the fishers, to consume less fish and seafood, thereby limiting their dietary protein intakes. Furthermore, the relationship between climate and coastal activities for both food and income is likely to affect all four components of food security, making the coastal communities even more vulnerable.

Food availability, accessibility and stability are threatened not only by climate variability but also by a number of development challenges, such as limited land and a small economy, and lack of irrigation facilities. Thus while addressing the community vulnerabilities associated with climate change and variability it is paramount to also manage other non-climatic factors that compound vulnerability to climate change-related food insecurity.

Conflict of Interest

The authors have not declared any conflict of interest.

REFERENCES

Agarwal B (1990). Social-security and the family-coping with seasonality and calamity in rural India. J. Peasant Stud. 17:341-412.
Aggarwal PK, Baethegan WE, Cooper P, Gommes R, Lee B, Meinke H, Rathore LS, Sivakumar MVK (2010). Managing climatic risks to combat land degradation and enhance food security: Key Information Needs. Procedia Environ. Sci. 1:305-312.
Albert GJ, Marc-Metian T (2009). Fishing for feed or fishing for food: Increasing global competition for small pelagic forage fish. AMBIO: J. Hum. Environ. 38:294-302.
Balaghi R, Badjeck MC, Bakari D, De Pauw E, De Wit A, Defourny P, Donatog S, Gommes R, Jlibenea M, Ravelo AC, Sivakumar MVK, Telahigue N, Tychon B (2010). Managing climatic risks for enhanced food security: key information capabilities. Procedia Environ. Sci. 1:313-323.
Barnett J (2011). Dangerous climate change in the Pacific Islands: food production and food security. Reg. Environ. Change 11:229-237.
Barrett CB, Bezuneh M, Aboud A, (2001a). Income diversification, poverty traps and policy shocks in Côte d'Ivoire and Kenya. Food Policy 26:367-384.
Barrett CB, Reardon T, Webb P (2001b). Non-farm income diversification and household livelihood strategies in rural Africa: Concepts, dynamics and policy implications. Food Policy 26:315-331.
Boetekees S, Immink M (2008). Placing food security and nutrition on Zanzibar's development policy agenda, a case study of successful policy development. Available online at: http://www.fao.org/docs/up/easypol/805/zanzibar_country_case_study_262en.pdf . Accessed 23 March 2012.

Boko M, Niang I, Nyong A, Vogel C, Githeko A, Medany M, Osman-Elasha B, Tabo R, Yanda P (2007). Africa. Climate Change 2007. In: ML Parry, OF Canziani, JP Palutikof, PJ van der Linden, CE Hanson (Eds.), Impacts, adaptation and vulnerability. Contribution of Working Group II to the Fourth Assessment Report of the Intergovernmental Panel on Climate Change. Cambridge University Press, Cambridge UK. 433-467.

Bondad-Reantaso MG, Subasinghe RP, Josupeit H, Cai J, Zhou X (2012). The role of crustacean fisheries and aquaculture in global food security: past, present and future. J. Invertebr. Pathol. 110:158–165.

Chakrabortya S, Newton AC (2011). Climate change, plant diseases and food security: An overview. Plant Pathology 60:2–14.

Chowdhury NT (2010) Water management in Bangladesh: an analytical review. Water Policy 12:32–51

Cordell D, Drangert J, White S (2009). The story of phosphorus: Global food security and food for thought. Glob. Environ. Change 19:292–305.

Dai A (2011). Drought under global warming: A review. WIREs Climate Change 2:45-65.

Droogers P (2004). Adaptation to climate change to enhance food security and preserve environmental quality: Example for southern Sri Lanka. Agric. Water Manage. 66:15–33.

Ellis F, Mdoe N (2003). Livelihoods and rural poverty reduction in Tanzania. World Dev. 31:367–1384.

Ericksen PJ (2008). Conceptualizing food systems for global environmental change research. Glob. Environ. Change 18:234–245.

Hahn MB, Riederer AM, Foster SO (2009). The livelihood vulnerability index: A pragmatic approach to assessing risks from climate variability and change- a case study in Mozambique. Glob. Environ. Change 19: 74–88.

Hanjra MA, Qureshi ME (2010). Global water crisis and future food security in an era of climate change. Food Policy 35:365–377.

Hill N (2005). Livelihoods in an artisanal fishing community and the effect of ecotourism. A report submitted in partial fulfilment of the requirements for the MSc and/or the DIC. Imperial College London, Faculty of Life Sciences, Department of Environmental Science & Technology, University of London.

Hillbruner C, Egan R (2008). Seasonality, household food security, and nutritional status in Dinajpur, Bangladesh. Food Nutr. Bull. 29:221-231.

Kangalawe RYM (2012). Food security and health in the southern highlands of Tanzania: A multidisciplinary approach to evaluate the impact of climate change and other stress factors. Afr. J. Environ. Sci. Technol. 6:50-66.

Kangalawe RYM, Mwakalila S, Masolwa P (2011). Climate change impacts, local knowledge and coping strategies in the Great Ruaha River Catchment Area, Tanzania. Nat. Resourc. 2:212-223.

Kent G (1998). Fisheries, food security and the poor. Food policy 22:393-404.

Kuhnlein HV, Receveur O, Soueida R, Egeland GM (2004). Arctic indigenous people's experience: The nutrition transition with changing dietary patterns and obesity. J. Nutr. 124:1447–1453.

Lange GM, Jiddawi N (2009). Economic value of marine ecosystems services in Zanzibar: Implications for marine conservation and sustainable development. Ocean Coastal Manage. 52:521-532.

Lyimo JG, Kangalawe RYM (2010). Vulnerability and adaptive strategies to the impact of climate change and variability. The case of rural households in semiarid Tanzania. Environ. Econ. 1:88-96.

McDowell JZ, Hess JJ (2012). Accessing Adaptation: Multiple stressors on livelihoods in the Bolivia highlands under a changing climate. Glob. Environ. Change 22:342-352.

Makame MO (2013). Vulnerability and adaptation of Zanzibar east coast communities to climate variability and change and other interacting stressors. PhD Thesis, Rhodes University, South Africa.

Maxwell D, Ahiadekeb C, Levine C, Armar-Klemesud M, Zakariahd S, Lampteyd G (1999). Alternative food-security indicators: Revisiting the frequency and severity of coping strategies. Food Policy 24:411–429

Maxwell DG (1996). Measuring food insecurity: The frequency and severity of coping strategies. Food Policy 21:291-303.

Merino G, Barange M, Blanchard JL, Harle J, Holmes R, Allen I, Allison EH, Badjeck MC, Dulvy NK, Holt J, Jennings S, Mullon C, Rodwell LD (2012). Can marine fisheries and aquaculture meet fish demand from a growing human population in a changing climate? Global Environ. Change 22:795-806.

Mustelin J, Klein RG, Assaid B, Sitari T, Khamis M, Mzee A, Haji T (2010). Understanding current and future vulnerability in coastal settings: community perceptions and preferences for adaptation in Zanzibar, Tanzania. Popul. Environ. 31:371–398.

Mwisongo A, Borg J (2002). Proceedings of the Kagera Health Sector Reform Laboratory 2nd Annual Conference. Ministry of Health, United Republic of Tanzania.

Oldewage-Therona WH, Dicks EG, Napier CE (2006). Poverty, household food insecurity and nutrition: Coping strategies in an informal settlement in the Vaal Triangle, South Africa. Public Health 120:795-804.

Owusu V, Abdulai A, Abdul-Rahman S (2011). Non-farm work and food security among farm households in Northern Ghana. Food Policy 36:108–118.

Peng S, Huang J, Sheehy JE, Laza RC, Visperas RM, Zhong X, Cassman KG (2004). Rice yields decline with higher night temperature from global warming. Proc. Natl. Acad. Sci. U.S.A. 101:9971-9975.

Receveur O, Boulay M, Kuhnlein HV (1997). Decreasing traditional food use affects diet quality for adult in 16 communities of the Canadian Northwest territories. J. Nutr.127:2179-2186.

Said S (2011). Ukame, mabadiliko ya tabianchi na umaskini Pemba. Mwananchi, 5 May 2011. Available at: http://www.mwananchi.co.tz/magazines/26-jungukuu/11650-ukame-mabadiliko-ya-tabianchi-na-umaskini-pemba.html. Accessed on 10 April 2012.

Valipour M (2014a). A comprehensive study on irrigation management in Asia and Oceania. Arch. Agron. Soil Sci. 61:1-25.

Valipour M (2014b). Future of agricultural water management in Africa. Arch. Agron. Soil Sci. 61:1-21.

Valipour M (2014c). Future of the area equipped for irrigation. Arch. Agron. Soil Sci. 60:1641-1660.

Valipour M (2014d). Pressure on renewable water resources by irrigation to 2060. Acta Adv. Agric. Sci. 2:32-42.

Valipour M, Ziatabar AM, Raeini-Sarjaz M, Gholami SMA, Shahnazari A, Fazlola R, Darzi-Naftchali A (2014). Agricultural water management in the world during past half century. Arch. Agron. Soil Sci. 61:1-22.

Walsh MT (2009). The use of wild and cultivated plants as famine foods on Pemba Island, Zanzibar. Plantes Soc. Ocean Occident. 42/43, 217-241.

Wang J (2010). Food security, food prices and climate change in China: A dynamic panel data analysis. Agric. Agric. Sci. Proc. 1: 321-324.

Zanzibar Revolutionary Government (2009). The Status of Zanzibar Coastal Resources: Towards the Development of Integrated Coastal Management Strategies and Action Plan, Department of Environment through support from Marine and Coastal Environmental Management Project, Zanzibar.

Ziervogel G, Ericksen PJ (2010). Adapting to climate change to sustain food security. WIREs Climate Change 1:525–540.

22

Economic impact of climate change and benefit of adaptations for maize production: Case from Namibia, Zambezi region

Teweldemedhin M. Y. , Durand W., Crespo O., Beletse Y. G. and Nhemachena C.

Department of Natural Resources and Agricultural Sciences, School of Natural Resources and Spatial Sciences, Polytechnic of Namibia, Windhoek, Namibia.

The aim of this research is to examine the impact of climate change in maize farmers' livelihood in Zambezi region, Namibia and benefit of adaptation. Trade-off analysis–multidimensional (TOA-MD) model was presented as a method for evaluation with a combination of simulated baseline production and future simulated yield using Decision Support Systems for Agro-technology Transfer (DSSAT) in maize production system, under five different climate scenarios of Global Circulation Models (GCMs). Even though the magnitude and the impact of different GCMs differs, the projections shows to have a negative economic impact with the highest going up to 76% and lowest to be around 46% loss without any adaption strategies in the Zambezi region. Adaptation strategies and some policy options were tested. The analysis suggests that the introduction of an irrigation system may be sufficient to offset the negative effects of climate change. Since various assumptions and uncertainties are associated with using the proposed approach and results should be interpreted with caution. Despite these limitations, the methodology presented in this study shows the potential to yield new insights into the way that realistic adaptation strategies could improve the livelihoods of smallholder farmers. To safeguard the limited productive assets of rural Namibian's, the study suggested policy aim to target pro-poor disaster management and other adoption mechanism is very important. Apart from protecting productive resources of the rural population, policy should target the diversification of the rural economic environment and strengthen rural-urban linkages.

Key words: Climate change, trade-off analysis–multidimensional (TOA-MD), maize, Namibia, Zambezi.

INTRODUCTION

Although, agriculture sector in Namibia contributes only about 4.1% to the gross domestic product (GDP), however it is regarded as an important part of the economy because it employs 37% of the work force, and sustain 70% of Namibia's population fully, or to a large extent, depend on agriculture for their livelihoods (CBS, 2012). As a comparison, fishing and fish processing contributed 3.6%, while the mining and quarrying industry still remained the highest contributor at 12.4% in 2010 (CBS, 2012).

Crop farming takes place in communal and commercial areas, with the former highly dependent on the rainfall condition. The combination of long dry spells, floods and the persistence of swarms of red-billed quelea birds during critical stages of crop development led to depressed crop yields. In 2007, the total cultivated area was estimated at around 500000 hectares planted, yet there is a potential to increase the land under cultivation (MWAF, 2009).

Namibia is believed to be known as the most vulnerable countries to climate change in Sub Saharan Africa. As it is characterized by semi-arid to hyper-arid conditions and highly variable rainfall; though small stretches of the country (about 8%) are classified as semi-humid or sub-tropical (MWAF, 2009). Rainfall distribution across the country varies from an average of <25mm per year in some parts of the Namibia Desert to 700mm in some parts of the Caprivi Strip, in the North East. The potential implications of climate change in Namibian small holder agriculture have received more attention in the last decade and several efforts have been made to characterize the impact. However, the methods used to date to assess impacts of climate change on smallholder agriculture are less suited to assess socio-economic impacts. To date, integrated climate change impact assessment that consider climate, biophysical and economic models have not been established for small holder agriculture in Namibia.

Study on impact of climate by Desert Research Foundation of Namibia (DRFN)(2008) indicated detected that trends in rainfall is typically more difficult, especially in highly variable arid climates such as Namibia. Considerable spatial heterogeneity in the trends has been observed, but it appears as if the northern and central regions of Namibia are experiencing a later onset and earlier cessation of rains, resulting in shorter seasons in most vicinities.

Description of the study area

The Zambezi Region, until 2013 known as the Caprivi Region, is one of the 14 regions of Namibia, located in the extreme north-east of the country (Figure 1). It is largely concurrent with the Caprivi Strip and takes its name from the Zambezi River that runs along its border. Katima Mulilo is the capital (17.5000° S, 24.2667° E). The climate of the region is characterized by summer rainfall (October to March), with an average rainfall of about 700 mm per year. In summer, January is the month with the highest average maximum temperature (30°C), and winter in July has the minim lowest average temperature of around 2.5°C.

Zambazi region domintley consisting of varying from sand to clay, at one end of the spectrum are heavy soil with high content of clay in areas which are regulary

flooded, that is the hydromorphic and organic clay soils. Those areas flooded most frquently hold water for the longest period, and often have a high content of organic materal dervied from decmposed reeds, sedges and other plants that grow in the water. Eastern Zambazi larglery clay-loam and West part of the region more sandy type soil. Genearlly speaking, the region dominated by clay-loam soils (about 35% of the area) and sand (about 50%) (Mandleson, 2011), of all economic activities agricutlure is the most important source of liveslihood the region livelihood depend on farming (both crops and livestock). Large areas have been cleared to plant crops, the continous increasing number of livestock population in the area create heavy grazing on the enviroment.

Due to relatively high rainfall compared to the other regions; as it has been mentioned on the above rainfall, distribution across the country varies from an average of <25 mm per year in parts of the Namibia Desert to 700 mm in some parts of the Zambezi Strip. Secondly, due to existence of perennial rivers in the region, this provides potential for introduction of small scale irrigation systems in the area.

METHODOLOGY

Overall, the project has three crosscutting themes that emphasise on: uncertainty, aggregation across scales, and representative agricultural pathways (RAPs). The uncertainty explores component of the uncertainty cascade. The aggregation across scales connects local, regional, and global agricultural information (Antle, 2011b). The RAPs processes develop scenarios that connect the representative concentration pathways and the socio-economic pathways (SSPs) that are needed to be included in the model. In this integrated climate change impact assessment research, there are three core questions need to be answered:

1. What is the sensitivity of current agricultural production systems to climate change? Current production system (1980 to 2009 Climate) and future climate current production system (2040 to 2069 Climate), without any adaption and RAPs effect,
2. What is the impact of climate change on future agricultural production systems? (Current production system with future trend on prices and technology on the production system, in addition to considering the effect of RAPs,
3. What are the benefits of climate change adaptations? Future climate production system that includes trend on future climate-adapted production system.

Figure 2 presents the general description of the entire project data processing and methodological framework on the climate assessment: blue colour coded shows the economic component, red for the climate component, green data process for crop modelling and white colour combine both crop and economic modeling. For this report results from the economic modeling only reported indicated blue colour.

Climate data

Due to insufficient data observation from Zambezi region Katima Mulilo station, AgMERRA data were used from Rundu weather around 500 km distance from the study area (climate data was

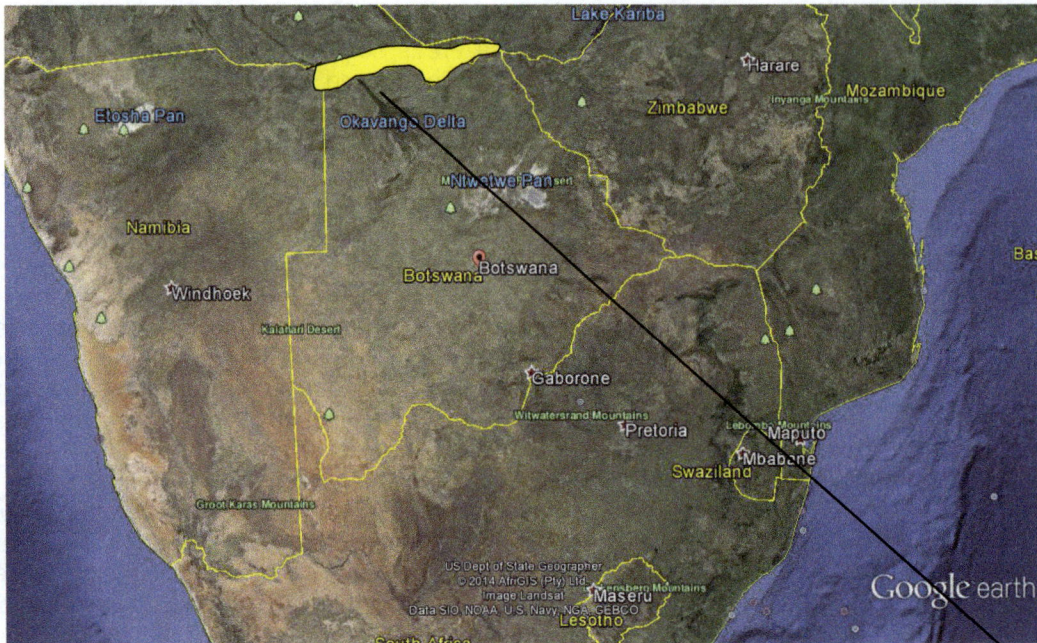

Study   areas   in   Southern

Figure 1. Geographical location of the study areas (Source: Google Earth (2014).

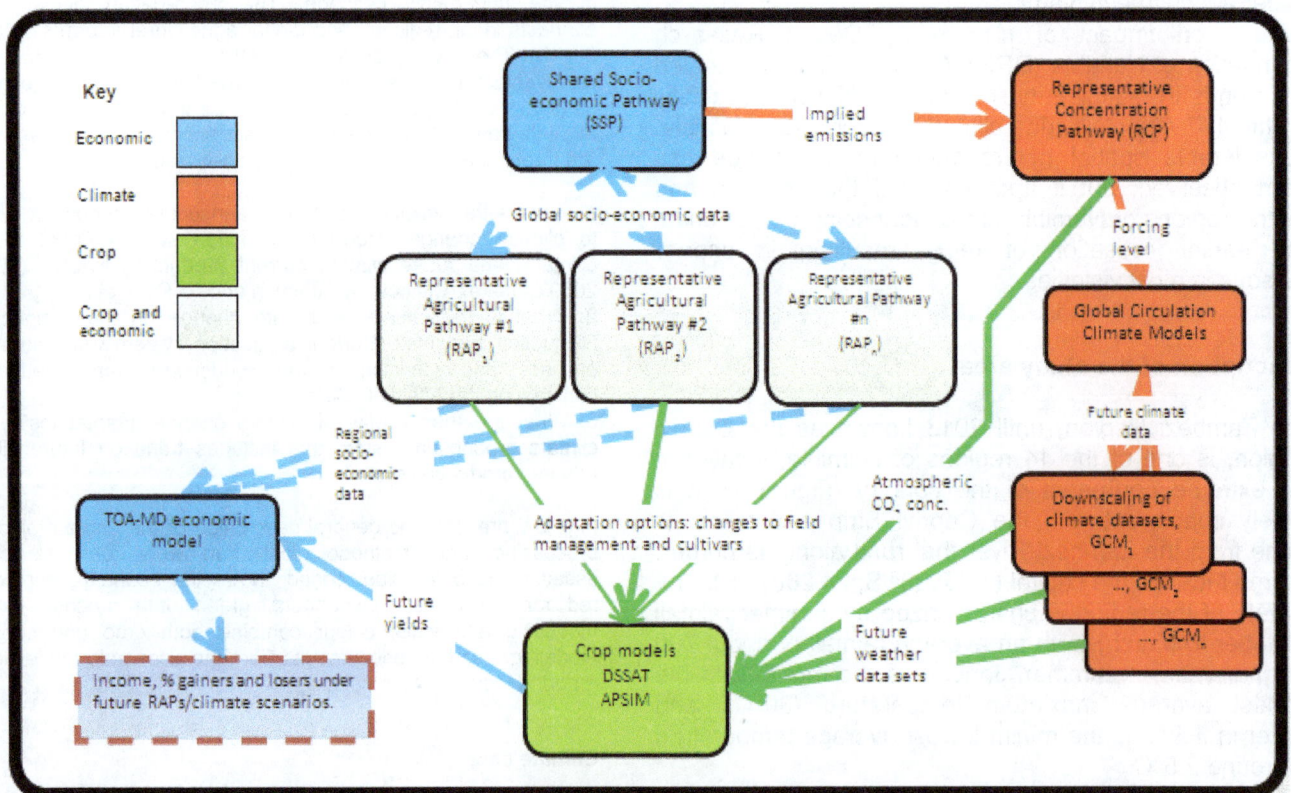

Figure 2. Integrated climate assessment methodological framework (Source: Developed by the research team).

collected by the Namibia weather center). At each location, changes from current climate (1980 to 2010) to near future (2010 to 2040), mid-century (2040 to 2070) and end of century (2070 to 2100) were computed for representative concentration pathways RCP4.5 and RCP8.5. The representative concentration pathways (RCPs) describe the heating effects of atmospheric greenhouse gases (GHGs) at the end of 2100. Twenty global circulation models (GCMs) were used to compute twenty delta changes in monthly temperatures and monthly rainfalls, hence producing 20 possible future weather scenarios per baseline per time period per RCP per station. However, as part of larger fast-track project objective, a first phase presented here consisted in using five of those GCMs only.

Crop data

From different source of literature review and through consultation of agricultural extension officers, the commonly used crop management practices of Zambezi region such as, planting data, soil depth, fertilizer application and harvesting date were used as an input for crop modeling. The physical and chemical properties of the dominant soils information were collected from the data base of Namibia Agricultural research center and some literature review (Mandleson, 2011). In this study, the decision support systems for agro-technology transfer (DSSAT) are used to model the day by day bio-physical growth of maize crops. The model is supported by data base management programs for soil, weather, crop management and experimental. Since the end result of this study was economic analysis, climate and crop results are not reported

Socio-economic data

The data for Zambezi region originate from the project 'Diversified Agriculture and Livelihood Support Options (DALSO) under the Red Cross initiative collected for which farm survey data were collected in 2012 (Mbai et al., 2013). For this analysis, a selection of 191 farms was extracted from the database for complete data (quantities and prices). For inputs (such as seeds, labor, fertilizer, and manure) assumptions were made for the number of families involved or employed during the season for labor; whereas all farmers used manure as their fertilizer. Livestock income calculated based on the potential hiring of oxen, that is the maximum farmer can hire out would be twenty oxen per season, as plough done in four pair.

Methodology for socio-economic impact

For the analysis of climate change economic impact and adaptation strategies, this study used the Tradeoff Analysis model for Multi-Dimensional Impact Assessment (TOA-MD). This model has been used for the analysis of technology adoption (Antle and Validivia, 2006; Nalukenge et al., 2006; Antle and Stoorvogel, 2006, 2008; Immerzeel et al., 2008; Claessens et al., 2008; Antle, 2011b; Antle and Valdivia, 2011) provides an overview of the methodology, and present a validation of the TOA-MD approach against more complex, spatially-explicit models of semi-subsistence agricultural systems.

In the TOA-MD model, farmers are assumed to be economically rational. This meant that they make decisions based on maximizing expected value and presented with a simple binary choice: they can continue to operate with production System 1, or they can switch to an alternative System 2 (Antle and Valdivia, 2011). The logic of the analysis is summarized as follows: farmers are initially operating a base technology with a base climate. This combination is defined as System 1. System 2 is defined as the case where farmers continue using the base technology under a perturbed climate. If

some farmers are worse off economically under the perturbed climate, they are said to be vulnerable to climate change. Overall vulnerability can be measured by the proportion of farmers made worse off, and can also be defined relative to some threshold, such as the poverty line, in which case it says how many more households are put into poverty by climate change (Antle and Valdivia, 2011).

The simulation model uses data on the spatial variability in economic returns to represent heterogeneity; such as heterogeneity in soils, climate, transportation costs, and the farm household's characteristics.

It is necessary to distinguish between three factors affecting the expected value of a production system: the production methods used, referred as the technology, and the physical environment in which the system is operating, for example, the climate, and the economic and social environment in which the system is operated. This is the socio-economic setting that we shall refer to as a Representative Agricultural Pathway (RAP) (Antle, 2011b).

$$\omega = \text{System 1 value} - \text{System 2 value}$$

$$= (P_1 Y_1 a_1 - C_1) - (P_2 Y_2 a_2 - C_2) \tag{1}$$

Where: P, price in System 1 and System 2 respectively; Y, production (Yield) System 1 and System 2 respectively; a, land use; C, Production cost in System 1 and System 2 respectively; ω, the difference between System 1 and System 2.

$$\omega = V_1 - V_2 = \text{losses from CC}$$

$$V_1 = \text{Value of CClim+XTech}$$

$$V_2 = \text{Value of FClim+XTech} \tag{2}$$

$$\mu_\omega = \mu_1 - \mu_2 \tag{3}$$

$$\sigma_\omega^2 = \sigma_1^2 + \sigma_2^2 - \sigma_1 \sigma_2 \sigma \beta_{12} \tag{4}$$

μ_1 & σ_1 from observed of System 1, but for System 2 derived using Random Relative Yield Model

$$C_2 = K_2 y_2 \tag{5}$$

So $V_2 = (P_2 y_2 a_2 - K_2) y_2 = \gamma y_2$

let $y_2 = y_1 + \varepsilon = (1 + \varepsilon / y_1) y_1 = R y_1$
define $R = y_2 / y_1 = $ relative yield
Then $v_2 = \gamma y_1 R$

It is important to take note "γ" is estimate from survey data, whereas, R is estimate using crop models. Since relative yield is assumed to be representative from the heterogonous population it is expected to be normally distributed.

Define: y_1 = actual crop yield in current climate
$\quad s_1$ = simulated crop yield with current climate = $b_1 y_1$
$\quad s_2$ = simulated crop yield with changed climate = $b_2 y_2$

Since we do not know y_2 so we use crop sim models to estimate it!
Assume $b_1 = b_2$ then

Table 1. Summary statistics of the data used for the TOA-MD model.

EAST Caprivi	System 1	System 2	RAP 1 (%)
Farm characteristics	**Mean (STD)**	**Mean (STD)**	
House hold size (persons)	4.89 (2.30)	4.89 (2.30)	
Non-Ag income (Rs.)/year	8937 (4085)	8937 (4085)	
Farm size (ha)	4.15 (4.78)	4.15 (4.78)	
Total farm size	382.2	382.2	
Population of farmers	69200	69200	
Poverty level/year	4536	4536	
Crops/maize			
Yield/ha (kg)	350.54 (126.5)	238.65 (84.84)	
Gross revenue/ha (Rs.)	1400 (455.6)	950.2 (468.07)	
Variable cost/ha (Rs.)	466.22 (151.5)	317.40 (112.85)	
Net return/ha (Rs)	935.95 (316.35)	637.2 (226.5)	
Price (Rs./kg)	4	4	
RAPS			
Land size			70
House hold size			60
None Agric income			40
Price			130
Variable cost			140
Herd size			50%

$R = y_2 / y_1 = s_2 / s_1$ (estimated from crop models!)

$y_2 = R\, y_1$

data for y_1 and R at a representative sample of sites, then

y_2 = climate perturbed yields = $R \times y_1$

Furthermore, Antle (2011a) show that in an economic adaptation analysis, accurate measurement of the economic, environmental and social impacts of technology adoption must take into account the statistical correlation between factors affecting adoption (For example, economic returns) and the other outcomes of interest. The TOA-MD model is designed to incorporate these correlations into the simulation of impacts on farm income and income-based poverty. In climate change assessment, the TOA-MD model implies that not all farms are affected in the same way – in most cases, some farms lose and some farms gain from climate change. Similarly, some farms may be willing to adopt technologies that facilitate adaptation to climate change, while others will not. The TOA-MD model allows researchers to simulate the impacts of the full range of adoption rates from zero to 100% (Claessens et al., 2008).

RESULTS AND DISCUSSION

Table 1 presents the Namibia case study, the farm systems characterizing for CCSM4 of the Global Circulation Models (GCMs) as an example. On average, those five climate scenarios is predicted to be hotter in the future (+2.0 to +3.5°C), with greater variability in rainfall. Future rainfall/precipitation projections are less consistent, with different climate models revealing different projections in the Southern Africa region.

Question 1: What is the sensitivity of current agricultural production systems to climate change?

The results of the sensitivity of current production systems to future climate change are presented in Table 2. The results show that future climate change is projected to be detrimental to crop production in the Zambezi region (Caprivi) in Namibia. Crop yields are expected to decrease by 11 to 23% due to expected changes in climate 71 to 77% of the farmers in the Zambezi region are expected to lose. Furthermore, Table 3 presents the predication of the model to the farmers' net welfare. The model predicts the net crop revenue would drop ranges from 38 to 108%, this would yield impacts on mean return would be range from -35 to -60%. Whereas analysis on per capita income (PCI) shows decreases of 38 to 98% due to climate change, while poverty analysis shows that all the farmers below the poverty line would increase ranges from 18 to 46%. The results imply that current crop production systems are sensitive to the effects of climate change.

Question 2: What is the impact of climate change on future agricultural production systems?

Table 4 presents the impact of climate change on future

Table 2. Sensitivity of current agricultural production systems to climate change.

Stratum 1	CCMS4		GFDL		HadGEM_2ES		MIROC-5		MPI-ESM	
	East	West	East	West	East	West	East	West	East	West
Observed mean yield (maize) (kg/ha)*	350.54	359.79	350.54	359.79	350.54	359.79	350.54	359.79	350.54	359.79
Mean yield change (crop name) (%) [defined as: (mean relative yield -1)*100]	-11	-13	-23	-19	-8	-7	-23	-12	-14.7	-17
Losers (%)	74.15	75.88	76.27	76.5	71.96	73.65	75.74	75.7	76.31	76.7
Gains (% mean net returns) - old	2.32	2.37	2.35	2.57	2.25	2.34	2.58	2.32	2.29	2.32
Losses (% mean net returns) - old	-39.55	-42.77	-48.75	-50.38	-32.36	-35.05	-53.75	-41.13	-47.26	-45.07
Gains (% mean net returns) - corrected	8.97	9.84	9.92	10.96	8.04	8.87	10.63	9.54	9.68	9.98
Losses (% mean net returns) - corrected	-53.34	-56.37	-63.92	-65.85	-44.96	-47.59	-70.96	-54.33	-61.93	-58.76
Observed net returns without climate change (NAD/ha)	5,614.64	5026.002	5,598.73	5054.545	5,483.08	3081.273	5646.119	5007.621	5582.315	4999.697
Observed net returns with climate change (NAD/ha)	60.68	-110.245	-430.34	-328.511	805.17	1920.942	-479.7	-36.0319	-394.48	-237.014
Observed per-capita income without climate change (NAD/Person/Year)	3,630.90	3089.925	3,626.93	3097.158	3598.072	4991.862	1819.377	3085.267	1811.417	3083.259
Observed per-capita income with climate change (NAD/Person/Year)	2,245.08	1788.308	2,122.56	1732.995	2430.845	413.1373	1055.122	1807.115	1065.755	1756.132
Projected poverty rate without climate change (%) **	25.66	33.51	25.61	33.29	26.37	33.8	12.69	33.63	12.85	33.64
Projected poverty rate with climate change (%) **	56.31	76.63	60.51	79.82	50.65	70.15	30.55	75.6	30.08	78.17

* Normalised. ** Poverty line: NAD2454 per capita per year (exchange rate against USD 1$ equivalent to NAD10 (Namibian Dollar)).

Table 3. Impact of climate change to return per capita income and poverty line.

Impact	CCMS4		GFDL		HadGEM_2ES		MIROC-5		MPI-ESM	
	East	West	East	West	East	West	East	West	East	West
Net impact (% mean net returns)	44.36	46.53	54.00	54.89	36.93	38.73	60.33	44.79	52.25	48.78
per capita income (PCI)	38.17	42.12	41.48	44.05	32.44	91.72	42.01	41.43	41.16	43.04
Poverty	30.64	43.12	34.89	46.54	24.28	36.36	17.86	41.97	17.23	44.53
Net revenue	98.92	102.19	107.69	106.50	85.32	37.66	108.50	100.72	107.07	104.74

crop production systems in the Zambezi region (Caprivi) region. The results show that about 38 to 65% of the farmers will lose as a result of climate change. Also, future climate change with RAPs and global trend expected to results in decreases in mean yield decrease from -3 to 17%, this would impact net revenues mixed impact for some climate scenarios provided positive and negative net impact. From example, GFDL-west, HadGEM_2ES (west and east), MIROC-5 (East) and MPI-ESM (East) projected to be positive net revenue; whereas, the remaining scenarios would be projected to be negative impact. As indicated in Table 5 with regards to welfare analysis that includes Per Capita Income (PCI) poverty line indicated that climate change will adversely affect livelihoods of Zambezi (Caprivi) substance farmers. For example poverty is expected to reduce marginally in GFDL (west), HadGEM_2ES (west) and MIROC-5 (west) by 1.78, 4.96 and 0.97% respectively. Whereas, for the remaining climate in the model project, there would be adverse effect, especially, GFDL (East) showed hard hit which is estimated to be around 49%;

Table 4. The impact of climate change on future agricultural production systems.

Stratum 1	CCMS4		GFDL		HadGEM_2ES		MIROC-5		MPI-ESM	
	East	West	East	West	East	West	East	West	East	West
Projected mean yield (maize) (kg/ha)	350.54	359.79	350.54	359.79	350.54	359.79	350.54	359.79	350.54	359.79
Mean yield change (crop name) (%) [defined as: (mean relative yield -1)*100]	-11.00	-13.00	-23.00	-19.00	-7.81	-3.00	-23.00	-12.00	-15.00	-17.00
Losers (%)	53.63	53.70	64.65	46.43	56.50	54.32	37.63	52.10	41.90	58.29
Gains (% mean net returns) - old	5.50	6.65	3.87	7.24	5.09	5.78	9.63	5.99	8.70	4.81
Losses (% mean net returns) - old	16.86	19.04	24.29	13.13	18.74	17.19	10.64	15.50	12.71	18.44
Gains (% mean net returns) - corrected	11.86	14.36	10.94	13.52	11.71	12.65	15.44	12.50	14.98	11.53
Losses (% mean net returns) - corrected	31.43	35.44	37.57	28.27	33.16	31.65	28.28	29.75	30.33	31.64
Projected net returns without climate change (NAD/ha)	8,271.68	6,967.43	8,715.93	6,548.04	6,560.45	5,986.71	5,661.81	6,476.75	5,364.18	7,121.20
Projected net returns with climate change (NAD/ha)	5,621.76	5,970.61	3,283.36	6,612.08	7,527.96	6,878.76	8,879.11	6,361.17	9,087.60	5,080.61
Projected per-capita income without climate change (NAD/Person/Year)	4,293.88	3,581.92	4,404.73	3,475.64	3,866.90	3,333.38	3,642.67	3,457.57	3,568.40	3,620.89
Projected per-capita income with climate change (NAD/Person/Year)	2,294.71	3,732.60	1,711.24	4,003.53	2,770.35	4,116.16	3,107.48	3,897.56	3,159.51	3,356.69
Projected poverty rate without climate change (%) **	21.80	28.48	20.24	30.21	26.80	32.47	29.76	30.38	30.75	27.86
Projected poverty rate with climate change (%) **	53.71	31.00	69.39	28.43	43.37	27.51	37.27	29.41	36.40	35.61

* Normalised. ** Poverty line: NAD2454 per capita per year (exchange rate against USD 1$ equivalent to NAD10 (Namibian Dollar)).

Table 5. The impact of climate change on future agricultural production systems.

Impact	CCMS4		GFDL		HadGEM_2ES		MIROC-5		MPI-ESM	
	East	West	East	West	East	West	East	West	East	West
Net impact	19.57	21.08	26.62	14.74	21.45	19.01	12.84	17.25	15.35	20.11
PCI	46.56	4.21	61.15	15.19	28.36	23.48	14.69	12.73	11.46	7.30
Poverty	31.92	2.53	49.15	1.78	16.58	4.96	7.52	0.97	5.65	7.75
Net revenue	32.04	14.31	62.33	0.98	14.75	14.90	56.82	1.78	69.41	28.66

Table 6. The benefits of adoption of climate change adaptations on future agricultural production systems.

Stratum 1	CCMS4		GFDL		HadGEM_2ES		MIROC-5		MPI-ESM	
	East	West	East	West	East	West	East	West	East	West
Mean yield change (crop name) (%) [defined as: (mean relative yield -1)*100]	249	196	218	199	322	-3	263	178	248	160
% adoption rate	74	69	70	54	70	77	74	70	75	67
Projected net returns without adaptation (ZAR/ha)	-40	1004	545	3349	-374	126	-410	1082	-217	1891
Projected net returns with adaptation (NAD/ha)	8508	6749	7953	5080	10117	7216	8046	6595	8702	5983
Projected per-capita income without adaptation (NAD/Person/Year)	2220	2071	2366	2665	2137	1848	2128	2090	2176	2295
Projected per-capita income with adaptation (NAD/Person/Year)	3015	4061	2876	3356	3416	4258	2900	3996	3063	3738
Projected poverty rate without climate change (%) **	57	63	52	44	59	71	60	62	58	55
Projected poverty rate with climate change (%) **	37	24	40	33	30	23	39	25	36	27

* Normalised. ** Poverty line: NAD2454 per capita per year (exchange rate against USD 1$ equivalent to NAD10 (Namibian Dollar)).

while PCI which is also in the model provided mixed results.

Question 3: What are the benefits of climate change adaptations?

Table 6 shows the benefits of adoption of climate change adaptations on future crop production systems in Namibia (Zambezi region). The adaptation package analysed for this study included the introduction of irrigation as adaptation measures and also RAPs included in the model.

The results show adoption ranging from 54 to 77% of the adapted crop production system under climate change. In addition, the mean yield changes shows an increase ranging from 160 to 322% increase (with exception HadGEM_2ES_West shows 3% reduction). This shows the option of irrigation usage, even over a

much smaller land area, would lead maize production to increase at least five-fold while also providing an opportunity for different crop varieties to be grown throughout the year. The overall effect would be to uplift the livelihoods and food security of those living within the study area. From this study, it can be concluded that the introduction of an irrigation system would compensate for the negative effects of climate change. Furthermore, net returns per farm increases by 18 to 29% as a result of adopting the adaptation package.

The results also show that poverty levels decreases by about 12% minimum and 39% maximum when farmers adopt the adaptation package and PCI increases by ranges from 22 to 130%. Generally, the adoption of the adaptation package helps to reduce the negative impacts of climate change of crop production systems in the Zambezi region in Namibia. However, further

analysis would be required to test different adaptation packages and RAPS on future crop production system (Table 7).

In summary, Figure 3 presented the impact of climate change on the net impact of farmers return for those three different core questions for WEST Zambezi. As shown in the figure for core question-1 it shows the change in climate for future (but without RAPs and adoption), under this scenario the net impact would be a loss of range from 35 to 46% for the East part of the region under those five different GCMs. For core question-2 that is with changed climate in future, without any adaptation measures, but with some policy change and impact of global market trend. Under this consideration the impact of climate change on the farmers' net return projected to be a loss of up to 6% on their net return (with only FGDL climate scenario yields 3% gain on the net return). When considering the adaptation for

Table 7. Impact of climate change with adaptation strategies to return, per capita income and poverty line.

Impact	CCMS4		GFDL		HadGEM_2ES		MIROC-5		MPI-ESM	
	East	West	East	West	East	West	East	West	East	West
Mean yield change	260	209	241	218	329	0	286	190	263	177
Per capita income (PCI)	36	96	22	26	60	130	36	91	41	63
Poverty	-20	-39	-13	-12	-29	-49	-21	-37	-22	-28
Net returns	-21141	572	1360	52	-2806	5636	-2060	509	-4101	216

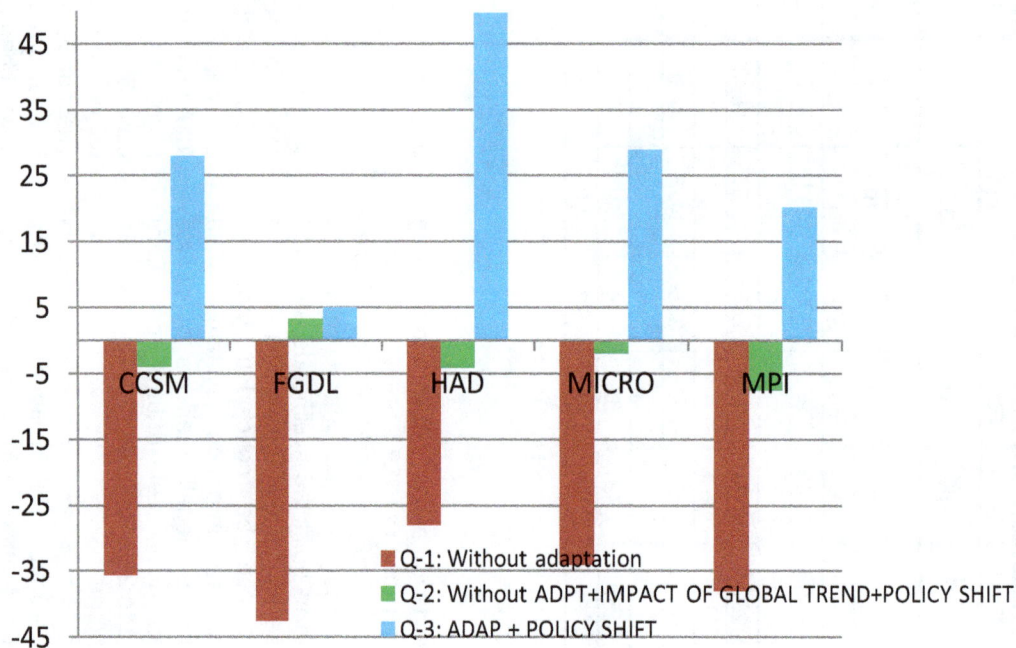

Figure 3. Percentage of net impact of change (for three different core questions) under different five GCMs for Zambezi region (West Zambezi).

inclusive policy shift which yield a positive and potential offset in the impact of climate change, farmers would gain up 45% of those different climate scenarios.

Similar study from IPCC (2014) reported that, there would be an increase in temperatures and changes in precipitation are very likely to reduce cereal crop productivity in Africa (specifically worst in the South-West of Southern Africa). This will have strong adverse effects on food security. New evidence is also emerging that high-value perennial crops could also be adversely affected by temperature rise. Pest, weed and disease pressure on crops and livestock is expected to increase as a result of climate change combined with other factors. Moreover, new challenges to food security are emerging as a result of strong urbanization trends on the continent of Africa and increasingly globalized food chains, which require better understanding of the multi-stressor context of food and livelihood security in both urban and rural contexts in Africa.

Figure 4 presented the impact of climate change on the net impact of farmers return for those three different core

questions for EAST Zambezi. As shown in the figure, it is different from the WEST presented earlier.

Conclusion

Vulnerability and adaptation assessments, particularly at the local level, face limited knowledge about exactly what to adapt to Namibia's natural variability and only exacerbates the shortcomings of global and regional climate models which allow only for broad statements of change. With a view of current technology and future climate change challenges the comprehensive climate assessment was done.

In this study, the TOA-MD model was presented as a method to evaluate the impacts of climate change and the economic viability of adaptation strategies using the kinds of data that are typically available for semi-subsistence systems are important. The method was applied to the maize production systems of the Zambezi region, in Namibia. With a combination of simulated

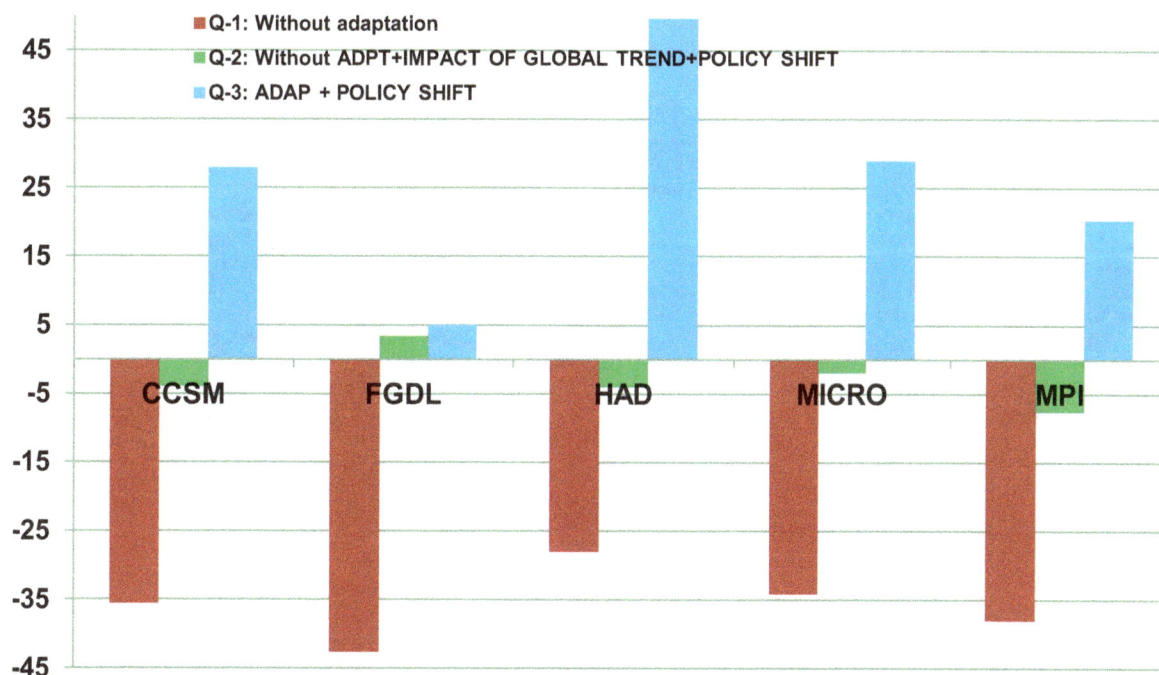

Figure 4. Percentage of net impact of change (for three different core questions) under different five GCMs for Zambezi region (East Zambezi).

baseline production and future simulated yield using DSSAT in maize production system, under five different climate scenarios to achieve those three core questions of the study. The Economic impacts of climate change to 2050 were analyzed. Even though the magnitude of climate impact differs under different GCMs climate change is projected to have a negative economic impact with the highest going up to 76% and lowest to be around 46% loss without any RAPs and adaption in the Zambezi region. Adaptation strategies were tested for the introduction of irrigation system and by introducing socio-economic scenarios based on Representative Agricultural Pathways.

Highly variable climatic conditions and the risk of extreme events: it is important that policy be developed to safeguard the limited productive assets of rural Namibian's by means of targeted, pro-poor disaster insurance schemes. Apart from protecting productive resources of the rural population, policy should target at the diversification of the rural economic environment and strengthen rural-urban linkages. These policy directions should receive adequate attention during the formulation of a rural development policy and strategy, which is currently lacking in Namibia's policy framework. A national debate to clarify the expectations of the agricultural sector to national development, also in lieu of climate change, should be initiated to streamline policies aimed at the sector. Outright conflicting goals prevail which further undermine the potential of this vulnerable sector as well as the sustainable use of the environment.

Conflict of interest

The authors have not declared any conflict of interests.

ACKNOWLEDGEMENT

This research was sponsored by the USDA and UK-AID, further we would like to acknowledge the coordination and leadership of Agricultural Model Inter-comparison and Improvement project (Agmip) from Colombia University and also the technical support of NASA.

REFERENCES

Antle JM, Valdivia RO (2006). Modelling the supply of ecosystem services from agriculture: a minimum-data approach. Austr. J. Agric. Resour. Econ. 50(2):1–15.

Antle JM (2011a). Parsimonious multi-dimensional impact assessment. Am. J. Agric. Econ. 93(5):1292–1311.

Antle JM (2011b). Representative Agricultural Pathways for Model Inter-comparison and Impact Assessment. Available at www.agmip.org/RAPs

Antle JM, Stoorvogel JJ (2006). Predicting the supply of ecosystem services from agriculture. Am. J. Agric. Econ. 88(5):1174–1180.

Antle JM, Stoorvogel JJ (2008). Agricultural carbon sequestration, poverty and sustainability. Environ. Dev. Econ. 13:327–352.

Antle JM, Valdivia R (2011). Methods for Assessing Economic, Environmental and Social Impacts of aquaculture technologies:

adoption of integrated agriculture-aquaculture in Malawi. Department of Agricultural and Resource Economics, Oregon State University, Corvallis OR 97331 USA.

Barnes J, MacGregor J, Alberts M (2012). Expected climate change impacts on land and natural resource use in Namibia: exploring economically efficient responses. Pastoralism: Research, Policy and Practice 2012, 2:22 available: http://www.pastoralismjournal.com/content/2/1/22

Central Bureau of Statistics (CBS) (2011). Namibian Central Bureau of Statistics (CBS). Available from: http://www.npc.gov.na

Claessens L, Stoorvogel JJ, Antle JM (2008). Ex ante assessment of dual-purpose sweet potato in the crop-livestock system of western Kenya: a minimum-data approach. Agric. Syst. 99(1):13–22.

Desert Research Foundation of Namibia (DRFN) (2008). Climate change vulnerability & adaptation assessment Namibia, Windhoek, Namibia.

Immerzeel W, Stoorvogel JJ, Antle JM (2008). Can payments for ecosystem services secure the water tower of Tibet? Agric. Syst. 96(3):52–63.

IPCC (2014). Climate Change 2014: Impacts, Adaptation, and Vulnerability. IPCC WGII AR5 Summary for Policymakers.

Jill EC, Jose C, Zaidi PH, Grudloyma P, Sanchez C, Jose LA, Suriphat T, Makumbi D, Cosmos M, Marianne B, Abebe M, Sarah H, Gary NA (2013). Identification of drought, heat, and combined drought and heat tolerant donors in maize (*Zea mays* L.). J. Crop Sci. 15(5):22-36.

Liao KJ, Tagaris E, Russell AG, Amar P, Manomaiphiboon K, Woo JH (2010). Cost analysis of impacts of climate change on regional air quality. J. Air Waste Manage. Assoc. 60(2):195-203.

Ministry of Agriculture, Water and Forestry (2009). Agricultural Survey. Directorate for Veterinary Services, Windhoek.

Reid H, Sahlen L, Stage J, MacGregor J (2007). The economic impact of climate change in Namibia: How climate change will affect the contribution of Namibia's natural resources to its economy. Discussion Paper International Institute for Environment and Development (IIED). Available at: http://www.google.com.na/url?sa=t&rct=j&q=&esrc=s&frm=1&source=web&cd=19&ved=0CHAQFjAIOAo&url=http%3A%2F%2Fpubs.iied.org%2Fpdfs%2F15509IIED.pdf&ei=V1SXUoO2IeW20wXm04HgAQ&usg=AFQjCNHv0k8QuQGlU6xeozgp9omOGC021g

Permissions

The contributors of this book come from diverse backgrounds, making this book a truly international effort. This book will bring forth new frontiers with its revolutionizing research information and detailed analysis of the nascent developments around the world.

We would like to thank all the contributing authors for lending their expertise to make the book truly unique. They have played a crucial role in the development of this book. Without their invaluable contributions this book wouldn't have been possible. They have made vital efforts to compile up to date information on the varied aspects of this subject to make this book a valuable addition to the collection of many professionals and students.

This book was conceptualized with the vision of imparting up-to-date information and advanced data in this field. To ensure the same, a matchless editorial board was set up. Every individual on the board went through rigorous rounds of assessment to prove their worth. After which they invested a large part of their time researching and compiling the most relevant data for our readers.

The editorial board has been involved in producing this book since its inception. They have spent rigorous hours researching and exploring the diverse topics which have resulted in the successful publishing of this book. They have passed on their knowledge of decades through this book. To expedite this challenging task, the publisher supported the team at every step. A small team of assistant editors was also appointed to further simplify the editing procedure and attain best results for the readers.

Apart from the editorial board, the designing team has also invested a significant amount of their time in understanding the subject and creating the most relevant covers. They scrutinized every image to scout for the most suitable representation of the subject and create an appropriate cover for the book.

The publishing team has been an ardent support to the editorial, designing and production team. Their endless efforts to recruit the best for this project, has resulted in the accomplishment of this book. They are a veteran in the field of academics and their pool of knowledge is as vast as their experience in printing. Their expertise and guidance has proved useful at every step. Their uncompromising quality standards have made this book an exceptional effort. Their encouragement from time to time has been an inspiration for everyone.

The publisher and the editorial board hope that this book will prove to be a valuable piece of knowledge for researchers, students, practitioners and scholars across the globe.

List of Contributors

R. B. Radin Firdaus
School of Social Sciences, Universiti Sains Malaysia, Pulau Pinang, Malaysia

Ismail Abdul Latiff
Faculty of Agriculture, Universiti Putra Malaysia, Serdang, Malaysia

P. Borkotoky
Faculty of Agriculture, Universiti Putra Malaysia, Serdang, Malaysia

K. I. Etonihu
Faculty of Agriculture, Nasarawa State University, Keffi, Nigeria

S. A. Rahman
Faculty of Agriculture, Nasarawa State University, Keffi, Nigeria

S. Usman
National Agricultural Extension and Research Liaison Services, Ahmadu Bello University, Zaria, Nigeria

Simbarashe Ndhleve
Department of Agricultural Economics and Extension, University of Fort Hare, South Africa

Ajuruchukwu Obi
Department of Agricultural Economics and Extension, University of Fort Hare, South Africa

Degye Goshu
School of Agricultural Economics and Agribusiness, Haramaya University, P.O. Box: 05, Haramaya University, Ethiopia

Belay Kassa
School of Agricultural Economics and Agribusiness, Haramaya University, P.O. Box: 05, Haramaya University, Ethiopia

Mengistu Ketema
School of Agricultural Economics and Agribusiness, Haramaya University, P.O. Box: 05, Haramaya University, Ethiopia

Abimbola O. Adepoju
Department of Agricultural Economics, University of Ibadan, Oyo State, Nigeria

Oluwakemi A. Obayelu
Department of Agricultural Economics, University of Ibadan, Oyo State, Nigeria

I. F. Ayanda
Department of Agricultural Economics and Extension Services, Kwara State University, Malete, Nigeria

A. A. Akinola
Department of Agricultural Economics, Obafemi Awolowo University, Ile- Ife, Osun State, Nigeria

R. Adeyemo
Department of Agricultural Economics, Obafemi Awolowo University, Ile- Ife, Osun State, Nigeria

Richard Bwalya
Institute of Economic and Social Research, University of Zambia, Lusaka, Zambia

Johnny Mugisha
Department of Agribusiness and Natural Resource Economics, School of Agricultural Sciences, Makerere University, P. O. Box 7062, Kampala Uganda

Theodora Hyuha
Department of Agribusiness and Natural Resource Economics, School of Agricultural Sciences, Makerere University, P. O. Box 7062, Kampala Uganda

Maria De Salvo
Department of Business Administration, University of Verona, via Della Pieve, 70, San Pietro in Cariano, Verona, Italy

Diego Begalli
Department of Business Administration, University of Verona, via Della Pieve, 70, San Pietro in Cariano, Verona, Italy

Giovanni Signorello
Department of Agri-food and Environmental System Management, University of Catania, via Santa Sofia, 100, Catania, Italy

A. G. Adekola
Department of Agricultural Economics and Extension, Igbinedion University, Okada, Edo State, Nigeria

F. O. Adereti
Department of Agricultural Extension and Rural Development, Obafemi Awolowo University, Ile Ife, Osun State, Nigeria

G. F. Koledoye
Department of Agricultural Extension and Rural Development, Obafemi Awolowo University, Ile Ife, Osun State, Nigeria

P. T. Owombo
Department of Agricultural Economics, Obafemi Awolowo University, Ile Ife, Osun State, Nigeria

Micah B. Masuku
Department of Agricultural Economics and Management, University of Swaziland, P. O. Luyengo, M205, Luyengo Swaziland

Ivy Drafor
Economics Department, Methodist University College Ghana, P. O. Box DC 940, Dansoman, Accra

Byron Zamasiya
International Centre for Tropical Agriculture, (CIAT), P.O. Box MP228 Mt Pleasant, Harare, Zimbabwe

Nelson Mango
International Centre for Tropical Agriculture, (CIAT), P.O. Box MP228 Mt Pleasant, Harare, Zimbabwe

Kefasi Nyikahadzoi
Centre for Applied Social Sciences, University of Zimbabwe, P. O. Box MP167 Mt Pleasant, Harare, Zimbabwe

Shephard Siziba
Department of Agricultural Economics and Extension, University of Zimbabwe, P. O. Box MP167, Mt Pleasant, Harare, Zimbabwe

Misginaw Tamirat
Department of Agricultural Economics, Jimma University College of Agriculture and Veterinary Medicine, P. O. Box 307, Jimma Ethiopia

Tilksew Getahun Bimerew
Department of Rural Development and Agricultural Extension, Haramaya University, Haremaya, Ethiopia

Fekadu Beyene
Department of Rural Development and Agricultural Extension, Haramaya University, Haremaya, Ethiopia

Francis Addeah Darko
Department of Agricultural Economics, Purdue University, 403 West State Street, Krannert Bldg.West Lafayette, IN 47907, Indiana, United States

Benjamin Allen
Department of Agricultural Economics, Purdue University, 403 West State Street, Krannert Bldg.West Lafayette, IN 47907, Indiana, United States

John Mazunda
Department of Agricultural Economics, Purdue University, 403 West State Street, Krannert Bldg.West Lafayette, IN 47907, Indiana, United States

Rafiullah Rahimzai
Department of Agricultural Economics, Purdue University, 403 West State Street, Krannert Bldg.West Lafayette, IN 47907, Indiana, United States

Craig Dobbins
Department of Agricultural Economics, Purdue University, 403 West State Street, Krannert Bldg.West Lafayette, IN 47907, Indiana, United States

Melaku Berhe
Department of Resource Economics, Agricultural Extension and Development, College of Dryland Agriculture and Natural Resources, Mekelle University, P.O.Box 231, Mekelle, Ethiopia

Dana Hoag
Department of Agricultural and Resource Economics, Colorado State University, Fort Collins, CO 80523-1172, USA

B. P. Montle
Polytechnic of Namibia, Windhoek, Namibia

M. Y. Teweldemedhin
Polytechnic of Namibia, Windhoek, Namibia

Tagel Gebrehiwot
Laboratory for Social Interactions and Economic Behaviour (LSEB), University of Twente, P. O. Box 217, 7514 AE Enschede, Netherlands

Anne van der Veen
Faculty of Geo-Information Science and Earth Observation, University of Twente, P. O. Box 217, 7514 AE Enschede, Netherlands

Robert Aidoo
Department of Agricultural Economics, Agribusiness and Extension, Kwame Nkrumah University of Science and Technology (KNUST), Kumasi, Ghana

Rita A. Danfoku
Department of Agricultural Economics, Agribusiness and Extension, Kwame Nkrumah University of Science and Technology (KNUST), Kumasi, Ghana

James Osei Mensah
Department of Agricultural Economics, Agribusiness and Extension, Kwame Nkrumah University of Science and Technology (KNUST), Kumasi, Ghana

Makame O. Makame
School of Natural and Social Sciences, State University of Zanzibar, P. O. Box 146, Zanzibar-Tanzania

Richard Y. M. Kangalawe
Institute of Resource Assessment, University of Dar es Salaam, P. O. Box 35097, Dar es Salaam; Tanzania

Layla A. Salum
School of Natural and Social Sciences, State University of Zanzibar, P. O. Box 146, Zanzibar-Tanzania

M. Y. Teweldemedhin
Department of Natural Resources and Agricultural Sciences, School of Natural Resources and Spatial Sciences, Polytechnic of Namibia, Windhoek, Namibia

W. Durand
Department of Natural Resources and Agricultural Sciences, School of Natural Resources and Spatial Sciences, Polytechnic of Namibia, Windhoek, Namibia

O. Crespo
Department of Natural Resources and Agricultural Sciences, School of Natural Resources and Spatial Sciences, Polytechnic of Namibia, Windhoek, Namibia

Y. G. Beletse
Department of Natural Resources and Agricultural Sciences, School of Natural Resources and Spatial Sciences, Polytechnic of Namibia, Windhoek, Namibia

C. Nhemachena
Department of Natural Resources and Agricultural Sciences, School of Natural Resources and Spatial Sciences, Polytechnic of Namibia, Windhoek, Namibia

www.ingramcontent.com/pod-product-compliance
Lightning Source LLC
Chambersburg PA
CBHW050442200326
41458CB00014B/5031